T0313775

Industrial Data Analytics for Diagnosis and Prognosis

Industrial Data Analytics for Diagnosis and Prognosis

A Random Effects Modelling Approach

Shiyu Zhou
University of Wisconsin – Madison

Yong Chen
University of Iowa

Published by John Wiley & Sons, Inc., Hoboken, New Jersey.

Published simultaneously in Canada.

Library of Congress Cataloging-in-Publication Data:
Names: Zhou, Shiyu, 1970- author. | Chen, Yong (Professor of industrial and systems engineering), author.
Title: Industrial data analytics for diagnosis and prognosis : a random effects modelling approach / Shiyu Zhou, Yong Chen.
Description: Hoboken. NJ : John Wiley & Sons, Inc., 2021. | Includes bibliographical references and index.
Identifiers: LCCN 2021000379 (print) | LCCN 2021000380 (ebook) | ISBN 9781119666288 (hardback) | ISBN 9781119666295 (pdf) | ISBN 9781119666301 (epub) | ISBN 9781119666271 (ebook)
Subjects: LCSH: Industrial engineering--Statistical methods. | Industrial management--Mathematics. | Random data (Statistics) | Estimation theory.
Classification: LCC T57.35 .Z56 2021 (print) | LCC T57.35 (ebook) | DDC 658.0072/7--dc23
LC record available at https://lccn.loc.gov/2021000379
LC ebook record available at https://lccn.loc.gov/2021000380

Cover image: © monsitj/ iStock/Getty Images
Cover design by Wiley

Set in 9.5/12.5pt STIXTwoText by Integra Software Services, Pondicherry, India.

10 9 8 7 6 5 4 3 2 1

To our families:

Yifan and Laura
Jinghui, Jonathan, and Nathan

Contents

Preface

Today, we are facing a data rich world that is changing faster than ever before. The ubiquitous availability of data provides great opportunities for industrial enterprises to improve their process quality and productivity. Industrial data analytics is the process of collecting, exploring, and analyzing data generated from industrial operations and throughout the product life cycle in order to gain insights and improve decision-making. This book describes industrial data analytics approaches with an emphasis on diagnosis and prognosis of industrial processes and systems.

A large number of textbooks/research monographs exist on diagnosis and prognosis in the engineering field. Most of these engineering books focus on model-based diagnosis and prognosis problems in dynamic systems. The model-based approaches adopt a dynamic model for the system, often in the form of a state space model, as the basis for diagnosis and prognosis. Different from these existing books, this book focuses on *the concept of random effects and its applications in system diagnosis and prognosis*. The impetus for this book arose from the current digital revolution. In this digital age, the essential feature of a modern engineering system is that a large amount of data from multiple similar units/machines during their operations are collected in real time. This feature poses significant intellectual opportunities and challenges. As for opportunities, since we have observations from potentially a very large number of similar units, we can compare their operations, share the information, and extract common knowledge to enable accurate and tailored prediction and control at the individual level. As for challenges, because the data are collected in the field and not in a controlled environment, the data contain significant variation and heterogeneity due to the large variations in working/usage conditions for different units. This requires that the analytics approaches should be not only general (so that the common information can be learned and shared), but also flexible (so that the behavior of an individual unit can be captured and controlled). The random effects modeling approaches can exactly address these opportunities and challenges.

Random effects, as the name implies, refer to the underlying random factors in an industrial process or system that impact on the outcome of the process. In diagnosis and prognosis applications, random effects can be used to model the sources of variation in a process and the variation among individual characteristics of multiple heterogeneous units. Some excellent books are available in the industrial statistics area. However, these existing books mainly focus on population level behavior and fixed effects models. The goal of this book is to adapt and bring the theory and techniques of random effects to the application area of industrial system diagnosis and prognosis.

The book contains two main parts. The first part covers general statistical concepts and theory useful for describing and modeling variation, fixed effects, and random effects for both univariate and multivariate data, which provides the necessary background for the second part of the book. The second part covers advanced statistical methods for variation source diagnosis and system failure prognosis based on the random effects modeling approach. An appendix summarizing the basic results in linear spaces and matrix theory is also included at the end of the book for the sake of completeness.

This book is intended for students, engineers, and researchers who are interested in using modern statistical methods for variation modeling, analysis, and prediction in industrial systems. It can be used as a textbook for a graduate level or advanced undergraduate level course on industrial data analytics and/or quality and reliability engineering. We also include "Bibliographic Notes" at the end of each chapter that highlight relevant additional reading materials for interested readers. These bibliographic notes are not intended to provide a complete review of the topic. We apologize for missing literature that is relevant but not included in these notes.

Many of the materials of this book come from the authors' recent research works in variation modeling and analysis, variation source diagnosis, and system condition and failure prognosis for manufacturing systems and beyond. We hope this book can stimulate some new research and serve as a reference book for researchers in this area.

Shiyu Zhou
Madison, Wisconsin, USA
Yong Chen
Iowa City, Iowa, USA

Acknowledgments

We would like to thank the many people we collaborated with that have led up to the writing of this book. In particular, we would like to thank Jianjun Shi, our Ph.D. advisor at the University of Michigan (now at Georgia Tech.), for his continuous advice and encouragement. We are grateful for our colleagues Daniel Apley, Darek Ceglarek, Yu Ding, Jionghua Jin, Dharmaraj Veeramani, Yilu Zhang for their collaborations with us on the related research topics. Grateful thanks also go to Raed Kontar, Junbo Son, and Chao Wang who have helped with the book including computational code to create some of the illustrations and designing the exercise problems. Many students including Akash Deep, Salman Jahani, Jaesung Lee, and Congfang Huang read parts of the manuscript and helped with the exercise problems. We thank the National Science Foundation for the support of our research work related to the book.

Finally, a very special note of appreciation is extended to our families who have provided continuous support over the past years.

S.Z. and Y.C.

Acronyms

AIC	Akaike Information Criterion
BIC	Bayesian Information Criterion
CDF	Cumulative Distribution Function
EM	Expectation–Maximization
GP	Gaussian Process
i.i.d.	Independent and identically distributed
IoT	Internet of Things
IQR	Interquartile Range
KCC	Key Control Characteristic
KQC	Key Quality Characteristic
LME	Linear Mixed Effects
LRT	Likelihood Ratio Test
MLE	Maximum Likelihood Estimation
MINQUE	Minimum Norm Quadratic Unbiased Estimation
MOGP	Multiple Output Gaussian Process
PCA	Principal Component Analysis
pdf	probability density function
PH	Proportional Hazards
RREF	Reduced Row Echelon Form
REML	Restricted Maximum Likelihood
RUL	Remaining Useful Life
r.v.	random variable(s)
SNR	Signal-to-Noise Ratio

Table of Notation

We will follow the custom of notations listed below throughout the book.

Item	Notation	Examples
Covariance matrix (variance–covariance matrix)	$\mathrm{cov}(\cdot)$, $\boldsymbol{\Sigma}$	$\mathrm{cov}(\mathbf{Y})$, $\boldsymbol{\Sigma}_{\mathbf{y}}$
Cumulative distribution function	$F(\mathbf{y})$, $F(\mathbf{y};\theta)$, $F_\theta(\mathbf{y})$	
Defined as	$:=$	$\mathbf{S}_{\mathbf{y}} := \frac{1}{N}\sum_{i=1}^{N}(\mathbf{y}_i\mathbf{y}_i^T)$
Density function	$f(\mathbf{y})$, $f(\mathbf{y};\theta)$, $f_\theta(\mathbf{y})$	
Estimated/predicted value	accent ˆ	$\hat{\theta}$, $\hat{\lambda}$
Expectation operation	$E(\cdot)$	$E(Y)$, $E(\mathbf{Y})$
Identity matrix	$\mathbf{I}$, $\mathbf{I}_n$ (n by n identity matrix)	
Indicator function	$I(\cdot)$	$I(y > a)$
Likelihood function	$L(\cdot)$	$L(\boldsymbol{\theta} \mid \mathbf{y})$
Matrix	boldface uppercase	$\mathbf{X}$, $\mathbf{Y}$
Matrix or vector transpose	superscript T	$\mathbf{X}^T$, $\mathbf{Y}^T$
Mean of a random variable (vector)	μ, $\boldsymbol{\mu}$	$\boldsymbol{\mu}_{\mathbf{y}}$
Model parameter	lowercase Greek letter	θ, λ
Negative log likelihood function	$-l(\cdot)$	$-l(\boldsymbol{\theta} \mid \mathbf{y}) = -\ln L(\boldsymbol{\theta} \mid \mathbf{y})$
Normal distribution function	$\mathcal{N}(\cdot,\cdot)$	$\mathcal{N}(\mu,\sigma^2)$, $\mathcal{N}(\boldsymbol{\mu},\boldsymbol{\Sigma})$
Parameter space	uppercase script Greek letters	Θ, Ω
Probability of	$\Pr(\cdot)$	$\Pr(t_1 \le T \le t_2)$

(Continued)

Item	Notation	Examples
Trace of a matrix	$\text{tr}(\cdot)$	$\text{tr}(\boldsymbol{\Sigma}_{\mathbf{y}})$
Variance of a r.v.	$\text{var}(\cdot)$, σ^2	$\text{var}(Y)$, $\text{var}(\epsilon)$, $\sigma^2_{b_i}$, σ^2_{ϵ}
Vector	boldface lowercase	$\mathbf{x}$, $\mathbf{y}$
Vector or matrix of zeros	$\mathbf{0}$, $\mathbf{0}_n$, $\mathbf{0}_{n\times m}$	
Vector of ones	$\mathbf{1}$, $\mathbf{1}_n$ (n by 1 vector of ones)	

1
Introduction

1.1 Background and Motivation

Today, we are facing a data rich world that is changing faster than ever before. The ubiquitous availability of data provides great opportunities for industrial enterprises to improve their process quality and productivity. Indeed, the fast development of sensing, communication, and information technology has turned modern industrial systems into a data rich environment. For example, in a modern manufacturing process, it is now common to conduct a 100% inspection of product quality through automatic inspection stations. In addition, many modern manufacturing machines are numerically-controlled and equipped with many sensors and can provide various sensing data of the working conditions to the outside world.

One particularly important enabling technology in this trend is the Internet of Things (IoT) technology. IoT represents a network of physical devices, which enables ubiquitous data collection, communication, and sharing. One typical application of the IoT technology is the remote condition monitoring, diagnosis, and failure prognosis system for after-sales services. Such a system typically consists of three major components as shown in Figure 1.1: (i) the in-field units (e.g., cars on the road), (ii) the communication network, and (iii) the back-office/cloud data processing center. The sensors embedded in the in-field unit continuously generate data, which are transmitted through the communication network to the back office. The aggregated data are then processed and analyzed at the back-office to assess system status and produce prognosis. The analytics results and the service alerts are passed individually to the in-field unit. Such a remote monitoring system can effectively improve the user

Industrial Data Analytics for Diagnosis and Prognosis: A Random Effects Modelling Approach, First Edition. Shiyu Zhou and Yong Chen.

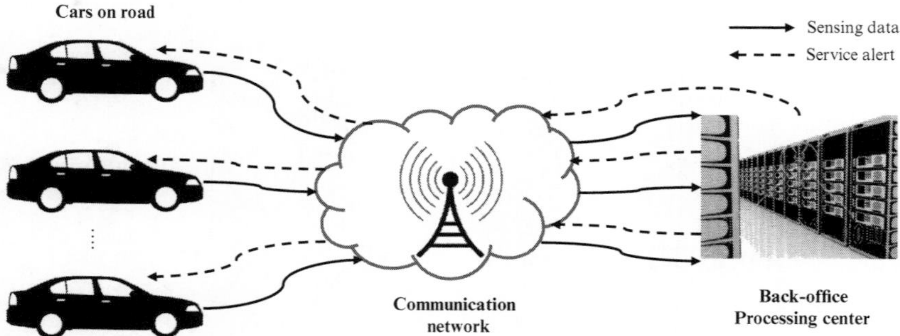

Figure 1.1 A diagram of an IoT enabled remote condition monitoring system.

experience, enhance the product safety, lower the ownership cost, and eventually gain competitive advantage for the manufacturer. Driven by the rapid development of information technology and the critical needs of providing fast and effective after-sales services to the products in a globalized market, the remote monitoring systems are becoming increasingly available.

The unprecedented data availability provides great opportunities for more precise and contextualized system condition monitoring, diagnosis, and prognosis, which are very challenging to achieve if only scarce data are available. Industrial data analytics is the process of collecting, exploring, and analyzing data generated from industrial operations throughout the product life cycle in order to gain insights and improve decision-making. Industrial data analytics encompasses a vast set of applied statistics and machine learning tools and techniques, including data visualization, data-driven process modeling, statistical process monitoring, root cause identification and diagnosis, predictive analytics, system reliability and robustness, and design of experiments, to name just a few. The focus of this book is industrial data analytics approaches that can take advantage of the unprecedented data availability. Particularly, we focus on the concept of random effects and its applications in system diagnosis and prognosis.

The terms *diagnosis* and *prognosis* were originally used in the medical field. Diagnosis is the identification of the disease that is responsible to the symptoms of the patient's illness, and prognosis is a forecast of the likely course of the disease. In the field of engineering, these terms have similar meanings: for an industrial system, diagnosis is the identification of the root cause of a system failure or abnormal working condition; and prognosis is the prediction of the system degradation status and the future failure or break down. Obviously, diagnosis and prognosis play a critical role in assuring smooth, efficient, and safe system operations. Indeed, diagnosis and prognosis have attracted ever-growing interest in recent years. This trend has been driven by the fact that capital goods manufacturers have been coming under increasing pressure to open up new sources of revenue and profit in recent years. Maintenance service costs constitute around 60–90% of the life-cycle costs of industrial machinery and installations. Systematic extension of the after-sales service business will be an increasingly important driver of profitable growth.

Due to the importance of diagnosis and prognosis in industrial system operations, a relatively large number of books/research monographs exist on this topic [Lewis et al., 2011, Niu, 2017, Wu et al., 2006, Talebi et al., 2009, Gertler, 1998, Chen and Patton, 2012, Witczak, 2007, Isermann, 2011, Ding, 2008, Si et al., 2017]. As implied by their titles, many of these books focus on model-based diagnosis and prognosis problems in dynamic systems. A model-based approach adopts a dynamic model, often in the form of a state space model, as the basis for diagnosis and prognosis. Then the difference between the observations and the model predictions, called residuals, are examined to achieve fault identification and diagnosis. For the prognosis, data-driven dynamic forecasting methods, such as time series modeling methods, are used to predict the future values of the interested system signals. The modeling and analysis of the system dynamics are the focus of the existing literature.

Different from the existing literature, this book focuses on *the concept of random effects and its applications in system diagnosis and prognosis*. Random effects, as the name implies, refer to the underlying random factors in an industrial process that impact on the outcome of the process. In diagnosis and prognosis applications, random effects can be used to model the sources of variation in a process and the variation among individual characteristics of multiple heterogeneous units. The following two examples illustrate the random effects in industrial processes.

Example 1.1 Random effects in automotive body sheet metal assembly processes

The concept of variation source is illustrated for an assembly operation in which two parts are welded together. In an automotive sheet metal assembly process, the sheet metals will be positioned and clamped on the fixture system through the matching of the locators (also called pins) on the fixture system and the holes on the sheet metals. Then the sheet metals will be welded together. Clearly, the accuracy of the positions of the locating pins and the tightness of the matching between the pins and the holes significantly influence the dimensional accuracy of the final assembly. Figure 1.2(a) shows the final product as designed. The assembly process is as follows: Part 1 is first located on the fixture and constrained by 4-way Pin L_1 and 2-way Pin L_2. A 4-way pin constrains the movement in two directions, while a 2-way pin only constrains the movement in one direction. Then, Part 2 is located by 4-way Pin L_3 and 2-way Pin L_4. The two parts are then welded together in a joining operation and released from the fixture.

If the position or diameter of Pin L_1 deviates from design nominal, then Part 1 will consequently not be in its designed nominal position, as shown in Figure 1.2(b). After joining Part 1 and Part 2, the dimensions of the final parts will deviate from the designed nominal values. One critical point that needs to be emphasized is that Figure 1.2(b) only shows one possible realization of produced assemblies. If we produce another assembly, the deviation of the position of Part 1 could be different. For instance, if the diameter of a pin is reduced due to pin wear, then the matching between the pin and the corresponding hole will be loose, which will lead to random wobble of the final position of part. This will in turn cause increased variation in the dimension of the produced final assemblies. As a result, mislocations of the pin can be manifested by either mean shift or variance change in the dimensional quality measurement such as M_1 and M_2 in the figure. In the case of mean shift error (for example due to a fixed position shift of the pin), the error can be compensated by process adjustment such as realignment of the locators. The variance change errors (for example due to a worn-out pin or the excessive looseness of a pin) cannot be easily compensated for in most cases. Also, note that each locator in the process is a potential source of the variance change errors, which is referred to as a *variation source*. The variation sources are random effects in the process that will impact on the final assembly quality. In most assembly processes, the pin wear is difficult to measure so the random effects are not directly observed. In a modern automotive body assembly process, hundreds of locators are used

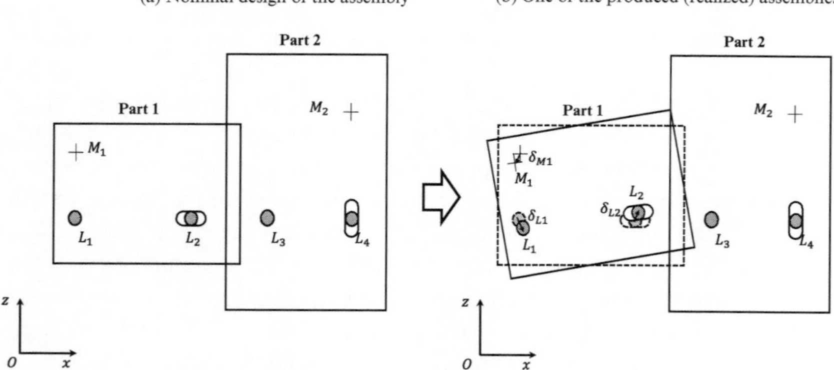

Figure 1.2 Random effects in an assembly operation.

to position a large number of parts and sub-assemblies. An important and challenging diagnosis problem is to estimate and identify the variation sources in the process based on the observed quality measurements.

Example 1.2 Random effects in battery degradation processes

In industrial applications, the reliability of a critical unit is crucial to guarantee the overall functional capabilities of the entire system. Failure of such a unit can be catastrophic. Turbine engines of airplanes, power supplies of computers, and batteries of automobiles are typical examples where failure of the unit would lead to breakdown of the entire system. For these reasons, the working condition of such critical units must be monitored and the remaining useful life (RUL) of such units should be predicted so that we can take preventive actions before catastrophic failure occurs. Many system failure mechanisms can be traced back to some underlying degradation processes. An important prognosis problem is to predict RUL based on the degradation signals collected, which are often strongly associated with the failure of the unit. For example, Figure 1.3 shows the evolution of the internal resistance signals of multiple automotive lead-acid batteries. The internal resistance measurement is known to be one of the best condition monitoring signals for the battery life prognosis [Eddahech et al., 2012]. As we can see from Figure 1.3, the internal resistance measurement generally increases with the service time of the battery, which indicates that the health status of the battery is deteriorating.

We can clearly see from Figure 1.3 that although similar, the progression paths of the internal resistance over time of different batteries are not identical. The difference is certainly expected due to many random factors in the material, manufacturing processes, and the working environment that vary

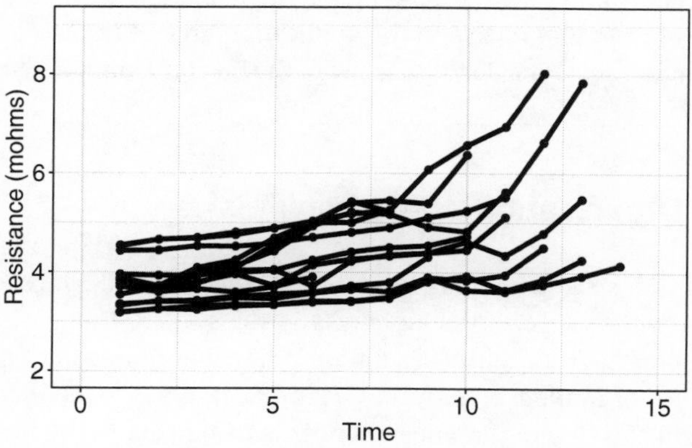

Figure 1.3 Internal resistance measures from multiple batteries over time.

from unit-to-unit. The random characteristics of degradation paths are random effects, which impact the observed degradation signals of multiple batteries.

The available data from multiple similar units/machines poses interesting intellectual opportunities and challenges for prognosis. As for opportunities, since we have observations from potentially a very large number of similar units, we can compare their operations/conditions, share the information, and extract common knowledge to enable accurate prediction and control at the individual level. As for challenges, because the data are collected in the field and not in a controlled environment, the data contain significant variation and heterogeneity due to the large variations in working conditions for different units. The data analytics approaches should not only be general (so that the common information can be learned and shared), but also flexible (so that the behavior of an individual subject can be captured and controlled).

Random effects always exist in industrial processes. The process variation caused by random effects is detrimental and thus random effects should be modeled, analyzed, and controlled, particularly in system diagnosis and prognosis. However, due to the limitation in the data availability, the data analytics approaches considering random effects have not been widely adopted in industrial practices. Indeed, before the significant advancement in communication and information technology, data collection in industries often occurs *locally* in very similar environments. With such limited data, the impact of random effects cannot be exposed and modeled easily. This situation has changed significantly in recent years due to the digital revolution as mentioned at the beginning of the section.

The statistical methods for random effects provide a powerful set of tools for us to model and analyze the random variation in an industrial process. The goal of this book is to provide a textbook for engineering students and a reference book for researchers and industrial practitioners to adapt and bring the theory and techniques of random effects to the application area of industrial system diagnosis and prognosis. The detailed scope of the book is summarized in the next section.

1.2 Scope and Organization of the Book

This book focuses on industrial data analytics methods for system diagnosis and prognosis with an emphasis on random effects in the system. Diagnosis concerns identification of the root cause of a failure or an abnormal working condition. In the context of random effects, the goal of diagnosis is to identify the variation sources in the system. Prognosis concerns using data to predict what will happen in the future. Regarding random effects, prognosis focuses on addressing unit-to-unit variation and making degradation/failure predictions for each individual unit considering the unique characteristic of the unit.

The book contains two main parts:

1. **Statistical Methods and Foundation for Industrial Data Analytics**
This part covers general statistical concepts, methods, and theory useful for describing and modelling the variation, the fixed effects, and the random effects for both univariate and multivariate data. This part provides necessary background for later chapters in part II. In part I, Chapter 2 introduces the basic statistical methods for visualizing and describing data variation. Chapter 3 introduces the concept of random vectors and multivariate normal distribution. Basic concepts in statistical modeling and inference will also be introduced. Chapter 4 focuses on the principal component analysis (PCA) method. PCA is a powerful method to expose and describe the variations in multivariate data. PCA has broad applications in variation source identification. Chapter 5 focuses on linear regression models, which are useful in modeling the fixed effects in a dataset. Statistical inference in linear regression including parameter estimation and hypothesis testing approaches will be discussed. Chapter 6 focuses on the basic theory of the linear mixed effects model, which captures both the fixed effects and the random effects in the data.
2. **Random Effects Approaches for Diagnosis and Prognosis**
This part covers the applications of the random effects modeling approach to diagnosis of variation sources and to failure prognosis in industrial processes/systems. Through industrial application examples, we will present variation pattern based variation source identification in Chapter 7. Variation source estimation methods based on the linear mixed effects model will be introduced in Chapter 8. A detailed performance comparison of different methods for practical applications is presented as well. In Chapter 9, the diagnosability issue for the variation source diagnosis problem will be studied. Chapter 10 introduces the mixed effects longitudinal modeling approach for forecasting system degradation and predicting remaining useful life based on the first time hitting probability. Some variations of the basic method such as the method considering mixture prior for unbalanced data in remaining useful life prediction are also presented. Chapter 11 introduces the concept of Gaussian processes as a nonparametric way for the modeling and analysis of multiple longitudinal signals. The application of the multi-output Gaussian process for failure prognosis will be presented as well. Chapter 12 introduces the method for failure prognosis combining the degradation signals and time-to-event data. The advanced joint prognosis model which integrates the survival regression model and the mixed effects regression model is presented.

1.3 How to Use This Book

This book is intended for students, engineers, and researchers who are interested in using modern statistical methods for variation modeling, diagnosis, and prediction in industrial systems.

This book can be used as a textbook for a graduate level or advanced undergraduate level courses on industrial data analytics. The book is fairly self-contained, although background in basic probability and statistics such as the concept of random variable, probability distribution, moments, and basic knowledge in linear algebra such as matrix operations and matrix decomposition would be useful. The appendix at the end of the book provides a summary of the necessary concepts and results in linear space and matrix theory. The materials in Part II of the book are relatively independent. So the instructor could combine selected chapters in Part II with Part I as the basic materials for different courses. For example, topics in Part I can be used for an advanced undergraduate level course on introduction to industrial data analytics. The materials in Part I and some selected chapters in Part II (e.g., Chapters 7, 8, and 9) can be used in a master's level statistical quality control course. Similarly, materials in Part I and selected later chapters in Part II (e.g., Chapters 10, 11, 12) can be used in a master's level course with emphasis on prognosis and reliability applications. Finally, Part II alone can be used as the textbook for an advanced graduate level course on diagnosis and prognosis.

One important feature of this book is that we provide detailed descriptions of software implementation for most of the methods and algorithms. We adopt the statistical programming language `R` in this book. `R` language is versatile and has a very large number of up-to-date packages implementing various statistical methods [R Core Team, 2020]. This feature makes this book fit well with the needs of practitioners in engineering fields to self study and implement the statistical modeling and analysis methods. All the `R` codes and data sets used in this book can be found at the book companion website.

Bibliographic Notes

Some examples of good books on system diagnosis and prognosis in engineering area are Lewis et al. [2011], Niu [2017], Wu et al. [2006], Talebi et al. [2009], Gertler [1998], Chen and Patton [2012], Witczak [2007], Isermann [2011], Ding [2008], Si et al. [2017]. Many good textbooks are available on industrial statistics. For example, Montgomery [2009], DeVor et al. [2007],

Colosimo and Del Castillo [2006], Wu and Hamada [2011] are on statistical monitoring and design. On the failure event analysis and prognosis, Meeker and Escobar [2014], Rausand et al. [2004], Elsayed [2012] are commonly cited references.

Part I

Statistical Methods and Foundation for Industrial Data Analytics

2

Introduction to Data Visualization and Characterization

Before making a chess move, an experienced chess player first explores the positions of the pieces on the chess board for noticeable patterns such as opponent's threats, special relationships between chess pieces, and the strengths and weaknesses of both sides, before digging into in-depth calculation of move sequences to find the optimal move. Similarly, a data scientist should also start with an exploration of the data set for noticeable patterns before conducting any in-depth analysis by building a sophisticated mathematical model or running a computationally intensive algorithm. Simple data exploration methods can help understand the basic data structure such as dimension and types of variables; discover initial patterns such as relationships among variables; identify missing values, outliers, and skewed distribution for the needs of data pre-processing and transformation. This chapter focuses on basic graphical and numerical methods for data description and exploration. We first look at a data set in the following example.

Example 2.1 (`auto_spec` data) The data set in `auto_spec.csv`, which is from the UCI Machine Learning Repository [Dua and Graff, 2017], contains the specifications of a sample of cars. The following `R` codes can be used to read the data file and obtain information on basic characteristics and structure of the data set.

```
# load data
auto.spec.df <- read.csv ("auto_spec.csv", header = T)
# show basic information of data set
dim (auto.spec.df)
names (auto.spec.df)
head(auto.spec.df)
summary(auto.spec.df)
```

Industrial Data Analytics for Diagnosis and Prognosis: A Random Effects Modelling Approach, First Edition. Shiyu Zhou and Yong Chen.

```
> dim(auto.spec.df)
[1] 205 23
> names(auto.spec.df)
 [1] "make"              "fuel.type"         "aspiration"
 [4] "num.of.doors"      "body.style"        "drive.wheels"
 [7] "engine.location"   "wheel.base"        "length"
[10] "width"             "height"            "curb.weight"
[13] "engine.type"       "num.of.cylinders"  "engine.size"
[16] "fuel.system"       "bore"              "stroke"
[19] "compression.ratio" "horsepower"        "peak.rpm"
[22] "city.mpg"          "highway.mpg"
> head(auto.spec.df)
        Make  Fuel.type Aspiration Num.of.doors  Body.style Drive wheels
1 Alfa-Romeo     Gas        Std            Two Convertible          Rwd
2 Alfa-Romeo     Gas        Std            Two Convertible          Rwd
3 Alfa-Romeo     Gas        Std            Two   Hatchback          Rwd
4       Audi     Gas        Std           Four       Sedan          Fwd
5       Audi     Gas        Std           Four       Sedan          Fwd
6       Audi     Gas        Std            Two       Sedan          Fwd
....
  Horsepower Peak.rpm City.mpg Highway.mpg
1        111     5000       21          27
2        111     5000       21          27
3        154     5000       19          26
4        102     5500       24          30
5        115     5500       18          22
6        110     5500       19          25
> summary(auto.spec.df)
        Make      Fuel.type  Aspiration  Num.of.doors     Body.style
Toyota    : 32  Diesel:  20  Std  :168     Four:114 Convertible: 6
Nissan    : 18  Gas   : 185  Turbo: 37     two : 89 Hardtop    : 8
Mazda     : 17                             NA's:  2 Hatchback  :70
Honda     : 13                                      Sedan      :96
Mitsubishi: 13                                      Wagon      :25
Subaru    : 12
(Other)   :100
....
```

```
     City.mpg          Highway.mpg
Min.    :13.00       Min.  :16.00
1st Qu.:19.00      1st Qu.:25.00
Median :24.00      Median :30.00
Mean    :25.22       Mean  :30.75
3rd Qu.:30.00      3rd Qu.:34.00
Max.    :49.00       Max.  :54.00
```

From the R outputs, we see that this data set contains 205 observations on 23 variables including manufacturer, fuel type, body style, dimension, horsepower, miles per gallon, and other specifications of a car. In statistics and data mining literature, an observation is also called a record, a data point, a case, a sample, an entity, an instance, or a subject, etc. The variables associated with an observation are also called attributes, fields, characteristics, or features, etc. The `summary()` function shows the basic summary information of each variable such as the mean, median, and range of values. From the summary information, it is obvious that there are two types of variable. A variable such as `fuel.type` and `body.style` has a finite number of possible values, and there is no numerical relationship among the values. Such a variable is referred to as a *categorical variable*. On the other hand, a variable such as `highway.mpg` and `horsepower` has continuous numerical values, and is referred to as a *numerical variable*. Beyond the basic data summary, graphical methods can be used to show more patterns of both types of variables, as discussed in the following subsection.

Note from the results of `summary()`, several variables in the `auto_spec` data set have *missing values*, which are represented by `NA`. Missing values are a common occurrence in real world data sets. There are various ways to handle the missing values in a data set. If the number of observations with missing values is small, those observations might be simply omitted. To do this, we can use the R function `na.omit()`. From the following R codes we can see that there are $205-197=8$ observations with missing values in this data set. So simply removing these observations is a reasonable way to handle the missing values for this data set.

```
> dim(na.omit(auto.spec.df))
[1]  197  23
```

If a significant number of observations in a data set have missing values, an alternative to simply removing observations with missing values is *imputation*, which is a process of replacing missing values with substituted values. A simple method of imputation is to replace missing values with a mean or median of the variable. More sophisticated procedures such as regression-based imputation do exist. These methods play important roles mainly in medical and scientific studies, where data collection from patients or subjects is often costly.

In most industrial data analytics applications where data are typically abundant, simpler methods of handling missing values are usually sufficient.

2.1 Data Visualization

Data visualization is used to represent the data using graphical methods. It is one of the most effective and intuitive ways to explore the important patterns in the data such as data distribution, relationship among variables, surprising clusters, and outliers. Data visualization is a fast-growing area and a large number and variety of tools have been developed. This section discusses some of the most basic and useful types of graphical methods or data plots for industrial data analytics applications.

2.1.1 Distribution Plots for a Single Variable

Bar charts can be used to display the distribution of a categorical variable, while histograms and box plots are useful tools to display the distribution of a numerical variable.

Distribution of A Categorical Variable – Bar Chart

In a bar chart, the horizontal axis corresponds to all possible values/categories of a categorical variable. The vertical axis shows the number of observations in each category. To draw a bar chart for a categorical variable in `R`, we need to first use the `table()` function to count the number of observations in each category. Then the `barplot()` function can be used to plot the calculated counts. For example, the following `R` codes plot the distribution of the `body.style` variable in the `auto_spec` data set.

```
bodystyle.freq <- table(auto.spec.df$body.style)
barplot(bodystyle.freq, xlab = "Body Style",
    ylim = c(0, 100))
```

The plotted bar chart is shown in Figure 2.1. From the bar chart, it is clear that most of the cars in the data are either sedans or hatchbacks.

Distribution of Numerical Variables – Histogram and Box Plot

A histogram can be used to approximately represent the distribution of a numerical variable with continuous values. A histogram can be considered as

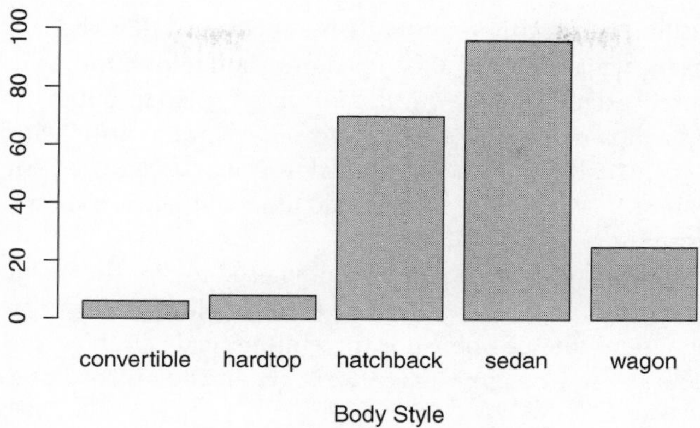

Figure 2.1 Bar chart of car body style.

a bar chart extended to continuous numerical variables. To draw a histogram, the entire range of the variable in the data set is divided into a number of consecutive equal sized intervals. Then a "bar" is shown for each interval to represent the number of observations in the interval.

Another commonly used plot that can represent distribution of a numerical variable is the box plot. We illustrate the basic elements of a box plot in Figure 2.2, which shows the box plot of the numerical variable `width` of the `auto_spec` data set. The bold line within the rectangle box represents the median value of the variable in the data set. The lower and upper bound of the box are corresponding to the first quartile (25th percentile) and the third quartile (75th percentile), respectively. The height of the box is the *interquartile range* (IQR), which is the distance between the first

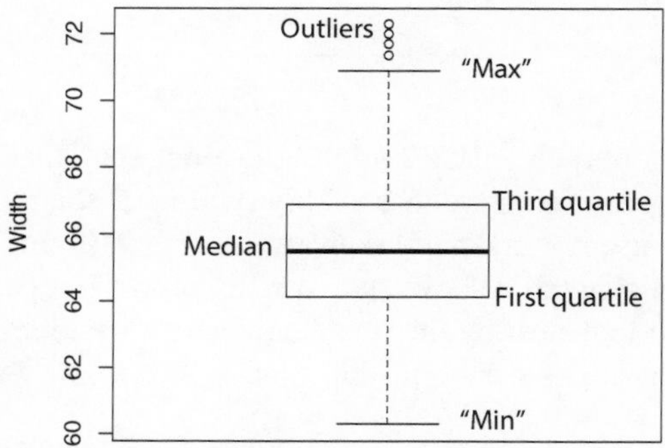

Figure 2.2 Elements of a box plot.

and the third quartile. The short horizontal lines above and below the box are called the *whiskers*, which represent the maximum and minimum of the values in the data set, excluding the "outliers". In box plots, an outlier is typically defined as a data point that is either above the third quartile with a distance greater than 1.5 times of the IQR or below the first quartile with a distance greater than 1.5 times of IQR. The individual outliers are shown by the open circles in the box plot in Figure 2.2.

The `R` functions `hist()` and `boxplot()` can be used to plot the histogram and box plot, respectively. The following `R` codes plot, as shown in Figure 2.3, the histograms and box plots for three numerical variables, the `length`, `horsepower`, and `compression.ratio`, in the `auto_spec` data set.

```
oldpar <- par(mfrow=c(2,3)) # split the plot into panels
hist(auto.spec.df$length, xlab = "Length",
     main = "Histogram of Length")
hist(auto.spec.df$horsepower, xlab = "Horsepower",
     main = "Histogram of Horsepower")
hist(auto.spec.df$compression.ratio,
   xlab = "Compression Ratio",
     main = "Histogram of Compression Ratio")
boxplot(auto.spec.df$length, ylab = "Length",
     main = "Boxplot of Length")
boxplot(auto.spec.df$horsepower, ylab = "Horsepower",
     main = "Boxplot of Horsepower")
boxplot(auto.spec.df$compression.ratio,
   ylab = "Compression Ratio",
     main = "Boxplot of Compression Ratio")
par(oldpar)
```

From the histogram and box plot of the variable `length`, it can be seen that the distribution of the car lengths in the data set has a fairly symmetric shape. In contrast, the distribution of horsepower is more skewed with a long (right) tail. The histogram of the compression ratios shows the existence of two groups or clusters of data, which is also indicated by the separate cluster of outliers with high compression ratios that can be seen in the box plot.

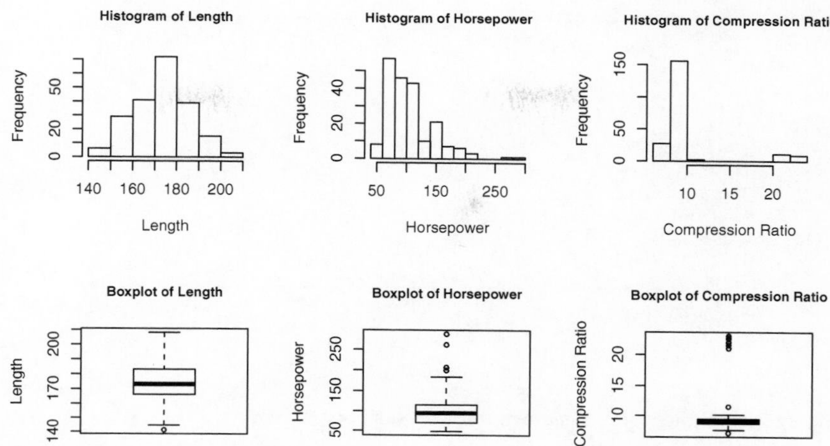

Figure 2.3 Histograms and box plots of three numerical variables.

2.1.2 Plots for Relationship Between Two Variables

The relationship between variables is one of the most useful patterns in industrial data analytics applications. For example, we are often interested in predicting a particular variable of interest, which is referred to as the *response variable*, based on available input information represented by a number of variables that are referred to as the *predictor variables*. In this situation, the relationship between the response variable and the predictor variables can help identify the most important predictors. Plotting of two variables can also be used to detect redundant variables and outliers in a data set. Depending on the types of variables being compared, different plots can be used to study the relationship between the variables.

Relationship Between Two Numerical Variables – Scatter Plot

In a scatter plot, each observation is represented by a point whose coordinates are the values for the two variables of this observation. The following `R` codes draw the scatter plot for two numerical variables, `horsepower` and `highway.mpg`, of the `auto_spec` data set.

```
plot(auto.spec.df$highway.mpg ~ auto.spec.df$horsepower,
     xlab = "Horsepower", ylab = "Highway MPG")
```

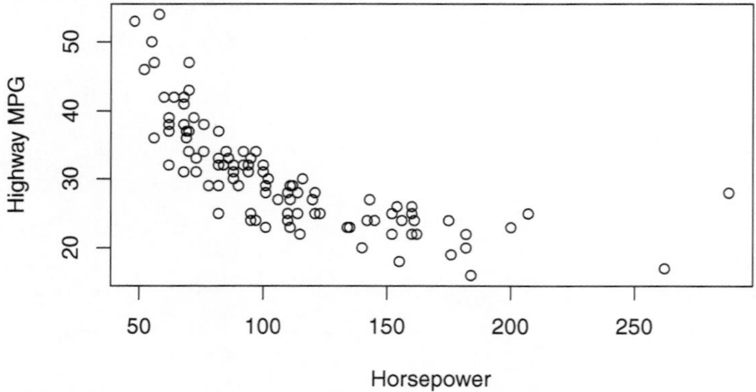

Figure 2.4 Scatter plot of highway MPG versus horsepower.

The obtained scatter plot is shown in Figure 2.4. It can be seen from the scatter plot that a general trend exists in the relationship between the highway MPG and the horsepower, where a car with higher horsepower is more likely to have a lower highway MPG.

Relationship Between A Numerical Variable and A Categorical Variable – Side-by-Side Box Plot

Side-by-side box plots can be used to show how the distribution of a numerical variable changes over different values of a categorical variable. The idea is to use a box plot to represent the distribution of the numerical variable at each value of the categorical variable. In Figure 2.5, we draw two side-by-side box plots for the `auto_spec` data set using the following `R` codes:

```
oldpar <- par(mfrow = c(1, 2))
boxplot(auto.spec.df$compression.ratio ~
        auto.spec.df$ fuel.type,
      xlab = "Fuel Type", ylab = "Compression Ratio")
boxplot(auto.spec.df$highway.mpg ~
        auto.spec.df$body.style,
      las = 2, xlab = "", ylab = "Highway MPG")
mtext("Body Style", side = 3, line = 1)
par(oldpar)
```

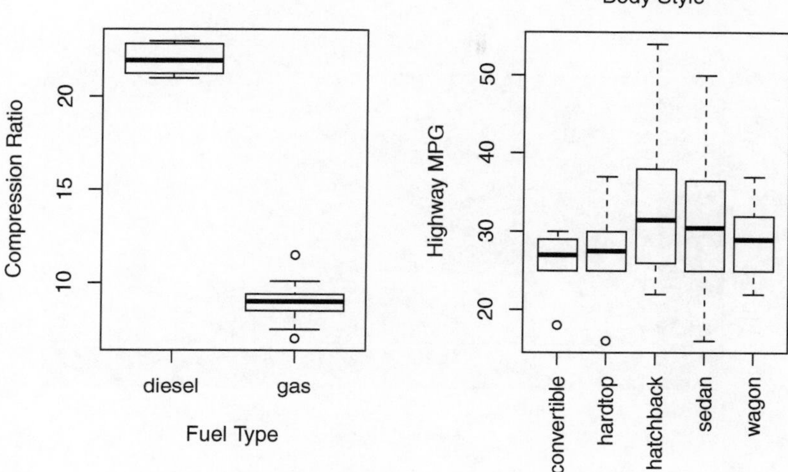

Figure 2.5 Side-by-side box plots.

The left panel of Figure 2.5 shows how the numerical variable `compression.ratio` is related to the two values (`diesel` and `gas`) of `fuel.type`. It is clear from the side-by-side box plot that a car with diesel fuel has a much higher compression ratio than a car with gas fuel. This also explains the separate cluster of outliers in the histogram and box plot of `compression.ratio` that is observed in Figure 2.3. The right panel of Figure 2.5 shows how `highway.mpg` is related to the five values of `body.style`. It can be seen that a hatchback car is more likely to have higher highway MPG while a convertible tends to have lower highway MPG.

Relationship Between Two Categorical Variables – Mosaic Plot

We can use a mosaic plot to see how values of two categorical variables are related to each other. Figure 2.6 shows a mosaic plot for `fuel.type` and `aspiration` of the `auto_spec` data set, which is drawn by the following `R` codes.

```
mosaicplot(fuel.type ~ aspiration, data = auto.spec.df,
        xlab = "Fuel Type", ylab = "Aspiration",
        color = c("green", "blue"),
        main = "Mosaic Plot")
```

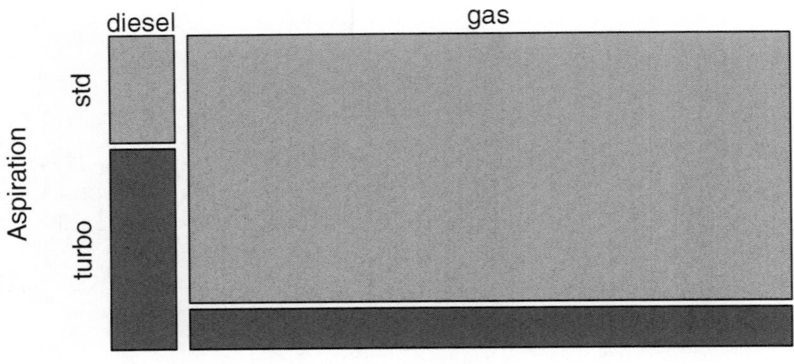

Figure 2.6 Mosaic plot for fuel type and aspiration.

In a mosaic plot, the height of a bar represents the *percentage* for each value of the variable in the vertical axis given a fixed value of the variable in the horizontal axis. For example, in Figure 2.6 the height of the bar corresponding to turbo aspiration is much higher when the fuel type is diesel than when it is gas, which means a higher percentage of diesel cars use turbo aspiration, while a lower percentage of gasoline cars use turbo aspiration. The width of a bar in a mosaic plot corresponds to the *frequency*, or the number of observations, for each value of the variable in the horizontal axis. For example, from Figure 2.6, the bars for gas fuel type is much wider than those for diesel fuel type, indicating that a much larger number of cars are gasoline cars in the data set.

2.1.3 Plots for More than Two Variables

It is very difficult to plot more than two variables in a two dimensional plot. This section introduces commonly used plots that show some aspects of how multiple variables are related to each other. In Chapter 4, we will study another technique called principal component analysis, which can also serve as a useful tool to visualize high dimensional data in a low dimensional space.

Color Coded Scatter Plot

We have seen that a scatter plot can effectively show the relationship between two numerical variables. By adding color coding to the points on a scatter plot of two numerical variables, we are able to study their relationship with a third variable. Typically, the third variable is a categorical variable, with each category represented by a different color. The color coded scatter plot is very useful in visualizing how some numerical variables can be used to predict a categorical variable. For the `auto_spec` data, we can use a color coded scatter plot to show how `fuel.type` is related to two of the numerical variables `horsepower` and `peak.rpm`. The color coded scatter plot is shown in Figure 2.7, which is created by the following `R` codes.

```
oldpar <- par(xpd = TRUE)
plot(auto.spec.df$peak.rpm ~ auto.spec.df$horsepower,
     xlab = "Horsepower", ylab = "Peak RPM",
     col = ifelse(auto.spec.df$fuel.type == "gas",
       "black", "gray"))
legend("topleft", inset = c(0, -0.2),
     legend = c("gas", "diesel"),
      col = c("black", "gray"), pch = 1, cex = 0.8)
par(oldpar)
```

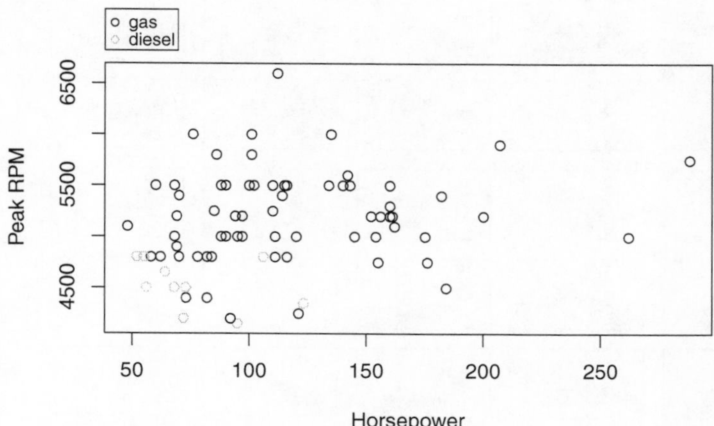

Figure 2.7 Scatter plot color coded by fuel type.

Although there is no clear relationship between the peak RPM and horsepower of a car from the scatter plot in Figure 2.7, it is obvious from the color coded plot that diesel cars tend to have low peak RPM and low horsepower.

Scatter Plot Matrix and Heatmap

The pairwise relationship of multiple numerical variables can be visualized simultaneously by using a matrix of scatter plots. The following R codes plot the scatter plot matrix for five of the numerical variables in the `auto_spec` data set: `wheel.base`, `height`, `curb.weight`, `city.mpg`, and `highway.mpg`. The column indices of the five variables are 8, 11, 12, 22, and 23, respectively.

```
var.idx <- c(8, 11, 12, 22, 23)
plot(auto.spec.df[, var.idx])
```

From the scatter plot matrix shown in Figure 2.8, there are different types of relationship among the variables. For example, there is a strong linear relationship between `city.mpg` and `highway.mpg`. Besides these two variables, `wheel.base`, `height`, and `curb.weight` are positively related to each other. And the `curb.weight` is negatively related to both `city.mpg` and `highway.mpg`.

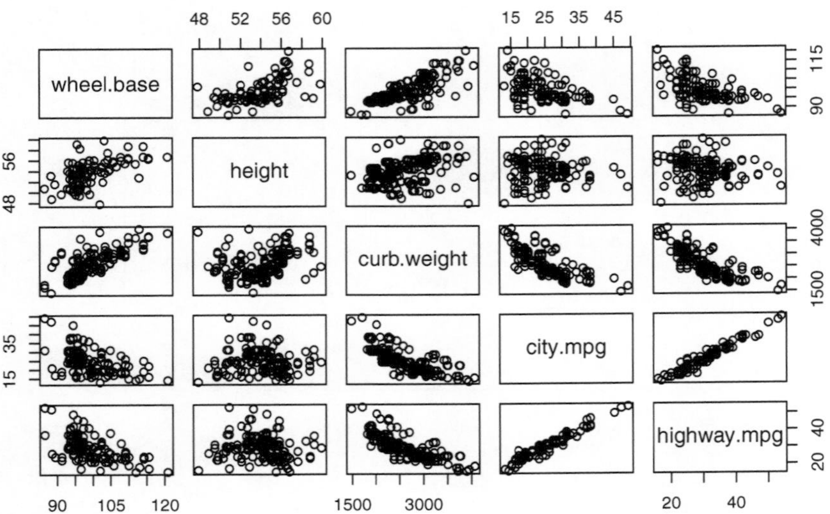

Figure 2.8 Scatter plot matrix for five numerical variables.

For a large number of numerical variables, it is difficult to visualize all pairwise scatter plots as in the scatter plot matrix. In this case, we can use a heatmap for pairwise correlations of the variables to quickly show the strength of the relationship. The heatmap uses different shades of colors to represent the values of the correlations so that the spots or regions of strong positive or negative relationship can be quickly detected. Detailed discussion of correlation is provided in Section 2.2. We draw the heatmap of correlations for all numerical variables in the `auto_spec` data set using the following R codes.

```
library(gplots)
var.idx <-c(8:12, 15, 17:23)
data.nomiss <- na.omit(auto.spec.df[, var.idx])
heatmap.2(cor(data.nomiss), Rowv = FALSE, Colv = FALSE,
          dendrogram = "none",
          cellnote = round(cor(data.nomiss),2),
          notecol = "black", key = FALSE,
          trace = 'none', margins=c(10,10))
```

In the above R codes, we use the `heatmap.2()` function from the `gplots` package to draw the heatmap. We first remove the observations with missing values using the `na.omit()` function. Then the heatmap is drawn for the pairwise correlations calculated by `cor()`. In the heatmap of all numerical variables, as shown in Figure 2.9, a lighter color indicates a strong positive (linear)

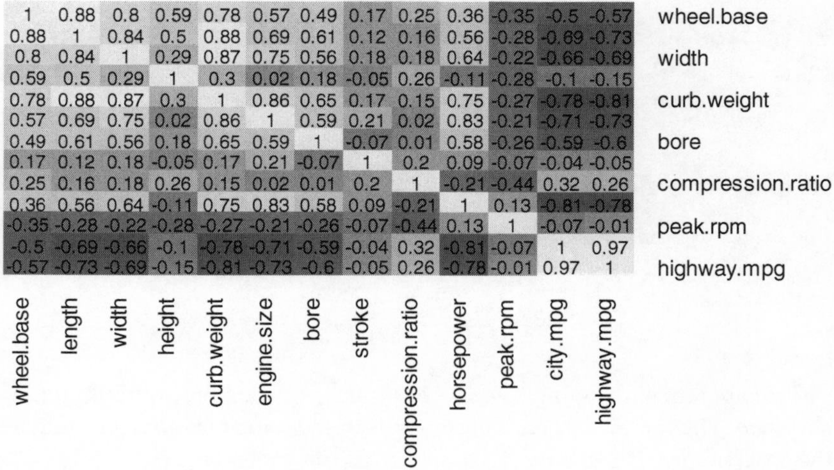

Figure 2.9 Heatmap of correlation for all numerical variables.

relationship between the variables and a darker color indicates a strong negative (linear) relationship. The correlation values are shown within each cell of the heatmap table. The diagonal cells have the lightest color because any variable has the strongest relationship to itself. From the heatmap in Figure 2.9, we can also see that the two MPG variables (`city.mpg` and `highway.mpg`) have strong negative relationships with many of the other numerical variables in the data set.

2.2 Summary Statistics

Data visualization is an effective and intuitive representation of the qualitative features of the data. Key characteristics of data can also be quantitatively summarized by numerical statistics. This section introduces common summary statistics for univariate and multivariate data.

2.2.1 Sample Mean, Variance, and Covariance

Sample Mean – Measure of Location

A *sample mean* or *sample average* provides a measure of location, or central tendency, of a variable in a data set. Consider a univariate data set, which is a data set with a single variable, that consists of a random sample of n observations $x_1, x_2, \dots, x_n$. The sample mean is simply the ordinary arithmetic average

$$\overline{x} = \frac{1}{n}\sum_{i=1}^{n} x_i.$$

For a data set $y_i, i = 1, 2, \dots, n$ obtained by multiplying each x_i by a constant a, i.e., $y_i = ax_i, i = 1, 2, \dots, n$, it is easy to see that

$$\overline{y} = a\overline{x}.$$

Sample Variance – Measure of Spread

The *sample variance* measures the spread of the data and is defined as

$$s^2 = \frac{\sum_{i=1}^{n}(x_i - \overline{x})^2}{n-1} = \frac{\sum_{i=1}^{n} x_i^2 - n\overline{x}^2}{n-1}. \tag{2.1}$$

The square root of the sample variance, $s = \sqrt{s^2}$, is called the *sample standard deviation*. The sample standard deviation is of the same measurement unit as the observations. For $y_i = ax_i, i = 1, 2, \dots, n$, its sample variance is

$$s_y^2 = a^2 s^2.$$

Sample Covariance and Correlation – Measure of Linear Association Between Two Variables

If each of the n observations of a data set is measured on two variables x_1 and x_2, let $(x_{11}, x_{21}, \dots, x_{n1})$ and $(x_{12}, x_{22}, \dots, x_{n2})$ denote the n observations on x_1 and x_2, respectively. The *sample covariance* of x_1 and x_2 is defined as

$$s_{12} = \frac{\sum_{i=1}^{n}(x_{i1} - \bar{x}_1)(x_{i2} - \bar{x}_2)}{n-1} = \frac{\sum_{i=1}^{n} x_{i1}x_{i2} - n\bar{x}_1\bar{x}_2}{n-1}, \tag{2.2}$$

where $\bar{x}_1$ and $\bar{x}_2$ are the sample means of x_1 and x_2, respectively. The value of sample covariance of two variables is affected by the linear association between them. From (2.2), if x_1 and x_2 have a strong positive linear association, they are usually both above their means or both below their means. Consequently, the product $(x_{i1} - \bar{x}_1)(x_{i2} - \bar{x}_2)$ will typically be positive and their sample covariance will have a large positive value. On the other hand, if x_1 and x_2 have a strong negative linear association, the product $(x_{i1} - \bar{x}_1)(x_{i2} - \bar{x}_2)$ will typically be negative and their sample covariance will have a negative value. If y_1 and y_2 are obtained by multiplying each measurement of x_1 and x_2 with a_1 and a_2, respectively, it is easy to see from (2.2) that the sample covariance of y_1 and y_2 is

$$s_{12}^{y} = a_1 a_2 s_{12}. \tag{2.3}$$

Equation (2.3) says that if the measurements are scaled, for example by changing measurement units, the sample covariance will be scaled correspondingly. The sample covariance's dependence on the measurement units makes it difficult to determine how large a sample covariance indicates a strong (linear) association between two variables. The *sample correlation* defined as follows is a measure of linear association that does not depend on the measurement units, or scaling of the variables

$$r_{12} = \frac{s_{12}}{s_1 s_2} = \frac{\sum_{i=1}^{n}(x_{i1} - \bar{x}_1)(x_{i2} - \bar{x}_2)}{\sqrt{\sum_{i=1}^{n}(x_{i1} - \bar{x}_1)^2 \sum_{i=1}^{n}(x_{i2} - \bar{x}_2)^2}}, \tag{2.4}$$

where s_1 and s_2 are the sample standard deviation of x_1 and x_2, respectively. The sample correlation ranges between -1 and 1, with values close to 1, -1, and 0 indicating a strong positive linear association, a strong negative linear association, and no linear association, respectively.

Example 2.2 To illustrate the calculation of summary statistics, we take a random sample of 10 observations, as shown in Table 2.1, from the `auto.`

`spec` data set on the variables `curb.weight`, `length`, and `width`. We use x_i, $i = 1,2,3$, to represent the three variables:

$$x_1 = \text{curb.weight}$$
$$x_2 = \text{length}$$
$$x_3 = \text{width}$$

Table 2.1 A random sample of 10 observations from the auto. spec data set.

x_1	x_2	x_3
3515	190.9	70.3
2300	168.7	64.0
2800	168.9	65.0
2122	166.3	64.4
2293	169.1	66.0
2765	176.8	64.8
2275	171.7	65.5
1890	159.1	64.2
2926	173.2	66.3
1909	158.8	63.6

To obtain the sample covariance for the variables `curb.weight` and `length` in the data set in Table 2.1, we first calculate the sample means $\bar{x}_1$, $\bar{x}_2$, and $\sum_{i=1}^{n} x_{i1}x_{i2}$ as:

$$\bar{x}_1 = \frac{3515 + 2300 \ + \cdots + 1909}{10} = 2479.5,$$
$$\bar{x}_2 = \frac{190.9 + 168.7 + \cdots + \ 158.8}{10} = 170.35,$$

$$\sum_{i=1}^{n} x_{i1}x_{i2} = (3515)(190.9) + (2300)(168.7) + \cdots + (1909)(158.8) = 4\,262\,679.$$

By (2.2), the sample covariance of the two variables can be obtained as

$$\begin{aligned} s_{12} &= \frac{\sum_{i=1}^{n} x_{i1}x_{i2} - n\bar{x}_1\bar{x}_2}{n-1} \\ &= \frac{4\,262\,679 - (10)(2479.5)(170.35)}{9} = 4316.8. \end{aligned}$$

The s_{12} value of 4316.8 itself cannot tell us whether the two variables have a strong or weak (linear) relationship. Such information can be provided by the correlation. To evaluate the sample correlation, we first need the sample variance of x_1 and x_2. By (2.1), we have

$$s_1^2 = \frac{\sum_{i=1}^{n} x_{i1}^2 - n\overline{x}_1^2}{n-1} = \frac{63\,844\,665 - (10)(2479.5)^2}{9} = 262\,829.2,$$

$$s_2^2 = \frac{\sum_{i=1}^{n} x_{i2}^2 - n\overline{x}_2^2}{n-1} = \frac{290\,947.8 - (10)(170.35)^2}{9} = 84.07.$$

By (2.4), we have

$$r_{12} = \frac{s_{12}}{s_1 s_2} = \frac{4316.8}{\sqrt{262\,829.2}\sqrt{84.07}} = 0.918,$$

which is close to 1 and corresponding to a strong positive linear association between the curb weight and length of cars.

Example 2.3 In R, the sample mean, variance, covariance, and correlation can be found using functions `mean()`, `var()`, `cov()`, and `cor()`, respectively. For example, the following R codes can be used to find the sample mean and sample variance of `curb.weight`, and the sample covariance and correlation between `curb.weight` and `length`, in the `auto.spec` data set.

```
mean(auto.spec.df$curb.weight)
var(auto.spec.df$curb.weight)
with(auto.spec.df, cov(curb.weight, length))
with(auto.spec.df, cor(curb.weight, length))
```

```
> mean(auto.spec.df$curb.weight)
[1] 2555.566
> var(auto.spec.df$curb.weight)
[1] 271107.9
> with(auto.spec.df, cov(curb.weight, length))
[1] 5638.336
> with(auto.spec.df, cor(curb.weight, length))
[1] 0.8777285
```

Note the results above are somewhat different from those in Example 2.2 because in this example we use the entire data set of `auto.spec`, instead of a small random subset of it as in Example 2.2.

2.2.2 Sample Mean Vector and Sample Covariance Matrix

A multivariate data set consists of n observations collected from n items or units and each observation contains measurements on p variables, $x_1, x_2, \ldots, x_p$. The measurement vector for the ith observation is denoted by

$$\mathbf{x}_i = \begin{pmatrix} x_{i1} \\ x_{i2} \\ \vdots \\ x_{ip} \end{pmatrix}.$$

The *sample mean vector* is the vector of sample means for the p variables, which is defined as

$$\overline{\mathbf{x}} = \begin{pmatrix} \overline{x}_1 \\ \overline{x}_2 \\ \vdots \\ \overline{x}_p \end{pmatrix} = \frac{1}{n} \sum_{i=1}^{n} \mathbf{x}_i,$$

where $\overline{x}_k$ is the sample mean of x_k, i.e., $\overline{x}_i = \frac{1}{n} \sum_{i=1}^{n} x_{ik}$, $k = 1, \ldots, p$.

The *sample covariance matrix* **S** is the matrix of sample variances and covariances of the p variables:

$$\mathbf{S} = \begin{pmatrix} s_{11} & s_{12} & \cdots & s_{1p} \\ s_{21} & s_{22} & \cdots & s_{2p} \\ \vdots & \vdots & & \vdots \\ s_{p1} & s_{p2} & \cdots & s_{pp} \end{pmatrix}.$$

The off-diagonal elements of **S** is the sample covariances of each pair of variables. For $j \neq k$,

$$s_{jk} = \frac{\sum_{i=1}^{n} (x_{ij} - \overline{x}_j)(x_{ik} - \overline{x}_k)}{n-1}. \tag{2.5}$$

The diagonal elements of **S**, s_{jj}, $j = 1, \ldots, p$ are the sample variance of the jth variable. It is easy to see that when $k = j$, the sample covariance in (2.5) is equal to s_j^2, the sample variance of the jth variable. So both notations s_{jj} and s_j^2

represent the sample variance of x_j. It is also obvious from (2.5) that $s_{jk} = s_{kj}$. So the sample covariance matrix $\mathbf{S}$ is a symmetric matrix. The sample covariance matrix $\mathbf{S}$ can also be written by the observation vector $\mathbf{x}_i$ as

$$\mathbf{S} = \frac{1}{n-1} \sum_{i=1}^{n} (\mathbf{x}_i - \overline{\mathbf{x}})(\mathbf{x}_i - \overline{\mathbf{x}})^T. \tag{2.6}$$

Similarly, we define the *sample correlation matrix* as

$$\mathbf{R} = \begin{pmatrix} 1 & r_{12} & \dots & r_{1p} \\ r_{21} & 1 & \dots & r_{2p} \\ \vdots & \vdots & & \vdots \\ r_{n1} & r_{n2} & \dots & 1 \end{pmatrix}.$$

The (j, k)th element of $\mathbf{R}$ is the sample correlation of the jth and kth variables:

$$r_{jk} = \frac{s_{jk}}{s_j s_k}.$$

The sample correlation between a variable and itself is equal to 1. So the diagonal elements of a sample correlation matrix are all equal to 1. The sample correlation matrix $\mathbf{R}$ is obviously symmetric since $r_{jk} = r_{kj}$.

Example 2.4 Consider the data set in Table 2.1. In Example 2.2, we found that $\overline{x}_1 = 2479.5$ and $\overline{x}_2 = 170.35$. Similarly, we can obtain $\overline{x}_3 = 65.41$. So the mean vector of $\mathbf{x} = (x_1 \ x_2 \ x_3)^T$ is given by

$$\overline{\mathbf{x}} = (2479.5 \ 170.35 \ 65.41)^T.$$

In Example 2.2, we calculated the sample variances, sample covariance, and sample correlation of x_1 and x_2. Similarly, we can obtain the sample variance of x_3 and its sample covariance and correlation with the other two variables as

$$s_3^2 = 3.71, \ s_{13} = 820.8, \ s_{23} = 15.56, \ r_{13} = 0.832, \ r_{23} = 0.881.$$

Note that while s_{23} is much smaller than s_{13}, r_{23} is greater than r_{13}, which indicates that the linear association between x_2 and x_3 is stronger than that of x_1 and x_3. This clearly shows that the magnitude of the covariance itself is not meaningful in characterizing how strong the relationship of two variables is. Combining all the sample variance, covariance, and correlation information, the sample covariance matrix and sample correlation matrix of $\mathbf{x} = (x_1 \ x_2 \ x_3)^T$ can be written as

$$\mathbf{S} = \begin{pmatrix} 262829.2 & 4316.8 & 820.8 \\ 4316.8 & 84.07 & 15.56 \\ 820.8 & 15.56 & 3.71 \end{pmatrix}, \quad \mathbf{R} = \begin{pmatrix} 1 & 0.918 & 0.832 \\ 0.918 & 1 & 0.881 \\ 0.832 & 0.881 & 1 \end{pmatrix}.$$

2.2.3 Linear Combination of Variables

We are often interested in some linear combinations of the variables $x_1, x_2, \ldots, x_p$. For example, for the `auto_spec` data set, two of the variables are `city.mpg` and `highway.mpg`. If you expect that 60% of the mileage for a car is on highway and 40% is on local roads, then the average MPG for a car can be estimated as $0.6 \times \text{highway.mpg} + 0.4 \times \text{city.mpg}$, which is a linear combination of `city.mpg` and `highway.mpg`. In general, let $c_1, c_2, \ldots, c_p$ be constants and consider the linear combination of the variables $x_1, x_2, \ldots, x_p$ given by

$$z = c_1 x_1 + c_2 x_2 + \cdots + c_p x_p.$$

For each observation of the data set, the corresponding value of the variable z can be found by

$$z_i = c_1 x_{i1} + c_2 x_{i2} + \cdots + c_p x_{ip} = \mathbf{c}^T \mathbf{x}_i, \ i = 1, \ldots, n,$$

where $\mathbf{c}^T = (c_1 \ c_2 \ \ldots \ c_p)$. It can be seen that the sample mean of z is

$$\begin{aligned} \bar{z} &= \frac{1}{n} \sum_{i=1}^{n} z_i \\ &= c_1 \left(\frac{1}{n} \sum_{i=1}^{n} x_{i1} \right) + \cdots + c_p \left(\frac{1}{n} \sum_{i=1}^{n} x_{ip} \right) \\ &= c_1 \bar{x}_1 + \cdots + c_p \bar{x}_p = \mathbf{c}^T \bar{\mathbf{x}}. \end{aligned} \tag{2.7}$$

The sample variance of z can be found as

$$s_z^2 = \frac{\sum_{i=1}^{n} (z_i - \bar{z})^2}{n-1} = \mathbf{c}^T \mathbf{S} \mathbf{c}. \tag{2.8}$$

Because sample variance is always non-negative, for any $\mathbf{c} \in \mathcal{R}^p$ we have $\mathbf{c}^T \mathbf{S} \mathbf{c} \geq 0$ from (2.8). Therefore, the sample covariance matrix $\mathbf{S}$ is always a positive semidefinite matrix.

In general, if we have q linear combinations of $x_1, x_2, \ldots, x_p$ defined by:

$$\begin{aligned} z_1 &= c_{11} x_1 + c_{12} x_2 + \cdots + c_{1p} x_p \\ z_2 &= c_{21} x_1 + c_{22} x_2 + \cdots + c_{2p} x_p \\ &\vdots \\ z_q &= c_{q1} x_1 + c_{q2} x_2 + \cdots + c_{qp} x_p \end{aligned}$$

or in matrix notation,

$$\mathbf{z} = \begin{pmatrix} z_1 \\ z_2 \\ \vdots \\ z_q \end{pmatrix} = \begin{pmatrix} c_{11} & c_{12} & \dots & c_{1p} \\ c_{21} & c_{22} & \dots & c_{2p} \\ \vdots & \vdots & & \vdots \\ c_{q1} & c_{q2} & \dots & c_{qp} \end{pmatrix} \begin{pmatrix} x_1 \\ x_2 \\ \vdots \\ x_p \end{pmatrix} = \mathbf{Cx}.$$

The sample mean vector and sample covariance matrix of

$$\mathbf{z}_i = \mathbf{Cx}_i, \quad i = 1,\ 2, \dots, n$$

are given by

$$\bar{\mathbf{z}} = \mathbf{C}\bar{\mathbf{x}} \tag{2.9}$$

$$\mathbf{S_z} = \mathbf{CSC}^T. \tag{2.10}$$

Obviously, (2.9) and (2.10) are generalizations of (2.7) and (2.8), respectively.

Example 2.5 For the `auto.spec` data set, using the `mean()` function of `R` the sample means of the variables `city.mpg` and `highway.mpg` can be found as 25.22 and 30.75, respectively. If we are interested in the overall MPG of a car, denoted by z, as the following weighted average of $x_1 =$ `city.mpg` and $x_2 =$ `highway.mpg`:

$$z = 0.4x_1 + 0.6x_2 = \mathbf{c}^T \begin{pmatrix} x_1 \\ x_2 \end{pmatrix},$$

where $\mathbf{c} = (0.4\ 0.6)^T$. Then by (2.7) the sample mean of the overall MPG in the data set is

$$\bar{z} = \mathbf{c}^T \bar{\mathbf{x}} = (0.4\ 0.6) \begin{pmatrix} 25.22 \\ 30.75 \end{pmatrix} = 28.54.$$

To find the sample variance of z, first we obtain the sample covariance matrix for `city.mpg` and `highway.mpg` using the `cov()` function of `R`:

```
cov(auto.spec.df[, c("city.mpg", "highway.mpg")])
cor(auto.spec.df[, c("city.mpg", "highway.mpg")])
```

The function `cor()` calculates the sample correlation matrix. Based on the output from the above `R` codes, we have

$$\mathbf{S} = \begin{pmatrix} 42.8 & 43.76 \\ 43.76 & 47.42 \end{pmatrix}, \quad \mathbf{R} = \begin{pmatrix} 1 & 0.971 \\ 0.971 & 1 \end{pmatrix}.$$

By (2.8), the sample variance of z is

$$s_z^2 = \mathbf{c}^T\mathbf{S}\mathbf{c} = (0.4\ 0.6)\begin{pmatrix} 42.8 & 43.76 \\ 43.76 & 47.42 \end{pmatrix}(0.4\ 0.6)^T = 44.9.$$

Bibliographic Notes

Data visualization methods are discussed in books in the data mining area, for example, Shmueli et al. [2017] and Williams [2011]. In this chapter, we mostly use the graphics functions from base `R`. A popular dedicated graphics package in `R` is the `ggplot2` package by Wickham [2016]. The `ggplot2` package provides more flexible and powerful graphics capability that can create presentation-quality visualization. However, it also comes with a significant learning curve to get familiar with the special technical language used in `ggplot2`. For those who use data visualizations on a regular basis, it is worth the time and effort to learn `ggplot2`.

Sample statistics such as sample mean vector and sample covariance matrix for multivariate observations are discussed in detail in many multivariate statistics books, for example, Johnson et al. [2002] and Rencher [2003].

Exercises

1. Consider the data in the following table with two numerical variables x_1 and x_2 and two categorical variables x_3 and x_4.

x_1	x_2	x_3	x_4
9	1	Yes	On
5	3	No	Off
1	2	Yes	Off
3	4	Yes	On
6	−1	No	On
3	3	Yes	On

(a) Manually sketch the scatter plot for x_1 and x_2.
(b) Manually sketch the mosaic plot for x_3 and x_4.

2. Consider the data set in Exercise 1. Manually calculate the sample mean vector, the sample covariance matrix, and the sample correlation matrix of $\mathbf{x} = (x_1\ x_2)^T$.
3. Consider the data in the following table with two numerical variables x_1 and x_2 and two categorical variables x_3 and x_4.

x_1	x_2	x_3	x_4
1	0	Yes	Working
4	6	No	Fail
2	2	Yes	Fail
0	3	No	Fail
3	4	No	Working
5	7	Yes	Working

   (a) Manually sketch the scatter plot for x_1 and x_2.
   (b) Manually sketch the mosaic plot for x_3 and x_4.
4. Consider the data set in Exercise 3. Manually calculate the sample mean vector, the sample covariance matrix, and the sample correlation matrix of $\mathbf{x} = (x_1\ x_2)^T$.
5. Consider the `auto_spec` data set in the file `auto_spec.csv`. Use `R` to draw appropriate plots to display the following information and comment on any patterns that can be found from the plots.
   (a) Distribution of the variables `fuel.type` and `aspiration`.
   (b) Distribution of each of the following three variables: `width`, `height`, and `highway.mpg`. Use two types of plots for each variable.
   (c) How does the `horsepower` affect the `city.mpg`?
   (d) The relationship between `horsepower` and `body.style`.
   (e) The relationship between `body.style` and `fuel.type`.
6. For the `auto_spec` data, use `R` to create a new variable named `cat.mpg`, which is equal to "high" if `highway.mpg` is at least 30, and "low" otherwise.
   (a) Using `R`, create a scatter plot of `horsepower` versus `curb.weight`, color-coded by the variable `cat.mpg`. Format the plot with appropriate labels and legend.
   (b) Use `R` to find the sample mean vector, the sample covariance matrix, and the sample correlation matrix of `highway.mpg` and `city.mpg`.
   (c) Assume that 75% of the mileage of a car is on a highway and 25% is on local roads, using the results from part (b), manually calculate the sample mean and sample variance of the overall average MPG of the cars in this data set.

(d) Use `R` to calculate the overall average MPG of each car in the data set based on the assumption in part (c). Then use `R` to find the sample mean and sample variance of the overall average MPG. Compare with the results in part (c).

7. Hot rolling is among the key steel-making processes that convert cast or semi-finished steel into finished products. A typical hot rolling process usually includes a melting division and a rolling division. The melting division is a continuous casting process that melts scrapped metals and solidifies the molten steel into semi-finished steel billet; the rolling division will further squeeze the steel billet by a sequence of stands. Each stand is composed of several rolls. The final long thin steel billet is coiled for transportation convenience and thus is often called a coil. Due to the recent development of computer and sensor technology, the whole hot rolling process is highly automated and monitored by a large number of sensors. Various types of sensors (optical sensor, temperature sensor, force sensor, etc.) are installed in the hot rolling process. The last rolling stands are equipped with some infrared sensors. These sensors take photos of the steel billets, and then the photos are processed to see if any defects are produced. We focus on two types of defect: *checkings* and *seams*.

(a) The file `hotrolling_defects.csv` contains the numbers of checkings and seams of 754 billets. Use `R` to generate two new variables corresponding to whether a billet has at least one checking defect and whether it has at least one seams defect, respectively. Use appropriate plots to visualize the distribution of each of these two new variables and the relationship between them.

(b) The file `stand_5_side_temp.csv` contains side temperature measurements when a steel billet is passing stand 5 of the rolling division. The side temperature is measured at 79 evenly spaced locations along the stand. Use `R` to draw a scatter plot matrix for the side temperature measurements at the first five locations of stand 5. Comment on noticeable patterns in relationship among the first five temperature variables.

(c) Use `R` to find the sample mean vector, the sample covariance matrix, and sample correlation matrix of the side temperature measurements at the first five locations of stand 5.

(d) Use `R` to draw a heatmap for the correlation of the side temperature measurements at the first 20 locations of stand 5. Which locations have the highest correlation in side temperature measurements?

3

Random Vectors and the Multivariate Normal Distribution

Informally, a random variable can be described as a variable whose value depends on the outcome of a random or chance phenomenon. Some examples of random variables are: the highway MPG of a new car randomly sampled from all cars on sale, the quality measurement of a product randomly sampled from a production line, the temperature measurement at a particular moment and location of a machine where temperature randomly varies over time. Due to the ubiquitous uncertainty and variation existing in industrial systems and processes, most variables of interest in industrial data analytics applications can be considered as random variables. Many industrial data analytics problems involve multiple random variables, which form a vector of random variables, also called as a random vector. In this chapter, we study the concept of random vectors and the multivariate normal distribution, the most commonly used model for a random vector.

3.1 Random Vectors

A *random vector* is a vector of random variables. Let $\mathbf{X} = (X_1 \, X_2 \ldots X_p)^T$ denote a random vector. The mean or expected value of a random vector is the vector of the mean values of each of its elements. The *mean vector* of $\mathbf{X}$ can be written as

$$\mu = E(\mathbf{X}) = \begin{pmatrix} E(X_1) \\ E(X_2) \\ \vdots \\ E(X_p) \end{pmatrix},$$

Industrial Data Analytics for Diagnosis and Prognosis: A Random Effects Modelling Approach, First Edition. Shiyu Zhou and Yong Chen.

where the univariate mean is defined as

$$\mu_i = E(X_i) = \begin{cases} \int_{-\infty}^{\infty} x_i f_i(x_i) dx_i & \text{if } X_i \text{ is a continuous random variable} \\ \sum_{x_i} x_i p_i(x_i) & \text{if } X_i \text{ is a discrete random variable} \end{cases},$$

where $f_i(x_i)$ is the probability density function of X_i if X_i is continuous and $p_i(x_i)$ is the probability mass function of X_i if X_i is discrete. The μ_i is also called the *population mean* of X_i because it is the mean of X_i over all possible values in the population. Similarly, the mean vector $\boldsymbol{\mu}$ is the population mean vector of $\mathbf{X}$.

To further explain the relationship and difference between the population mean and the sample mean introduced in Section 2.2, we first consider a univariate random variable X and its population mean μ. Consider a random sample of observations from the population, say, $X_1, X_2, \ldots, X_n$. The sample mean $\overline{X} = \frac{1}{n}\sum_{i=1}^{n} X_i$ is a random variable because the observations X_1, X_2, ..., X_n are all random variables with values varying from sample to sample. For example, let X represent the measured intensity of the current of a wafer produced by a semiconductor manufacturing process. Then we take a random sample of $n = 10$ wafers from this process and compute the sample mean of the measured intensities of the current and get the result $\overline{x} = 1.02$. Now we repeat this process, taking a second sample of $n = 10$ wafers from the same process and the resulting sample mean is 1.04. The sample means differ from sample to sample because they are random variables. Consequently, the sample mean, and any other function of the random observations, is a random variable. On the other hand, the population mean μ does not depend on the samples and is a (usually unknown) constant. When we take a sample with very large sample size n, the sample mean will be very close to the population mean μ with high probability. As the sample mean $\overline{X}$ is a random variable, we can evaluate its mean and variance. It is easy to see that $E(\overline{X}) = \mu$ and $\text{var}(\overline{X}) = \sigma^2 / n$, where σ^2 is the variance of X. An estimator of a parameter is called *unbiased* if its mean is equal to the true value of the parameter. $\overline{X}$ is a commonly used estimator of μ because it is unbiased and has a smaller variance for a larger sample size n.

This concept can be extended to a p-dimensional random vector $\mathbf{X}$ with mean vector $\boldsymbol{\mu}$. Consider a random sample $\mathbf{X}_1, \mathbf{X}_2, \ldots, \mathbf{X}_n$ from the population of $\mathbf{X}$. The sample mean vector $\overline{\mathbf{X}}$ is a random vector with population mean $E(\overline{\mathbf{X}}) = \boldsymbol{\mu}$ and population covariance matrix $\text{cov}(\overline{\mathbf{X}}) = \frac{1}{n}\boldsymbol{\Sigma}$, where $\boldsymbol{\Sigma}$ is the population covariance matrix of $\mathbf{X}$. The population covariance matrix is defined shortly. The sample mean vector $\overline{\mathbf{X}}$ is an unbiased estimator of the population mean vector $\boldsymbol{\mu}$.

The (population) *covariance matrix* of a random vector $\mathbf{X}$ is defined as

$$\boldsymbol{\Sigma} = \mathrm{cov}(\mathbf{X}) = \begin{pmatrix} \sigma_{11} & \sigma_{12} & \cdots & \sigma_{1p} \\ \sigma_{21} & \sigma_{22} & \cdots & \sigma_{2p} \\ \vdots & \vdots & & \vdots \\ \sigma_{p1} & \sigma_{p2} & \cdots & \sigma_{pp} \end{pmatrix}.$$

The ith diagonal element of $\boldsymbol{\Sigma}$ is the population variance of X_i:

$$\sigma_{ii} = \sigma_i^2 = \begin{cases} \int_{-\infty}^{\infty} (x_i - \mu_i)^2 f_i(x_i) dx_i & \text{if } X_i \text{ is a continuous random variable} \\ \sum_{x_i} (x_i - \mu_i)^2 p_i(x_i) & \text{if } X_i \text{ is a discrete random variable} \end{cases}.$$

The (j,k)th off-diagonal element of $\boldsymbol{\Sigma}$ is the population covariance between X_j and X_k:

$$\begin{aligned} \sigma_{jk} &= E(X_j - \mu_j)(X_k - \mu_k) \\ &= \begin{cases} \int_{-\infty}^{\infty}\int_{-\infty}^{\infty} (x_j - \mu_j)(x_k - \mu_k) f_{jk}(x_j, x_k) dx_j dx_k & \textit{if } X_j, X_k \text{ are continuous random variables} \\ \sum_{x_j}\sum_{x_k} (x_j - \mu_j)(x_k - \mu_k) p_{jk}(x_j, x_k) & \textit{if } X_j, X_k \text{ are discrete random variables} \end{cases}, \end{aligned}$$

where $f_{jk}(x_j, x_k)$ and $p_{jk}(x_j, x_k)$ are the joint density function and joint probability mass function, respectively, of X_j and X_k. The population covariance measures the linear association between the two random variables. It is clear that $\sigma_{jk} = \sigma_{kj}$ and the covariance matrix $\boldsymbol{\Sigma}$ is symmetric. The same as the sample covariance matrix, the population covariance matrix $\boldsymbol{\Sigma}$ is always positive semidefinite.

Similar to the population mean, the population variance and covariance can be estimated by the sample variance and covariance introduced in Section 2.2. The sample variance and covariance are both random variables, and are unbiased estimators of the population variance and covariance. Consequently, the sample covariance matrix $\mathbf{S}$ is an unbiased estimator of the population covariance matrix $\boldsymbol{\Sigma}$, that is, $E(\mathbf{S}) = \boldsymbol{\Sigma}$.

As for the sample covariance, the value of the population covariance of two random variables depends on the scaling, possibly due to the difference of measuring unit of the variables. A scaling-independent measure of the degree of linear association between the random variables X_j and X_k is given by the population correlation:

$$\rho_{jk} = \frac{\sigma_{jk}}{\sqrt{\sigma_{jj}}\sqrt{\sigma_{kk}}}.$$

It is clear that $\rho_{jk} = \rho_{kj}$. And the population *correlation matrix* of a random vector $\mathbf{X}$ is a symmetric matrix defined as

$$\text{cor}(\mathbf{X}) = \begin{pmatrix} 1 & \rho_{12} & \cdots & \rho_{1p} \\ \rho_{21} & 1 & \cdots & \rho_{2p} \\ \vdots & \vdots & & \vdots \\ \rho_{p1} & \rho_{p2} & \vdots & 1 \end{pmatrix}.$$

For univariate variables X and Y and a constant c, we have $E(X+Y) = E(X) + E(Y)$ and $E(cX) = cE(X)$. Similarly, for random vectors $\mathbf{X}$ and $\mathbf{Y}$ and a constant matrix $\mathbf{C}$, it can be seen that

$$E(\mathbf{X}+\mathbf{Y}) = E(\mathbf{X}) + E(\mathbf{Y}) \tag{3.1}$$

$$E(\mathbf{CX}) = \mathbf{C}(E(\mathbf{X})).$$

The covariance matrix of $\mathbf{Z} = \mathbf{CX}$ is

$$\mathbf{\Sigma}_{\mathbf{Z}} = \text{cov}(\mathbf{Z}) = \text{cov}(\mathbf{CX}) = \mathbf{C}\mathbf{\Sigma}_{\mathbf{x}}\mathbf{C}^T. \tag{3.2}$$

The similarity of (3.2) and (2.10) is pretty clear. When $\mathbf{C}$ is a row vector $\mathbf{c}^T = (c_1, c_2, \ldots, c_p)$, $\mathbf{CX} = \mathbf{c}^T\mathbf{X} = c_1X_1 + \cdots + c_pX_p$ and

$$E(\mathbf{c}^T\mathbf{X}) = \mathbf{c}^T\boldsymbol{\mu} \tag{3.3}$$

$$\text{var}(\mathbf{c}^T\mathbf{X}) = \mathbf{c}^T\mathbf{\Sigma}\mathbf{c}, \tag{3.4}$$

where $\boldsymbol{\mu}$ and $\mathbf{\Sigma}$ are the mean vector and covariance matrix of $\mathbf{X}$.

Let $\mathbf{X}_1$ and $\mathbf{X}_2$ denote two subvectors of $\mathbf{X}$, i.e., $\mathbf{X} = \begin{pmatrix} \mathbf{X}_1 \\ \mathbf{X}_2 \end{pmatrix}$. The mean vector and the covariance matrix of $\mathbf{X}$ can be partitioned as

$$\boldsymbol{\mu} = E(\mathbf{X}) = E\begin{pmatrix} \mathbf{X}_1 \\ \mathbf{X}_2 \end{pmatrix} = \begin{pmatrix} E(\mathbf{X}_1) \\ E(\mathbf{X}_2) \end{pmatrix} = \begin{pmatrix} \boldsymbol{\mu}_1 \\ \boldsymbol{\mu}_2 \end{pmatrix} \tag{3.5}$$

$$\mathbf{\Sigma} = \text{cov}\begin{pmatrix} \mathbf{X}_1 \\ \mathbf{X}_2 \end{pmatrix} = \begin{pmatrix} \mathbf{\Sigma}_{11} & \mathbf{\Sigma}_{12} \\ \mathbf{\Sigma}_{21} & \mathbf{\Sigma}_{22} \end{pmatrix}, \tag{3.6}$$

where $\mathbf{\Sigma}_{11} = \text{cov}(\mathbf{X}_1)$ and $\mathbf{\Sigma}_{22} = \text{cov}(\mathbf{X}_2)$. The matrix $\mathbf{\Sigma}_{12}$ contains the covariance of each component in $\mathbf{X}_1$ and each component in $\mathbf{X}_2$. Based on the symmetry of $\mathbf{\Sigma}$, we have $\mathbf{\Sigma}_{21} = \mathbf{\Sigma}_{12}^T$.

3.2 Density Function and Properties of Multivariate Normal Distribution

Normal distribution is the most commonly used distribution for continuous random variables. Many statistical models and inference methods are based on the univariate or multivariate normal distribution. One advantage of the normal distribution is its mathematical tractability. More importantly, the normal distribution turns out to be a good approximation to the "true" population distribution for many sample statistics and real-world data due to the *central limit theorem*, which says that the summation of a large number of independent observations from any population with the same mean and variance approximately follows a normal distribution.

Recall that a univariate random variable X with mean μ and variance σ^2 is normally distributed, which is denoted by $X \sim \mathcal{N}(\mu, \sigma^2)$, if it has the probability density function

$$f(x) = \frac{1}{\sqrt{2\pi\sigma^2}} e^{-(x-\mu)^2/2\sigma^2}, \quad -\infty < x < \infty. \tag{3.7}$$

The multivariate normal distribution is an extension of the univariate normal distribution. If a p-dimensional random vector $\mathbf{X}$ follows a multivariate normal distribution with mean vector $\boldsymbol{\mu}$ and covariance matrix $\boldsymbol{\Sigma}$, the probability density function of $\mathbf{X}$ has the form

$$f(\mathbf{x}) = \frac{1}{(2\pi)^{p/2}|\boldsymbol{\Sigma}|^{1/2}} e^{-(\mathbf{x}-\boldsymbol{\mu})^T\boldsymbol{\Sigma}^{-1}(\mathbf{x}-\boldsymbol{\mu})/2}. \tag{3.8}$$

We denote the p-dimensional normal distribution by $\mathcal{N}_p(\boldsymbol{\mu}, \boldsymbol{\Sigma})$.

From (3.8), the density of a p-dimensional normal distribution depends on $\mathbf{x}$ through the term $(\mathbf{x}-\boldsymbol{\mu})^T\boldsymbol{\Sigma}^{-1}(\mathbf{x}-\boldsymbol{\mu})$, which is the square of the distance from $\mathbf{x}$ to $\boldsymbol{\mu}$ standardized by the covariance matrix. Then it is clear that the set of $\mathbf{x}$ values yielding a constant height for the density form an ellipsoid. The set of points with the same height for the density is called a *contour*. The constant probability density contour of a p-dimensional normal distribution is:

$$\{\mathbf{x} \mid (\mathbf{x}-\boldsymbol{\mu})^T\boldsymbol{\Sigma}^{-1}(\mathbf{x}-\boldsymbol{\mu}) = c^2\},$$

which forms the surface of an ellipsoid centered at $\boldsymbol{\mu}$ with standardized distance between $\mathbf{x}$ and $\boldsymbol{\mu}$ equal to c. And the contour with larger distance c has a smaller height value for the density. It can be shown that the axes of the ellipsoid contours of constant density for the p-dimensional normal distribution are in the directions of the eigenvectors of $\boldsymbol{\Sigma}$ with lengths proportional to the square roots of the corresponding eigenvalues of $\boldsymbol{\Sigma}$.

Example 3.1: Consider a bivariate ($p = 2$) normally distributed random vector $\mathbf{X} = (X_1 \ X_2)^T$. Suppose the mean vector is $\boldsymbol{\mu} = (0 \ \ 0)^T$ and the covariance matrix is

$$\boldsymbol{\Sigma} = \begin{pmatrix} 1 & \rho \\ \rho & 1 \end{pmatrix}.$$

So the variance of both variables is equal to one and the covariance matrix coincides with the correlation matrix. The inverse of the covariance matrix is

$$\boldsymbol{\Sigma}^{-1} = \frac{1}{1-\rho^2}\begin{pmatrix} 1 & -\rho \\ -\rho & 1 \end{pmatrix}$$

and $|\boldsymbol{\Sigma}| = 1-\rho^2$. Substituting $\boldsymbol{\Sigma}^{-1}$ and $|\boldsymbol{\Sigma}|$ in (3.8), we have

$$f(x_1, x_2) = \frac{1}{2\pi\sqrt{1-\rho^2}} \exp\left\{-\frac{1}{2(1-\rho^2)}(x_1^2 + x_2^2 - 2\rho x_1 x_2)\right\} \tag{3.9}$$

From (3.9), if $\rho = 0$, the joint density can be written as $f(x_1, x_2) = f(x_1)f(x_2)$, where $f(x)$ is the univariate normal density as given in (3.7), with $\mu = 0$ and $\sigma = 1$. So in this case X_1 and X_2 are independent. This result is true for general multivariate normal distribution, as discussed later in this section.

By solving the characteristic equation $|\boldsymbol{\Sigma} - \lambda\mathbf{I}| = 0$, the two eigenvalues of $\boldsymbol{\Sigma}$ are $\lambda_1 = 1+\rho$ and $\lambda_2 = 1-\rho$. Based on $\boldsymbol{\Sigma}\mathbf{v} = \lambda\mathbf{v}$, the corresponding eigenvectors can be obtained as

$$\mathbf{v}_1 = \begin{pmatrix} \frac{\sqrt{2}}{2} \\ \frac{\sqrt{2}}{2} \end{pmatrix}, \ \mathbf{v}_2 = \begin{pmatrix} \frac{-\sqrt{2}}{2} \\ \frac{\sqrt{2}}{2} \end{pmatrix}.$$

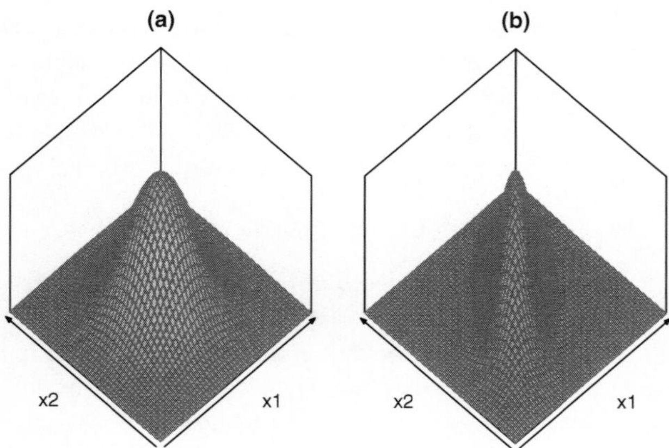

Figure 3.1 Two bivariate normal distributions, (a) $\rho = 0$ (b) $\rho = 0.75$

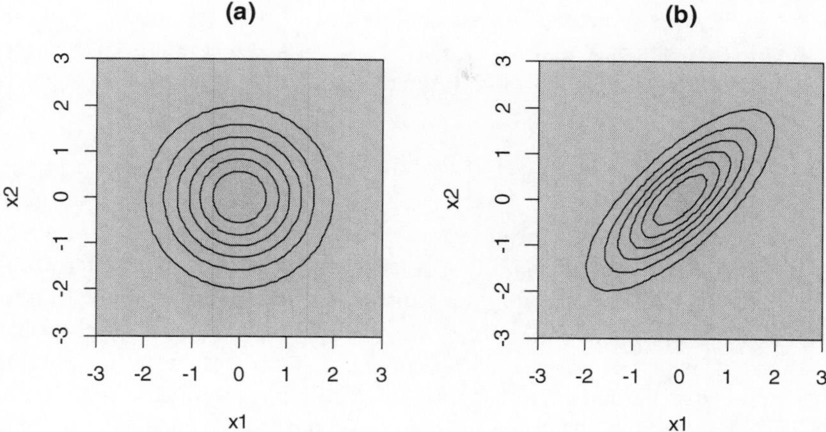

Figure 3.2 Contour plots for the distributions in Figure 3.1

So the major axis of the ellipse contour of constant density is along the line $x_1 = x_2$ and the minor axis is orthogonal to the major axis. The larger the correlation coefficient ρ, the more elongated the ellipse contour. As an example, two bivariate normal distributions with $\rho = 0$ and $\rho = 0.75$ are shown in Figure 3.1(a) and Figure 3.1(b), respectively. Notice how the presence of correlation causes the probability distribution to concentrate along the line $x_1 = x_2$. When $\rho = 0$, it is easy to see that the constant-density contour is a circle, as shown in Figure 3.2(a). For $\rho = 0.75$, the constant-density contour is an ellipse shown in Figure 3.2(b).

Properties of the Multivariate Normal Distribution

We list some of the most useful properties of the multivariate normal distribution. These properties make it convenient to manipulate normal distributions, which is one of the reasons for the popularity of the normal distribution. Suppose the random vector $\mathbf{X}$ follows a p-dimensional normal distribution $\mathcal{N}_p(\mu, \Sigma)$.

- Normality of linear combinations of the variables in $\mathbf{X}$. Let $\mathbf{c}$ be a vector of constants. From (3.3) and (3.4), we have $E(\mathbf{c}^T\mathbf{X}) = \mathbf{c}^T\mu$ and $\text{var}(\mathbf{c}^T\mathbf{X}) = \mathbf{c}^T\Sigma\mathbf{c}$. This is true for any random vector $\mathbf{X}$. When $\mathbf{X}$ follows a multivariate normal distribution, we have the additional property that $\mathbf{c}^T\mathbf{X}$ also follows a (univariate) normal distribution. That is, if $\mathbf{X} \sim \mathcal{N}_p(\mu, \Sigma)$, then $\mathbf{c}^T\mathbf{X} \sim \mathcal{N}(\mathbf{c}^T\mu, \mathbf{c}^T\Sigma\mathbf{c})$. In general, if $\mathbf{C}$ is a $q \times p$ matrix, $\mathbf{CX}$ still follows a multivariate normal distribution. From (3.1) and (3.2), we have $E(\mathbf{CX}) = \mathbf{C}\mu$ and $\text{cov}(\mathbf{CX}) = \mathbf{C}\Sigma\mathbf{C}^T$. So $\mathbf{CX} \sim \mathcal{N}_q(\mathbf{C}\mu, \mathbf{C}\Sigma\mathbf{C}^T)$.

- Normality of subvectors. Let $\mathbf{X}_1 = (X_1, X_2, \dots, X_q)$ be the subvector of the first q elements of $\mathbf{X}$ and $\mathbf{X}_2 = (X_{q+1}, X_{q+2}, \dots, X_p)$ be the subvector of the remaining $p-q$ elements of $\mathbf{X}$. From (3.5) and (3.6), $\boldsymbol{\mu}$ and $\boldsymbol{\Sigma}$ can be partitioned as

$$\boldsymbol{\mu} = \begin{pmatrix} \boldsymbol{\mu}_1 \\ \boldsymbol{\mu}_2 \end{pmatrix} \quad \boldsymbol{\Sigma} = \begin{pmatrix} \boldsymbol{\Sigma}_{11} & \boldsymbol{\Sigma}_{12} \\ \boldsymbol{\Sigma}_{21} & \boldsymbol{\Sigma}_{22} \end{pmatrix}, \tag{3.10}$$

where $\boldsymbol{\mu}_i$ and $\boldsymbol{\Sigma}_{ii}$ are the mean vector and covariance matrix of $\mathbf{X}_i$, for $i = 1, 2$. If $\mathbf{X}$ follows a multivariate normal distribution, we have the additional property that both $\mathbf{X}_1$ and $\mathbf{X}_2$ follow a multivariate normal distribution. That is, if $\mathbf{X} \sim \mathcal{N}_p(\boldsymbol{\mu}, \boldsymbol{\Sigma})$, then $\mathbf{X}_1 \sim \mathcal{N}_q(\boldsymbol{\mu}_1, \boldsymbol{\Sigma}_{11})$ and $\mathbf{X}_2 \sim \mathcal{N}_{p-q}(\boldsymbol{\mu}_2, \boldsymbol{\Sigma}_{22})$. A special case of this property is that each element of $\mathbf{X}$ also follows a (univariate) normal distribution. That is, if $\mathbf{X} \sim \mathcal{N}_p(\boldsymbol{\mu}, \boldsymbol{\Sigma})$, then $X_j \sim \mathcal{N}(\mu_j, \sigma_{jj})$, $j = 1, 2, \dots, p$. The converse of this result is not true. If each element of a random vector $\mathbf{X}$ follows a univariate normal distribution, $\mathbf{X}$ may not follow a multivariate normal distribution.
- Zero covariance implies independence. If $\mathbf{X} \sim \mathcal{N}_p(\boldsymbol{\mu}, \boldsymbol{\Sigma})$ and $\mathbf{X} = \begin{pmatrix} \mathbf{X}_1 \\ \mathbf{X}_2 \end{pmatrix}$, the mean vector and covariance matrix of $\mathbf{X}$ can be partitioned as in (3.10). The subvectors $\mathbf{X}_1$ and $\mathbf{X}_2$ are independent if and only if $\boldsymbol{\Sigma}_{12} = \mathbf{0}$. Specifically, for any two elements X_i and X_j of $\mathbf{X}$, X_i and X_j are independent if and only if $\sigma_{ij} = \text{cov}(X_i, X_j) = 0$. Note that if X_i and X_j do not follow joint normal distribution, and X_i and X_j are independent, we still have $\text{cov}(X_i, X_j) = 0$. However, the converse is not necessarily true. That is, if $\text{cov}(X_i, X_j) = 0$, X_i and X_j may not be independent.
- Conditional distributions are normal. Suppose $\mathbf{X} \sim \mathcal{N}_p(\boldsymbol{\mu}, \boldsymbol{\Sigma})$, $\mathbf{X} = \begin{pmatrix} \mathbf{X}_1 \\ \mathbf{X}_2 \end{pmatrix}$, and the mean vector and covariance matrix of $\mathbf{X}$ is given by (3.10). If $\mathbf{X}_1$ and $\mathbf{X}_2$ are not independent, we have $\boldsymbol{\Sigma}_{12} \neq \mathbf{0}$ and the conditional distribution of $\mathbf{X}_1$ given $\mathbf{X}_2 = \mathbf{x}_2$, is multivariate normal with

$$E(\mathbf{X}_1 \mid \mathbf{X}_2 = \mathbf{x}_2) = \boldsymbol{\mu}_1 + \boldsymbol{\Sigma}_{12}\, \boldsymbol{\Sigma}_{22}^{-1}(\mathbf{x}_2 - \boldsymbol{\mu}_2) \tag{3.11}$$

$$\text{cov}(\mathbf{X}_1 \mid \mathbf{X}_2 = \mathbf{x}_2) = \boldsymbol{\Sigma}_{11} - \boldsymbol{\Sigma}_{12}\boldsymbol{\Sigma}_{22}^{-1}\boldsymbol{\Sigma}_{21}. \tag{3.12}$$

Note that the mean vector of the conditional distribution is a linear function of $\mathbf{x}_2$. But the covariance matrix of the conditional distribution does not depend on $\mathbf{x}_2$. If $\mathbf{X}_1$ and $\mathbf{X}_2$ are independent, clearly the conditional distribution of $\mathbf{X}_1$ given $\mathbf{X}_2 = \mathbf{x}_2$ is simply $\mathcal{N}_q(\boldsymbol{\mu}_1, \boldsymbol{\Sigma}_{11})$, the unconditional distribution of $\mathbf{X}_1$.

3.3 Maximum Likelihood Estimation for Multivariate Normal Distributions

If the population distribution is assumed to be multivariate normal with mean vector $\boldsymbol{\mu}$ and covariance matrix $\boldsymbol{\Sigma}$. The parameters $\boldsymbol{\mu}$ and $\boldsymbol{\Sigma}$ can be estimated from a random sample of n observations $\mathbf{x}_1, \mathbf{x}_2, \ldots, \mathbf{x}_n$. A commonly used method for parameter estimation is the *maximum likelihood estimation* (MLE), and the estimated parameter values are called the *maximum likelihood estimates.* The idea of the maximum likelihood estimation is to find $\boldsymbol{\mu}$ and $\boldsymbol{\Sigma}$ that maximize the joint density of the $\mathbf{x}$'s, which is called the *likelihood function.* For multivariate normal distribution, the likelihood function is

$$\begin{aligned} L(\boldsymbol{\mu}, \boldsymbol{\Sigma}; \mathbf{x}_1, \mathbf{x}_2, \ldots, \mathbf{x}_n) &= \prod_{i=1}^{n} f(\mathbf{x}_i; \boldsymbol{\mu}, \boldsymbol{\Sigma}) \\ &= \prod_{i=1}^{n} \frac{1}{(2\pi)^{p/2} \left|\boldsymbol{\Sigma}\right|^{1/2}} e^{-(\mathbf{x}_i-\boldsymbol{\mu})^T \boldsymbol{\Sigma}^{-1} (\mathbf{x}_i-\boldsymbol{\mu})/2} \\ &= \frac{1}{(2\pi)^{np/2} \left|\boldsymbol{\Sigma}\right|^{n/2}} e^{-\sum_{i=1}^{n} (\mathbf{x}_i-\boldsymbol{\mu})^T \boldsymbol{\Sigma}^{-1} (\mathbf{x}_i-\boldsymbol{\mu})/2}. \end{aligned} \tag{3.13}$$

It is often easier to find the MLE by minimizing the negative *log likelihood function*, which is given by

$$\begin{aligned} -l(\boldsymbol{\mu}, \boldsymbol{\Sigma}) &= -\ln\left(L\left(\boldsymbol{\mu}, \boldsymbol{\Sigma}\right)\right) \\ &= \frac{np}{2} \ln\left(2\pi\right) + \frac{n}{2} \ln\left|\boldsymbol{\Sigma}\right| + \frac{1}{2} \sum_{i=1}^{n} \left(\mathbf{x}_i - \boldsymbol{\mu}\right)^T \boldsymbol{\Sigma}^{-1} \left(\mathbf{x}_i - \boldsymbol{\mu}\right). \end{aligned} \tag{3.14}$$

Taking the derivative of (3.14) with respect to $\boldsymbol{\mu}$, we have

$$\frac{\partial}{\partial \boldsymbol{\mu}} l\left(\boldsymbol{\mu}, \boldsymbol{\Sigma}\right) = -\sum_{i=1}^{n} \boldsymbol{\Sigma}^{-1}(\mathbf{x}_i - \boldsymbol{\mu}). \tag{3.15}$$

Setting the partial derivative in (3.15) to zero, the MLE of $\boldsymbol{\mu}$ is obtained as

$$\hat{\boldsymbol{\mu}} = \frac{1}{n} \sum_{i=1}^{n} \mathbf{x}_i, \tag{3.16}$$

which is the sample mean vector of the data set $\mathbf{x}_1, \mathbf{x}_2, \ldots, \mathbf{x}_n$. The derivation of the MLE of $\boldsymbol{\Sigma}$ is more involved and beyond the scope of this book. The result is given by

$$\hat{\boldsymbol{\Sigma}} = \frac{1}{n} \sum_{i=1}^{n} (\mathbf{x}_i - \bar{\mathbf{x}})(\mathbf{x}_i - \bar{\mathbf{x}})^T = \frac{n-1}{n} \mathbf{S}, \tag{3.17}$$

where $\mathbf{S}$ is the sample covariance matrix as given in (2.6). Since the MLE $\widehat{\boldsymbol{\Sigma}}$ uses n instead of $n-1$ in the denominator, it is a biased estimator. So the sample covariance matrix $\mathbf{S}$ is more commonly used to estimate $\boldsymbol{\Sigma}$, especially when n is small.

One useful property of MLE is the *invariance property*. In general, let $\hat{\boldsymbol{\theta}}$ denote the MLE of the parameter vector $\boldsymbol{\theta}$. Then the MLE of a function of $\boldsymbol{\theta}$, denoted by $h(\boldsymbol{\theta})$, is given by $h(\hat{\boldsymbol{\theta}})$. This result makes it very convenient to find the MLE of any function of a parameter, given the MLE of the parameter. For example, based on (3.17), it is easy to see that the MLE of the variance of X_j, the jth element of $\mathbf{X}$, is given by

$$\hat{\sigma}_{jj} = \frac{1}{n}\sum_{i=1}^{n}(X_{ij} - \overline{X}_j)^2.$$

Then based on the invariance property, the MLE of the standard deviation $\sqrt{\sigma_{jj}}$ is $\sqrt{\hat{\sigma}_{jj}}$.

The MLE has some good asymptotic properties and usually performs well for data sets of large sample sizes. For example, under mild regularity conditions, MLE satisfies the property of consistency, which guarantees that the estimator converges to the true value of the parameter as the sample size becomes infinite. In addition, under certain regularity conditions, the MLE is asymptotically normal and efficient. That is, as the sample size becomes infinite, the distribution of MLE will converge to a normal distribution with variance equal to the optimal asymptotic variance. The details of the regularity conditions are beyond the scope of this book. But these conditions are quite general and often satisfied in common circumstances.

3.4 Hypothesis Testing on Mean Vectors

In this section, we study how to determine if the population mean $\boldsymbol{\mu}$ is equal to a specific value $\boldsymbol{\mu}_0$ when the observations follow a normal distribution. We start by reviewing the hypothesis testing results for univariate data. Suppose $X_1, X_2, \ldots, X_n$ are a random sample of independent univariate observations following the normal distribution $\mathcal{N}(\mu, \sigma^2)$. The test on μ is formulated as

$$H_0 : \mu = \mu_0 \quad \text{vs.} \quad H_1 : \mu \neq \mu_0,$$

where H_0 is the null hypothesis and H_1 is the (two-sided) alternative hypothesis. For this test, we use the following test statistic:

$$t = \frac{\overline{X} - \mu_0}{s/\sqrt{n}}, \tag{3.18}$$

where $\overline{X}$ is the sample mean $\overline{X} = \frac{1}{n}\sum_{i=1}^{n} X_i$ and s^2 is the sample variance $s^2 = \frac{1}{n-1}\sum_{i=1}^{n}(X_i - \overline{X})^2$. The sample mean $\overline{X}$ follows $\mathcal{N}(\mu, \sigma^2/n)$ and $(n-1)s^2/\sigma^2$ follows a χ^2 distribution with $n-1$ degrees of freedom. Consequently, under H_0 the t statistic in (3.18) follows a Student's t-distribution with $n-1$ degrees of freedom. We reject H_0 at significance level α and conclude that μ is not equal to μ_0 if $|t| > t_{\alpha/2,\,n-1}$, where $t_{\alpha/2,\,n-1}$ denotes the upper $100(\alpha/2)$th percentile of the t-distribution with $n-1$ degrees of freedom. Intuitively, $|t| > t_{\alpha/2,\,n-1}$ indicates that we only have a small probability to observe $|t|$ if we sample from the Student's t-distribution with $n-1$ degrees of freedom. Thus, it is very likely the null hypothesis H_0 is not correct and we should reject H_0.

The test based on a fixed significance level α, say $\alpha = 0.05$, has the disadvantage that it gives the decision maker no idea about whether the observed value of the test statistic is just barely in the rejection region or if it is far into the region. Instead, the *p*-value can be used to indicate how strong the evidence is in rejecting the null hypothesis H_0. The *p*-value is the probability that the test statistic will take on a value that is at least as extreme as the observed value when the null hypothesis is true. The smaller the *p*-value, the stronger the evidence we have in rejecting H_0. If the *p*-value is smaller than α, H_0 will be rejected at the significance level of α. The *p*-value based on the t statistic in (3.18) can be found as

$$P = 2\Pr(T(n-1) > |t|),$$

where $T(n-1)$ denotes a random variable following a t distribution with $n-1$ degrees of freedom.

We can define the $100(1-\alpha)\%$ confidence interval for μ as

$$\left[\overline{x} - t_{\alpha/2,n-1}\frac{s}{\sqrt{n}},\ \overline{x} + t_{\alpha/2,\,n-1}\frac{s}{\sqrt{n}}\right].$$

It is easy to see that the null hypothesis H_0 is not rejected at level α if and only if μ_0 is in the $100(1-\alpha)\%$ confidence interval for μ. So the confidence interval consists of all those "plausible" values of μ_0 that would not be rejected by the test of H_0 at level α.

To see the link to the test statistic used for a multivariate normal distribution, we consider an equivalent rule to reject H_0, which is based on the square of the t statistic:

$$t^2 = \frac{(\overline{X} - \mu_0)^2}{s^2/n} = n(\overline{X} - \mu_0)(s^2)^{-1}(\overline{X} - \mu_0). \tag{3.19}$$

We reject H_0 at significance level α if $t^2 > (t_{\alpha/2,\,n-1})^2$.

For a multivariate distribution with unknown mean $\boldsymbol{\mu}$ and known $\boldsymbol{\Sigma}$, we consider testing the following hypotheses:

$$H_0 : \boldsymbol{\mu} = \boldsymbol{\mu_0} \quad \text{vs.} \quad H_1 : \boldsymbol{\mu} \neq \boldsymbol{\mu_0}. \tag{3.20}$$

Let $\mathbf{X}_1, \mathbf{X}_2, \ldots, \mathbf{X}_n$ denote a random sample from a multivariate normal population. The test statistic in (3.19) can be naturally generalized to the multivariate distribution as

$$T^2 = n\left(\overline{\mathbf{X}} - \boldsymbol{\mu_0}\right)^T \mathbf{S}^{-1}\left(\overline{\mathbf{X}} - \boldsymbol{\mu_0}\right), \tag{3.21}$$

where $\overline{\mathbf{X}}$ and $\mathbf{S}$ are the sample mean vector and the sample covariance matrix of $\mathbf{X}_1, \mathbf{X}_2, \ldots, \mathbf{X}_n$. The T^2 statistic in (3.19) is called *Hotelling's* T^2 in honor of Harold Hotelling who first obtained its distribution. Assuming H_0 is true, we have the following result about the distribution of the T^2-statistic:

$$\frac{n-p}{(n-1)p} T^2 \sim F_{p,\,n-p},$$

where $F_{p,n-p}$ denotes the F-distribution with p and $n-p$ degrees of freedom. Based on the results on the distribution of T^2, we reject H_0 at the significance level of α if

$$T^2 > \frac{(n-1)p}{n-p} F_{\alpha,p,n-p}, \tag{3.22}$$

where $F_{\alpha,p,n-p}$ denotes the upper (100α)th percentile of the F-distribution with p and $n-p$ degrees of freedom. The p-value of the test based on the T^2-statistic is

$$P = \Pr\left(F(p, n-p) > \frac{n-p}{(n-1)p} T^2\right),$$

where $F(p, n-p)$ denotes a random variable distributed as $F_{p,n-p}$.

The T^2 statistic can also be written as

$$T^2 = (\overline{\mathbf{X}} - \boldsymbol{\mu_0})^T \left(\frac{1}{n}\mathbf{S}\right)^{-1} (\overline{\mathbf{X}} - \boldsymbol{\mu_0}),$$

which can be interpreted as the standardized distance between the sample mean $\overline{\mathbf{X}}$ and $\boldsymbol{\mu_0}$. The distance is standardized by $\mathbf{S}/n$, which is equal to the sample covariance matrix of $\overline{\mathbf{X}}$. When the standardized distance between $\overline{\mathbf{X}}$

and $\boldsymbol{\mu}_0$ is beyond the critical value given in the right-hand side of (3.22), the true mean is not likely equal to be $\boldsymbol{\mu}_0$ and we reject H_0.

The concept of univariate confidence interval can be extended to multivariate confidence region. For p-dimensional normal distribution, the $100\left(1-\alpha\right)\%$ confidence region for $\boldsymbol{\mu}$ is defined as

$$\left\{\boldsymbol{\mu} \mid n(\overline{\mathbf{x}}-\boldsymbol{\mu})^T \mathbf{S}^{-1}(\overline{\mathbf{x}}-\boldsymbol{\mu}) \leq \frac{(n-1)p}{n-p} F_{\alpha,p,n-p}\right\}.$$

It is clear that the confidence region for $\boldsymbol{\mu}$ is an ellipsoid centered at $\overline{\mathbf{x}}$. Similar to the univariate case, the null hypothesis $H_0 : \boldsymbol{\mu} = \boldsymbol{\mu}_0$ is not rejected at level α if and only if $\boldsymbol{\mu}_0$ is in the $100(1-\alpha)\%$ confidence region for $\boldsymbol{\mu}$.

The T^2-statistic can also be derived as the *likelihood ratio test* of the hypotheses in (3.20). The likelihood ratio test is a general principle of constructing statistical test procedures and having several optimal properties for reasonably large samples. The detailed study of the likelihood ratio test theory is beyond the scope of this book.

Substituting the MLE of $\boldsymbol{\mu}$ and $\boldsymbol{\Sigma}$ in (3.16) and (3.17), respectively, into the likelihood function in (3.13), it is easy to see

$$\max_{\boldsymbol{\mu},\boldsymbol{\Sigma}} L\left(\boldsymbol{\mu},\boldsymbol{\Sigma}\right) = \frac{1}{\left(2\pi\right)^{np/2}\left|\widehat{\boldsymbol{\Sigma}}\right|^{n/2}} e^{-np/2},$$

where $\widehat{\boldsymbol{\Sigma}}$ is the MLE of $\boldsymbol{\Sigma}$ given in (3.17). Under the null hypothesis $H_0 : \boldsymbol{\mu} = \boldsymbol{\mu}_0$, the MLE of $\boldsymbol{\Sigma}$ with $\boldsymbol{\mu} = \boldsymbol{\mu}_0$ fixed can be obtained as

$$\widehat{\boldsymbol{\Sigma}}_0 = \frac{1}{n}\sum_{i=1}^{n}(\mathbf{x}_i - \boldsymbol{\mu}_0)(\mathbf{x}_i - \boldsymbol{\mu}_0)^T.$$

It can be seen that $\widehat{\boldsymbol{\Sigma}}_0$ is the same as $\widehat{\boldsymbol{\Sigma}}$ except that $\overline{\mathbf{X}}$ is replaced by $\boldsymbol{\mu}_0$.

The likelihood ratio test statistic is the ratio of the maximum likelihood over the subset of the parameter space specified by H_0 and the maximum likelihood over the entire parameter space. Specifically, the likelihood ratio test statistic of $H_0 : \boldsymbol{\mu} = \boldsymbol{\mu}_0$ is

$$LR = \frac{\max_{\boldsymbol{\Sigma}} L(\boldsymbol{\mu}_0, \boldsymbol{\Sigma})}{\max_{\boldsymbol{\mu},\boldsymbol{\Sigma}} L(\boldsymbol{\mu}, \boldsymbol{\Sigma})} = \left(\frac{\left|\widehat{\boldsymbol{\Sigma}}\right|}{\left|\widehat{\boldsymbol{\Sigma}}_0\right|}\right)^{n/2}. \tag{3.23}$$

The test based on the T^2-statistic in (3.21) and the likelihood ratio test is equivalent because it can be shown that

$$LR = \left(1 + \frac{T^2}{n-1}\right)^{-n/2}. \tag{3.24}$$

Example 3.2: Hot rolling is among the key steel-making processes that convert cast or semi-finished steel into finished products. A typical hot rolling process usually includes a melting division and a rolling division. The melting division is a continuous casting process that melts scrapped metals and solidifies the molten steel into semi-finished steel billet; the rolling division will further squeeze the steel billet by a sequence of stands in the hot rolling process. Each stand is composed of several rolls. The `side_temp_defect` data set contains the side temperature measurements on 139 defective steel billets at Stand 5 of a hot rolling process where the side temperatures are measured at 79 equally spaced locations spread along the stand. In this example, we focus on the three measurements at locations 2, 40, and 78, which correspond to locations close to the middle and the two ends of the stands. The nominal mean temperature values at the three locations are 1926, 1851, and 1872, which are obtained based on a large sample of billets without defects. We want to check if the defective billets have significantly different mean side temperature from the nominal values. We can, therefore, test the hypothesis

$$H_0 : \mu = \begin{pmatrix} 1926 \\ 1851 \\ 1872 \end{pmatrix}$$

The following `R` codes calculate the sample mean, sample covariance matrix, and the T^2-statistic for the three side temperature measurements.

```
side.temp.defect <- read.csv("side_temp_defect.csv",
      header = F)
X <- side.temp.defect[, c(2, 40, 78)]
mu0 <- c(1926, 1851, 1872)
x.bar <- apply(X, 2, mean) # sample mean
S <- cov(X) # sample var-cov matrix
n <- nrow(X)
p <- ncol(X)
alpha = 0.05
T2 <- n*t(x.bar-mu0)%*%solve(S)%*%(x.bar -mu0)
F0 <- (n-1)*p/(n-p)*qf(1-alpha, p, n-p)
p.value <- 1 - pf((n-p)/((n-1)*p)*T2, p, n-p)
```

Using the above `R` codes, the sample mean and sample covariance matrix are obtained as

$$\bar{\mathbf{x}} = \begin{pmatrix} 1930 \\ 1848 \\ 1864 \end{pmatrix}, \quad \mathbf{S} = \begin{pmatrix} 2547.4 & -111.0 & 133.7 \\ -111.0 & 533.1 & 300.7 \\ 133.7 & 300.7 & 562.5 \end{pmatrix}$$

The T^2-statistic is obtained by (3.21) as $T^2 = 19.71$. The right-hand side of (3.22) at $\alpha = 0.05$ is obtained as $F_0 = 8.13$. Since the observed value of T^2 exceeds the critical value F_0, we reject the null hypothesis H_0 and conclude that the mean vector of the three side temperatures of the defective billets is significantly different from the nominal mean vector. In addition, the *p*-value is $0.0004 < \alpha = 0.05$, which further confirms that H_0 should be rejected.

3.5 Bayesian Inference for Normal Distribution

Let $\mathcal{D} = \{\mathbf{x}_1, \mathbf{x}_2, \ldots, \mathbf{x}_n\}$ denote the observed data set. In the maximum likelihood estimation, the distribution parameters are considered as fixed. The estimation errors are obtained by considering the random distribution of possible data sets $\mathcal{D}$. By contrast, in Bayesian inference, we treat the observed data set $\mathcal{D}$ as the only data set. The uncertainty in the parameters is characterized through a probability distribution over the parameters.

In this subsection, we focus on Bayesian inference of normal distribution when the mean $\boldsymbol{\mu}$ is unknown and the covariance matrix $\boldsymbol{\Sigma}$ is assumed as known. The Bayesian inference is based on the Bayes' theorem. In general, the Bayes' theorem is about the conditional probability of an event A given that an event B occurs:

$$\Pr(A \mid B) = \frac{\Pr(B \mid A)\Pr(A)}{\Pr(B)}.$$

Applying Bayes' theorem for Bayesian inference of $\boldsymbol{\mu}$, we have

$$f(\boldsymbol{\mu} \mid \mathcal{D}) = \frac{f(\mathcal{D} \mid \boldsymbol{\mu}) g(\boldsymbol{\mu})}{f(\mathcal{D})}, \tag{3.25}$$

where $g(\boldsymbol{\mu})$ is the *prior distribution* of $\boldsymbol{\mu}$, which is the distribution *before* observing the data, and $f(\boldsymbol{\mu} \mid \mathcal{D})$ is called as the *posterior distribution*, which is the distribution *after* we have observed $\mathcal{D}$. The function $f(\mathcal{D} \mid \boldsymbol{\mu})$ on the right-hand side of (3.25) is the density function for the observed data set $\mathcal{D}$. If it is viewed as a function of the unknown parameter $\boldsymbol{\mu}$, $f(\mathcal{D} \mid \boldsymbol{\mu})$ is exactly the likelihood function of $\boldsymbol{\mu}$. Therefore the Bayes' theorem can be stated in words as

$$\text{posterior} \propto \text{likelihood} \times \text{prior}, \tag{3.26}$$

where $\propto$ stands for "is proportional to". Note the denominator $p(\mathcal{D})$ in the right-hand side of (3.25) is a constant which does not depend on the parameter $\boldsymbol{\mu}$. It plays the normalization role to ensure the left-hand side is a valid probability density function and integrates to one. Taking the integral of the right-hand side of (3.25) with respect to $\boldsymbol{\mu}$ and setting it to be equal to one, it is easy to see that

$$f(\mathcal{D}) = \int f(\mathcal{D} \mid \boldsymbol{\mu}) g(\boldsymbol{\mu}) d\boldsymbol{\mu}.$$

A point estimate of $\boldsymbol{\mu}$ can be obtained by maximizing the posterior distribution. This method is called the *maximum a posteriori* (MAP) estimate. The MAP estimate of $\boldsymbol{\mu}$ can be written as

$$\hat{\boldsymbol{\mu}}_{MAP} = \underset{\boldsymbol{\mu}}{\operatorname{argmax}}\, f(\boldsymbol{\mu} \mid \mathcal{D}) = \underset{\boldsymbol{\mu}}{\operatorname{argmax}}\, f(\mathcal{D} \mid \boldsymbol{\mu}) g(\boldsymbol{\mu}). \tag{3.27}$$

From (3.27), it can be seen that the MAP estimate is closely related to MLE. Without the prior $g(\boldsymbol{\mu})$, the MAP is the same as the MLE. So if the prior follows a uniform distribution, the MAP and MLE will be equivalent. Following this argument, if the prior distribution has a flat shape, we expect that the MAP and MLE are similar.

We first consider a simple case where the data follow a univariate normal distribution with unknown mean μ and known variance σ^2. The likelihood function based on a random sample of independent observations $\mathcal{D} = \{x_1, x_2, \ldots, x_n\}$ is given by

$$f(\mathcal{D} \mid \mu) = \prod_{i=1}^{n} f\left(x_i \mid \mu\right) = \frac{1}{(2\pi\sigma^2)^{n/2}} e^{-\frac{1}{2\sigma^2}\sum_{i=1}^{n}(x_i-\mu)^2}.$$

Based on (3.26), we have

$$f(\mu \mid \mathcal{D}) \propto f(\mathcal{D} \mid \mu) g(\mu),$$

where $g(\mu)$ is the probability density function of the prior distribution. We choose a normal distribution $\mathcal{N}(\mu_0, \sigma_0^2)$ as the prior for μ. This prior is a *conjugate prior* because the resulting posterior distribution will also be normal. By completing the square in the exponent of the likelihood and prior, the posterior distribution can be obtained as

$$\mu \mid \mathcal{D} \sim \mathcal{N}(\mu_n, \sigma_n^2),$$

where

$$\mu_n = \frac{\sigma^2}{n\sigma_0^2 + \sigma^2}\mu_0 + \frac{n\sigma_0^2}{n\sigma_0^2 + \sigma^2}\overline{x} \tag{3.28}$$

$$\frac{1}{\sigma_n^2} = \frac{1}{\sigma_0^2} + \frac{n}{\sigma^2}. \tag{3.29}$$

The posterior mean given in (3.28) can be understood as a weighted average of the prior mean μ_0 and the sample mean $\overline{x}$, which is the MLE of μ. When the sample size n is very large, the weight for $\overline{x}$ is close to one and the weight for μ_0 is close to 0, and the posterior mean is very close to the MLE, or the sample mean. On the other hand, when n is very small, the posterior mean is very close the prior mean μ_0. Similarly, if the prior variance σ_0^2 is very large, the prior distribution has a flat shape and the posterior mean is close to the MLE. Note that because the mode of a normal distribution is equal to the mean, the MAP of μ is exactly μ_n. Consequently, when n is very large, or when the prior is flat, the MAP is close to the MLE.

Equation (3.29) shows the relationship between the posterior variance and the prior variance. It is easier to understand the relationship if we consider the inverse of the variance, which is called the *precision*. A high (low) precision corresponds to a low (high) variance. Equation (3.29) basically says that the posterior precision is equal to the prior precision with an added precision contribution proportional to n. Each observation adds a contribution of $\frac{1}{\sigma^2}$, the

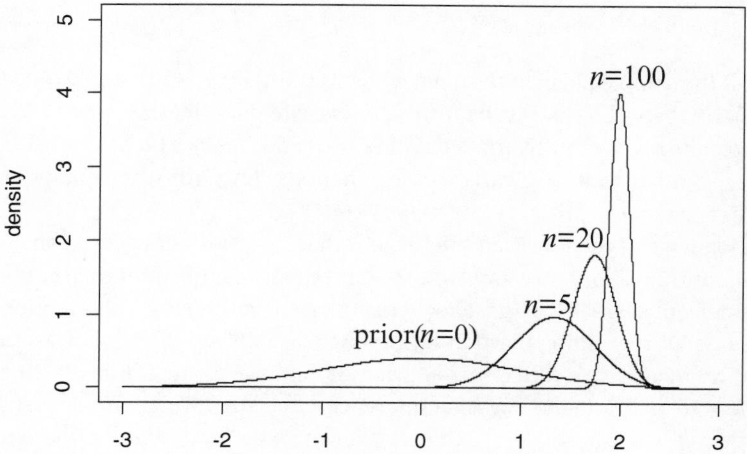

Figure 3.3 Posterior distribution of the mean with various sample sizes

precision of x_i, to the posterior precision. When n is very large, the posterior precision becomes very high, or equivalently the posterior variance becomes very small. On the other hand, when n is very small, the posterior precision and variance will be very close to the prior precision and variance. Specifically, when $n=0$, the posterior distribution is the same as the prior distribution. We illustrate the posterior distribution of the mean with known variance under various sample sizes in Figure 3.3, where the data are generated from $\mathcal{N}(2,1)$ and the prior distribution of the mean is $\mathcal{N}(0,1)$. It is clear from Figure 3.3 that with sample size n getting larger, the posterior distribution of the mean becomes more and more concentrated at the true mean.

When the data follow a p-dimensional multivariate normal distribution with unknown mean $\boldsymbol{\mu}$ and known covariance matrix $\boldsymbol{\Sigma}$, the posterior distribution based on a random sample of independent observations $\mathcal{D}=\{\mathbf{x}_1, \mathbf{x}_2, \dots, \mathbf{x}_n\}$ is given by

$$f(\boldsymbol{\mu} \mid \mathcal{D}) \propto f(\mathcal{D} \mid \boldsymbol{\mu}) g(\boldsymbol{\mu}) = \prod_{i=1}^{n} f(\mathbf{x}_i \mid \boldsymbol{\mu}) g(\boldsymbol{\mu}),$$

where $g(\boldsymbol{\mu})$ is the density of the conjugate prior distribution $\mathcal{N}_p(\boldsymbol{\mu}_0, \boldsymbol{\Sigma}_0)$. Similar to the univariate case, the posterior distribution of $\boldsymbol{\mu}$ can be obtained as

$$\boldsymbol{\mu} \mid \mathcal{D} \sim \mathcal{N}_p(\boldsymbol{\mu}_n, \boldsymbol{\Sigma}_n),$$

where

$$\boldsymbol{\mu}_n = (\boldsymbol{\Sigma}_0^{-1} + n\boldsymbol{\Sigma}^{-1})^{-1}(\boldsymbol{\Sigma}_0^{-1}\boldsymbol{\mu}_0 + n\boldsymbol{\Sigma}^{-1}\overline{\mathbf{x}}) \tag{3.30}$$

$$\boldsymbol{\Sigma}_n^{-1} = \boldsymbol{\Sigma}_0^{-1} + n\boldsymbol{\Sigma}^{-1}, \tag{3.31}$$

where $\overline{\mathbf{x}}$ is the sample mean of the data, which is the MLE of $\boldsymbol{\mu}$. It is easy to see the similarity between the results for the univariate data in (3.28) and (3.29) and the results for the multivariate data in (3.30) and (3.31). The MAP of $\boldsymbol{\mu}$ is exactly $\boldsymbol{\mu}_n$. Similar to the univariate case, when n is large, or when the prior distribution is flat, the MAP is close to the MLE.

One advantage of the Bayesian inference is that the prior knowledge can be included naturally. Suppose, for example, a randomly sampled product turns out to be defective. A MLE of the defective rate based on this single observation would be equal to 1, implying that all products are defective. By contrast, a Bayesian approach with a reasonable prior should give a much less extreme conclusion. In addition, the Bayesian inference can be performed in a sequential manner very naturally. To see this, we can write the posterior distribution of $\boldsymbol{\mu}$ with the contribution from the last data point x_n separated out as

$$f(\boldsymbol{\mu} \mid \mathcal{D}) \propto \left[g(\boldsymbol{\mu}) \prod_{i=1}^{n-1} f(\mathbf{x}_i \mid \boldsymbol{\mu}) \right] f(\mathbf{x}_n \mid \boldsymbol{\mu}). \tag{3.32}$$

Equation (3.32) can be viewed as the posterior distribution given a single observation $\mathbf{x}_n$ with the term in the square bracket treated as the prior. Note that the term in the square brackets is just the posterior distribution (up to a normalization constant) after observing $n-1$ data points. Equation (3.32) says that we can treat the posterior based on the first $n-1$ observations as the prior and update the posterior based on the next observation using the Bayes' theorem. This process can be repeated sequentially for each new observation. The sequential update of posterior under the Bayesian framework is very useful when observations are collected sequentially over time.

Example 3.3: For the `side_temp_defect` data set from a hot rolling process, suppose the true covariance matrix of the side temperatures measured at location 2, 40, and 78 of Stand 5 is known and given by

$$\mathbf{S} = \begin{pmatrix} 2547.4 & -111.0 & 133.7 \\ -111.0 & 533.1 & 300.7 \\ 133.7 & 300.7 & 562.5 \end{pmatrix}.$$

We use the nominal mean temperatures as given in Example 3.2 as the mean of the prior distribution and a diagonal matrix with variance equal to 100 for each temperature variable as its covariance matrix:

$$\boldsymbol{\mu}_0 = \begin{pmatrix} 1926 \\ 1851 \\ 1872 \end{pmatrix}, \quad \boldsymbol{\Sigma}_0 = \begin{pmatrix} 100 & 0 & 0 \\ 0 & 100 & 0 \\ 0 & 0 & 100 \end{pmatrix}.$$

Based on (3.30) and (3.31), the following `R` codes calculate the posterior mean and covariance matrix for $\boldsymbol{\mu}$ using the first five ($n = 5$) observations in the data set.

```
Sigma <- matrix(c(2547.4, -111.0, 133.7,
                  -111.0, 533.1, 300.7,
                  133.7, 300.7, 562.5),
                nrow = 3, ncol = 3, byrow = T)
Precision <- solve(Sigma)
Sigma0 <- diag(rep(100, 3))
Precision0 <- solve(Sigma0)
mu0 <- c(1926, 1851, 1872)
```

```
n <- 5
X.n <- side.temp.defect[1:n, c(2, 40, 78)]
x.bar <- apply(X.n, 2, mean)
mu.n <- solve(Precision0+n*Precision)%*%
   (Precision0%*%mu0+n*Precision%*%x.bar)
Sigma.n <- solve(Precision0 + n*Precision)
```

The posterior mean and covariance matrix are obtained as

$$\boldsymbol{\mu}_5 = \begin{pmatrix} 1930 \\ 1856 \\ 1854 \end{pmatrix}, \quad \boldsymbol{\Sigma}_5 = \begin{pmatrix} 83.37 & -2.61 & 2.83 \\ -2.61 & 46.85 & 15.37 \\ 2.83 & 15.37 & 48.235 \end{pmatrix}.$$

Compared to the sample mean of the first five observations, which is (1943 1850 1838)^T, the posterior mean has some deviations from both the sample mean and the prior mean $\boldsymbol{\mu}_0$. Now we use the first 100 ($n = 100$) observations to find the posterior mean by changing `n` in the `R` codes from 5 to 100. The posterior mean and covariance matrix are

$$\boldsymbol{\mu}_{100} = \begin{pmatrix} 1940 \\ 1849 \\ 1865 \end{pmatrix}, \quad \boldsymbol{\Sigma}_{100} = \begin{pmatrix} 20.28 & -0.87 & 1.03 \\ -0.87 & 4.97 & 2.72 \\ 1.03 & 2.72 & 5.235 \end{pmatrix}.$$

Compared to the sample mean vector of the first 100 observations, which is (1944 1849 1865)^T, the posterior mean with $n = 100$ observations is very close to the sample mean, while the influence of the prior mean is very small. In addition, the posterior variance for the mean temperature at each of the three locations is much smaller for $n = 100$ than for $n = 5$.

Bibliographic Notes

Multivariate normal distribution and its inference are thoroughly discussed in multivariate statistics books, for example, Johnson et al. [2002], Rencher [2003], and Anderson [2003]. Particularly, proofs of many theoretical results and properties can be found in Anderson [2003].

Exercises

1. Consider two discrete random variables X and Y with joint probability mass function $p(x, y)$ given in the following table:

x	y	$p(x, y)$
−1	−1	0.24
−1	1	0.06
0	−1	0.16
0	1	0.14
1	−1	0.40
1	1	0.00

Find the mean vector, the covariance matrix, and the correlation matrix of the random vector $(X\ Y)^T$.

2. A random vector $\mathbf{X} = (X_1\ X_2\ X_3\ X_4)^T$ has mean vector and covariance matrix given as

$$\mu = \begin{pmatrix} 1 \\ 2 \\ 3 \\ 4 \end{pmatrix} \quad \text{and} \quad \mathbf{\Sigma} = \begin{pmatrix} 3 & 0 & 0 & 3 \\ 0 & 3 & 2 & -1 \\ 0 & 2 & 2 & -1 \\ 3 & -1 & -1 & 5 \end{pmatrix}.$$

Let $\mathbf{X}_1 = \begin{pmatrix} X_1 \\ X_2 \end{pmatrix}$, $\mathbf{X}_2 = \begin{pmatrix} X_3 \\ X_4 \end{pmatrix}$, $\mathbf{A} = \begin{pmatrix} 1 \\ 2 \end{pmatrix}^T$, and $\mathbf{B} = \begin{pmatrix} 1 & -2 \\ 2 & -1 \end{pmatrix}$. Please find

(a) $E(\mathbf{X}_1)$
(b) $E(\mathbf{AX}_1)$
(c) $\text{cov}(\mathbf{X}_1)$
(d) $\text{var}(\mathbf{AX}_1)$
(e) $E(\mathbf{X}_2)$
(f) $E(\mathbf{BX}_2)$
(g) $\text{cov}(\mathbf{X}_2)$
(h) $\text{cov}(\mathbf{BX}_2)$

3. Repeat Exercise 2, but with $\mathbf{A}$ and $\mathbf{B}$ replaced by

$$\mathbf{A} = \begin{pmatrix} 1 \\ -1 \end{pmatrix}^T \quad \text{and} \quad \mathbf{B} = \begin{pmatrix} 2 & -1 \\ 0 & 2 \end{pmatrix}.$$

4. Let $\mathbf{X} = (X_1\ X_2\ X_3)^T$ be a random vector with $\mathbf{X} \sim \mathcal{N}(\boldsymbol{\mu}, \boldsymbol{\Sigma})$ with

$$\boldsymbol{\mu} = \begin{pmatrix} -2 \\ 1 \\ 4 \end{pmatrix}, \quad \boldsymbol{\Sigma} = \begin{pmatrix} 1 & -2 & 0 \\ -2 & 5 & 0 \\ 0 & 0 & 4 \end{pmatrix}.$$

Which of the following random variables are independent? Please explain.
(a) X_1 and X_2
(b) X_2 and X_3
(c) X_1 and X_3
(d) (X_1, X_2) and X_3
(e) $\dfrac{2X_1 + X_2}{3}$ and X_3

5. Consider the random vector $\mathbf{X}$ in Exercise 2.
   (a) Find the distribution of $X_2 + X_3 + X_4$.
   (b) Find the distribution of $3X_2 - 2X_3 + X_4$.
   (c) Find the joint distribution of $X_2 + X_3 + X_4$ and $3X_2 - 2X_3 + X_4$.
   (d) Find the distribution of $X_1 - X_2 + 2X_3 + X_4$
   (e) Find a 2×1 vector $\mathbf{c}$ such that X_2 and $X_2 + \mathbf{c}^T \begin{pmatrix} X_3 \\ X_4 \end{pmatrix}$ are independent.
   (f) Find a 2×1 vector $\mathbf{c}$ such that X_2 and $X_4 + \mathbf{c}^T \begin{pmatrix} X_2 \\ X_3 \end{pmatrix}$ are independent.
6. Consider the random vector in Exercise 4.
   (a) Find the conditional distribution of X_1, given that $X_3 = x_3$.
   (b) Find the conditional distribution of X_1, given that $X_2 = x_2$.
7. Consider the random vector $\mathbf{X}$ in Exercise 2.
   (a) Find the conditional distribution of X_1, given that $X_2 = x_2$ and $X_3 = x_3$.
   (b) Find the conditional distribution of X_2, given that $X_3 = x_3$ and $X_4 = x_4$.
   (c) Find the conditional distribution of X_3, given that $X_2 = x_2$ and $X_4 = x_4$.
   (d) Find the conditional distribution of $(X_2\ X_3)^T$, given that $X_4 = x_4$.
8. Calculate by hand the maximum likelihood estimates of the mean vector $\boldsymbol{\mu}$ and the covariance matrix $\boldsymbol{\Sigma}$ of $(X_2\ X_3)^T$ based on the first five observations of the last two variables in Table 2.1, assuming the observations are from a bivariate normal population.

9. Consider a random sample of size $n = 3$ from a bivariate normal population as shown in the following table.

x_1	x_2
5	8
9	5
7	2

Evaluate the T^2-statistic used to test $H_0 : \boldsymbol{\mu} = \boldsymbol{\mu}_0$ based on this data set, where $\boldsymbol{\mu}_0 = (8\ 4)^T$. What is the distribution of the T^2-statistic in this case?

10. Consider the data from a bivariate normal population in the following table:

x_1	x_2
4	14
10	11
8	11
10	12

(a) Evaluate the T^2-statistic for testing $H_0 : \boldsymbol{\mu} = \begin{pmatrix} 9 & 13 \end{pmatrix}^T$ using the data.
(b) Specify the distribution of the T^2-statistic from (a).
(c) Using (a) and (b), test H_0 at the $\alpha = 0.05$ level. What conclusion do you reach?

11. Use the data in Exercise 10 to calculate the likelihood ratio test statistic LR using (3.23). Verify the correctness of (3.24) for this data.

12. Consider the hot rolling process as described in Example 3.2. Check if the mean side temperatures for the defective billets at the following locations along Stand 5 deviate significantly from the nominal values:
(a) locations 10 and 15 with nominal mean temperatures equal to 1852.6 and 1872.4, respectively.
(b) locations 6, 7, and 8 with nominal mean temperatures equal to 1878.0, 1868.5, and 1860.6, respectively.
(c) locations 17, 18, 19, and 20 with nominal mean temperatures equal to 1876.7, 1875.7, 1872.7, and 1868.5, respectively.

13. Perform Bayesian inference for the mean of side temperatures at locations 6, 7, and 8 based on the data set `side_temp_defect`. Please use the sample covariance of all the data at these three locations as the true covariance matrix and assume it is known. The mean and covariance matrix of the prior distribution is:

$$\boldsymbol{\mu}_0 = \begin{pmatrix} 1878.0 \\ 1868.5 \\ 1860.6 \end{pmatrix}, \quad \boldsymbol{\Sigma}_0 = \begin{pmatrix} 100 & 0 & 0 \\ 0 & 100 & 0 \\ 0 & 0 & 100 \end{pmatrix}.$$

Please find the posterior distribution of the mean temperatures at locations 6, 7, and 8 based on the first five ($n = 5$) observations, and the posterior distribution based on the first 100 ($n = 100$) observations, respectively. Comment on how the posterior distributions are different for different sample sizes. And compare the MAP estimate with the MLE.

4

Explaining Covariance Structure: Principal Components

4.1 Introduction to Principal Component Analysis

For data with a number of correlated variables, most of the variation in the data can often be explained by just a few new variables, which are called principal components. In principal component analysis (PCA), each principal component is a linear combination of the original variables. PCA can be used in dimension reduction when the majority of the variation in the data can be represented by a small number of principal components. In industrial applications, each principal component may correspond to a source of data variation. Consequentially PCA is a useful tool to identify variation sources such as process faults. Such applications are discussed in detail in Chapter 7.

Example 4.1 Principal component analysis on two variables of the `Auto` data set. We will use the `Auto` data set from the `ISLR` package of `R`, which contains information for cars such as gas mileage, horsepower, and weight, to illustrate the use of PCA. We first consider two variables in the `Auto` data set: `displacement` and `horsepower`. From the scatter plot in Figure 4.1 it is clear that there is a strong positive correlation between the two variables. Two orthogonal directions, $\mathbf{v}_1$ and $\mathbf{v}_2$ are marked on the scatter plot. It can be seen that most of the data variation is distributed along the direction $\mathbf{v}_1$. The remaining variation not captured by the $\mathbf{v}_1$ direction is distributed along $\mathbf{v}_2$. The direction $\mathbf{v}_1$ corresponds to the first principal component of the data, which is the direction in which the data exhibit the largest variation. The direction $\mathbf{v}_2$ corresponds to the second principal component, which is orthogonal to $\mathbf{v}_1$ and captures the remaining variation in the data not captured by $\mathbf{v}_1$.

Industrial Data Analytics for Diagnosis and Prognosis: A Random Effects Modelling Approach, First Edition. Shiyu Zhou and Yong Chen.

The principal components are defined as linear combinations of the centered variables, with the mean of each variable subtracted. The data are centered because the principal components are used to explain only the covariance structure of the data. For the `Auto` data set shown in Figure 4.1, the two principal components are defined as:

$$b_i = v_{i1}(\text{displacement} - 194.4) + v_{i2}(\text{horsepower} - 104.5), i = 1, 2 \quad (4.1)$$

where 194.4 and 104.5 is the mean of `displacement` and `horsepower`, respectively, and the coefficients of the linear combination, v_{i1} and v_{i2}, are called the *loadings* of the *i*th principal component. The loading vectors $\mathbf{v}_i = (v_{i1}\ v_{i2})^T$, $i = 1,2$, correspond exactly to the two orthogonal directions marked in Figure 4.1.

The principal components of a data set can be obtained using the `R` function `princomp()`. The following `R` codes are used to find b_1 and b_2 for the data in Figure 4.1.

```
Auto.df <- read.csv("Auto.csv", header = T)
names(Auto.df)
Auto.2pc <- princomp(Auto.df[, 3:4])
Auto.2pc$loadings
```

The `loadings` value of `princomp()` gives the matrix of loadings, with each column corresponding to the loadings of a principal component:

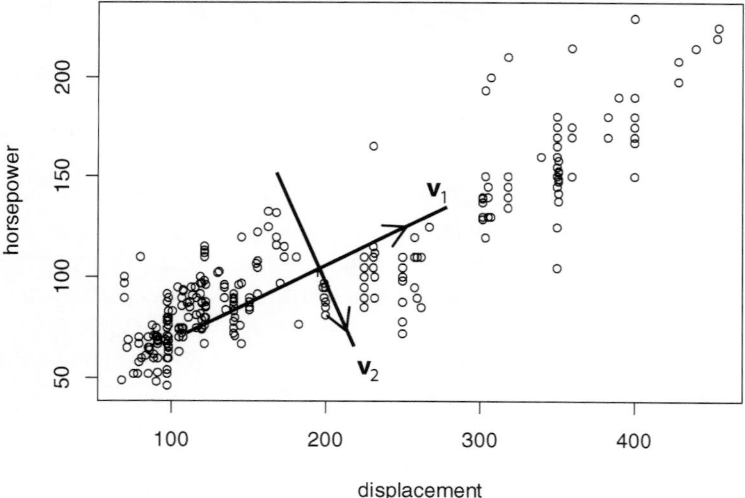

Figure 4.1 Scatter plot of horsepower and displacement.

```
Loadings:
              Comp.1   Comp.2
displacement   0.947    0.320
horsepower     0.320   -0.947
```

Based on the outputs above, the first principal component on the two variables displacement and horsepower is:

$$b_1 = (0.947)(\text{displacement} - 194.4) + (0.320)(\text{horsepower} - 104.5) \qquad (4.2)$$

and the second principal component is:

$$b_2 = (0.320)(\text{displacement} - 194.4) + (-0.947)(\text{horsepower} - 104.5).$$

Since the first principal component has positive loading weights on displacement and horsepower, and knowing that engine displacement is a determining factor of the power and torque that an engine produces, we can interpret the first principal component as a measure of the overall power of a car. The second principal component has loading weights of opposite signs on displacement and horsepower, indicating that it represents the differences between displacement and horsepower.

Note that the loading vectors, $\begin{pmatrix} 0.947 \\ 0.320 \end{pmatrix}$ and $\begin{pmatrix} 0.320 \\ -0.947 \end{pmatrix}$, of the first and second principal component, respectively, point to exactly the directions of $\mathbf{v}_1$ and $\mathbf{v}_2$ in Figure 4.1. We have seen from Figure 4.1 that the first principal component b_1 captures most of the variation in the data. The exact variation captured by each of the principal components can be obtained using summary() as:

```
> summary(Auto.2pc)
Importance of components:
                          Comp.1      Comp.2
Standard deviation      110.1865     16.0982
Proportion of variance    0.9791      0.0209
Cumulative proportion     0.9791      1.0000
```

From the outputs of summary(), we see that the first principal component explains 97.91% of the total variance in the data, while the second principal component explains only 2.09% of the total variance. This confirms that the first principal component b_1 accounts for most of the variation in this data set.

For a given observation and its displacement and horsepower values, (4.1) and (4.2) can be used to calculate its corresponding b_1 and b_2 values, which are called the *principal component scores*. The principal component scores are equal to the projected values of the (centered) observation onto the direction of the corresponding principal components. The principal component scores can be found by the `scores` value of the `princomp()` function as:

```
> dim(Auto.2pc$scores)
[1]  392  2
> head(Auto.2pc$scores)
     Comp.1   Comp.2
[1,] 114.8  11.8708
[2,] 166.8  -7.5149
[3,] 131.7  -3.5531
[4,] 118.4  -8.0366
[5,] 113.3   0.7962
[6,] 252.2 -13.4768
```

Based on the above outputs, each of the 392 observations in this data set has two principal component scores. The scores of the first six observations are printed. As an example, the first observation in the `Auto` data set is a Chevrolet Chevelle Malibu, with `displacement=307` and `horsepower=130`. Its first principal component score is then $(0.947)(307 - 194.4) + (0.320)(130 - 104.5) = 114.8$, matching the `R` output above. As we found from `summary()`, the first principal component scores capture 97.9% of the total variance in the data. If we would like to reduce the dimension of this two-variable data set to one, we can just keep the first principal component scores, which will retain 97.9% of the variation in the data in a one-dimensional space.

4.1.1 Principal Components for More Than Two Variables

The concept of principal components illustrated using the two variable example of the `Auto` data can be generalized to more than two variables. Generally, the ith principal component of a data set of p numerical variables is defined as:

$$b_i = v_{i1}(x_1 - \bar{x}_1) + v_{i2}(x_2 - \bar{x}_2) + \cdots + v_{ip}(x_p - \bar{x}_p), i = 1, 2, \ldots, p. \quad (4.3)$$

The vector of loadings for the ith principal component, $(v_{i1}\ v_{i2}\ \cdots\ v_{ip})^T$, is the direction of the ith principal component in p-dimensional space of all variables. The directions of the p principal components are orthogonal to each other. The first principal component captures most of the variation in the data. And the ith principal component captures most of the remaining variation in the data that are not captured by the first $(i - 1)$ components, for $i = 2, \ldots, p$. For each p-dimensional observation in a data set, its *score* for the ith principal component can be calculated using (4.3).

Example 4.2 Principal component analysis on five variables of the `Auto` data set. Revisiting the `Auto` data set, now we study the covariance structure of the $p = 5$ numerical variables: `mpg`, `displacement`, `horsepower`, `weight`, and `acceleration`. The following `R` codes are used to perform PCA for these five variables:

```
#column indices of numerical variables
cont.var <- c(1, 3:6)
# PCA for unscaled data
Auto.pc <- princomp(Auto.df[, cont.var])
summary(Auto.pc)
Auto.pc$loadings
```

Based on the results from `summary()`, the total variance of the data is dominated by the first principal component, which accounts for 99.76% of the total variance. The principal component loadings are:

```
Loadings:
              Comp.1  Comp.2   Comp.3   Comp.4   Comp.5
mpg                                      0.999
displacement  -0.114  -0.946    0.303
horsepower            -0.298   -0.949
weight        -0.993   0.121
acceleration                                     0.996
```

In the loadings output, the blanks represent small loading values. The `R` code `print(Auto.pc$loadings, cutoff = 0)` can be used to show the complete loading matrix. The `weight` variable has much higher (absolute) loading weight than the other variables in the first principal component. Does this provide any interesting insight of the covariance structure of the data? Not really. The only reason that `weight` has much larger loading weight than the other variables is that it has much larger variance than the other variables, which can be seen from the following `R` outputs:

```
> apply(Auto.df[, cont.var], 2, var)
mpg       displacement  horsepower  weight    acceleration
6.092e+01  1.095e+04    1.482e+03 7.215e+05   7.611e+00
```

The `weight` variable contributes to almost all the total variance of the data. However, this is not particularly interesting because the dominance of `weight` in variance is completely determined by its measurement unit, which is the *pound* in this data set. Suppose the measurement unit of `weight` is changed to *ton*, its variance will be reduced by a factor of 4×10^6 and its loading weight in the first principal component will be very small. Because it is undesirable for the principal components obtained to depend on an arbitrary choice of units, we typically normalize each variable to have standard deviation equal to one before we perform PCA, as discussed next.

4.1.2 PCA with Data Normalization

When the variables have common measurement units and their scales contain important information, such as dimensional deviations at different locations of a product or temperatures at different points on a part, it may be better to perform PCA without normalizing the variables. However, if the measurement units of different variables are not commensurate, or the scales do not reflect importance, it is generally advisable to perform PCA with normalization, i.e., dividing each (centered) variable by its standard deviation. Since the covariance matrix of normalized variables is the correlation matrix of the original variables, PCA with normalization is equivalent to perform PCA on the correlation matrix. In `princomp()`, we can set `cor = T` to perform PCA with normalization. The following `R` codes perform PCA on the (five) normalized variables in Example 4.2 for the `Auto` data set:

```
Auto_cor.pc <- princomp(Auto.df[, cont.var], cor = T)
```

The proportion of variance explained and the loadings of each principal component based on the normalized variables are:

```
> summary(Auto_cor.pc)
Importance of components:
                   Comp.1 Comp.2  Comp.3   Comp.4   Comp.5
Standard
deviation          1.9816 0.8438  0.47500  0.28788  0.22966
Proportion of
Variance           0.7854 0.1424  0.04512  0.01658  0.01055
Cumulative
Proportion         0.7854 0.9278  0.97288  0.98945  1.00000

> print(Auto_cor.pc$loadings, cutoff = 0)

Loadings:
               Comp.1   Comp.2   Comp.3   Comp.4   Comp.5
mpg             0.444    0.304    0.839    0.020    0.075
displacement-0.483   -0.135    0.371   -0.476   -0.620
horsepower   -0.484    0.124    0.206    0.826   -0.160
weight       -0.471   -0.326    0.305   -0.159    0.744
acceleration 0.335   -0.876    0.150    0.257   -0.178
```

Using normalized variables, the first principal component accounts for 78.5% of total variance. The first two principal components, collectively, explain more than 90% of total variance. Consequently, the data variation is summarized very well by the first two principal components.

Thanks to normalization, the principal component loading weights are not determined by the scaling of individual variables. From the loading matrix, the first principal component in terms of the normalized variables can be written as:

$$\begin{aligned} b_1 = 0.444\ \text{mpg} - 0.483\ \text{displacement} - 0.484\ \text{horsepower} \\ - 0.471\ \text{weight} + 0.335\ \text{acceleration} \end{aligned} \tag{4.4}$$

The first principal component has positive weights in `mpg` and `acceleration` and negative weights in the other three variables. All five weights have similar magnitude. This principal component can be interpreted as a measure of "compactness" of a car since a larger value of b_1 corresponds to lower power and weight, higher fuel efficiency, and inferior acceleration performance (note that a smaller value of `acceleration` corresponds to better performance since it measures the time to reach a certain speed). These are all typical characteristic of a compact size car. In the second principal component, `acceleration` has a negative weight of magnitude much larger than that of the other variables. So the second principal component relates more to a high performance in acceleration. Similarly, the third principal component relates more to the fuel efficiency of a car.

Sometimes it is helpful to flip the sign of a principal component loading vector when interpreting the principal components. A principal component is not affected by a sign flip. As shown in Figure 4.1 the principal component corresponds to a line that extends in either direction. Flipping its sign would not change the direction. If we flip the sign of the first principal component in (4.4), we can interpret it as an overall size index since after the sign flip a large value of the principal component would correspond to the characteristics of a large size car. Two different software packages will yield the same principal component loading vectors for the same data set, although the signs of those loading vectors may differ.

4.1.3 Visualization of Principal Components

A *biplot* can be used to show the data in the space of the first two principal components based on the principal component scores. The loading weights of each variable are also shown on a biplot using lines with arrows. Figure 4.2 shows a biplot of the principal components for the `Auto` data. The following `R` codes are used to draw the biplot in Figure 4.2:

```
data_labels <- rep("+", dim(Auto.df)[1])
data_labels[59] <- "VW Type3"
biplot(Auto_cor.pc, scale = 0,
       xlabs = data_labels,
       xlab = "First Principal Component",
       ylab = "Second Principal Component")
abline(h = 0, v = 0, lty = 2)
```

The R function `biplot()` is used to draw a biplot for PCA. We use the symbol "+" to represent the principal component scores of each observation, except for the 59th observation (a "VolksWagen type 3" car) where "VW Type3" is used. The data labels on a biplot can be set by "`xlabs=`" in `biplot()`. The `scale=0` argument to `biplot()` ensures that the principal component scores and loadings are scaled appropriately. The `abline()` function is used to draw the horizontal and vertical dashed lines passing the origin. From the biplot in Figure 4.2, the `VM Type3` car has a positive score in the first principal component and a negative score in the second principal component, which means that it is a compact car with slow acceleration. The plot of the variables using the arrows in Figure 4.2 show consistent patterns as to what we see from the principal component loading vectors. For example, the `acceleration` has the greatest influence (with negative weight) on the second principal component among the five variables. Overall, we see that the three power/weight variables (`horsepower`, `displacement`, and `weight`) are located close to each other on the biplot, indicating that the power/weight variables are correlated with each other—cars with more weights tend to have higher power.

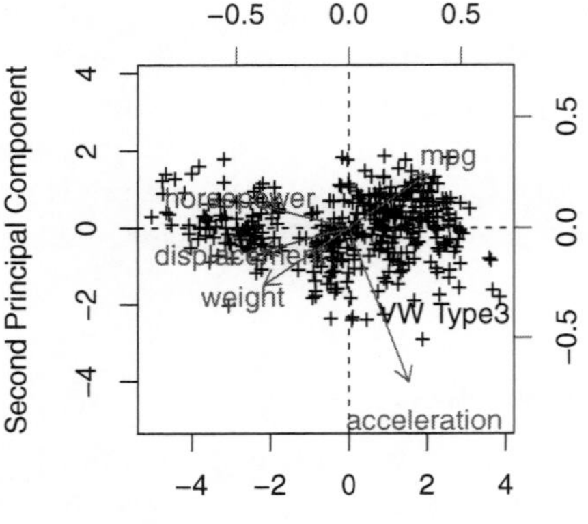

Figure 4.2 Biplot of the first two principal components for the Auto data.

4.1.4 Number of Principal Components to Retain

When PCA is used for dimension reduction, we would like to retain a small subset of principal components that can summarize most variation in the data. Typically the first few principal components can explain the majority of the variation in the data. An important question is how many of those principal components we should retain. A commonly used tool to help decide the number of principal components is the so-called *scree plot*. It is a plot of the variance explained by each of the principal component, which always follows a decreasing trend. The scree plot for the Auto data with five variables is shown in Figure 4.3. The R code used to draw this scree plot is:

```
screeplot(Auto_cor.pc, type = "lines",
main = "Scree Plot for PCA of Auto Data")
```

Typically the variance explained by the principal components drops quickly for the first few principal components. After a certain principal component, the variance explained stays at a sustained low level, forming an "elbow" shape in the scree plot. For the scree plot in Figure 4.3, an elbow can be identified at the second, and maybe also the third principal component. After identifying an elbow, we can retain the principal components *before* the elbow since they explain much larger variations in the data than those in the flat segment. Following this idea, for the Auto data with five variables, we can retain the first or the first two principal components. It should be noted that this type of visual analysis is inherently ad hoc. Unfortunately, there is no well-accepted objective method to decide the number of principal components. In practice,

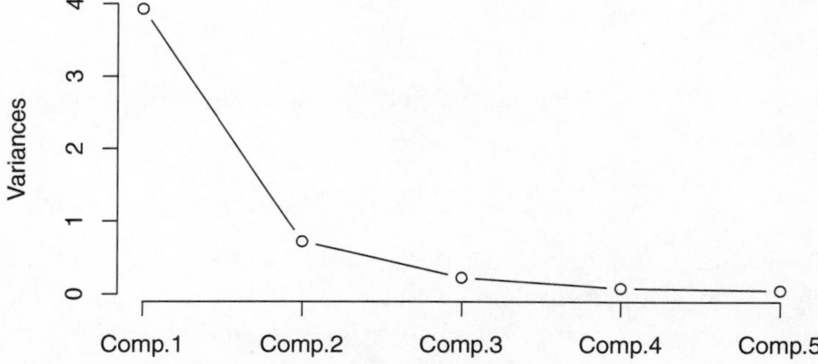

Figure 4.3 Scree plot for the Auto data with five variables.

other factors such as interpretability of the selected principal components in a specific application should also be considered.

4.2 Mathematical Formulation of Principal Components

Consider a data set consisting of n p-dimensional observation vectors, $\mathbf{x}_1, \mathbf{x}_2, \ldots, \mathbf{x}_n$, where $\mathbf{x}_i = (x_{i1}\ x_{i2}\ \ldots\ x_{ip})^T$. From Section 4.1, the data should be first centered by subtracting the sample mean of each variable before performing PCA. Starting from this section, we assume all the observations $\mathbf{x}_i$ are already centered with the sample mean of each variable equal to zero. As illustrated in Figure 4.1, the first principal component corresponds to the direction that captures the most variation in the data. To find the first principal component, consider a linear combination of $x_{i1}, x_{i2}, \ldots, x_{ip}$, denoted by:

$$b_{i1} = \mathbf{v}_1^T \mathbf{x}_i = v_{11}x_{i1} + v_{12}x_{i2} + \cdots + v_{1p}x_{ip},$$

where $\mathbf{v}_1 = (v_{11}\ v_{12}\ \ldots\ v_{1p})^T$. We want to find $\mathbf{v}_1$ that maximizes the sample variance of $b_{i1}, i = 1, 2, \ldots, n$. From (2.8), the sample variance of $b_{i1} = \mathbf{v}_1^T \mathbf{x}_i$ is equal to $\mathbf{v}_1^T \mathbf{S} \mathbf{v}_1$, where $\mathbf{S}$ is the sample covariance matrix of $\mathbf{x}_1, \mathbf{x}_2, \ldots, \mathbf{x}_n$. Obviously $\mathbf{v}_1^T \mathbf{S} \mathbf{v}_1$ can be increased unboundedly if we allow $\| \mathbf{v}_1 \| \to \infty$. Therefore, a well-defined optimization problem to find $\mathbf{v}_1$ is to maximize $\mathbf{v}_1^T \mathbf{S} \mathbf{v}_1$ subject to the constraint that $\mathbf{v}_1^T \mathbf{v}_1 = 1$. The optimal solution of this problem should satisfy the condition that it is a stationary point of the Lagrange function:

$$\mathbf{v}_1^T \mathbf{S} \mathbf{v}_1 - \lambda_1 (\mathbf{v}_1^T \mathbf{v}_1 - 1),$$

where λ_1 is the Lagrange multiplier. By setting the first derivative of the Lagrange function with respect to $\mathbf{v}_1$ to be equal to zero, we have

$$\mathbf{S} \mathbf{v}_1 = \lambda_1 \mathbf{v}_1,$$

which indicates that $\mathbf{v}_1$ must be an eigenvector of $\mathbf{S}$ and λ_1 is the corresponding eigenvalue. Thus, we have

$$\mathbf{v}_1^T \mathbf{S} \mathbf{v}_1 = \mathbf{v}_1^T (\lambda_1 \mathbf{v}_1) = \lambda_1. \tag{4.5}$$

To maximize $\mathbf{v}_1^T \mathbf{S} \mathbf{v}_1$, λ_1 should be the largest eigenvalue of $\mathbf{S}$ and the optimal $\mathbf{v}_1$ is the eigenvector with eigenvalue λ_1. And $\mathbf{v}_1$ is the loading vector of the first principal component and the corresponding linear combination $b_{i1} = \mathbf{v}_1^T \mathbf{x}_i$, $i = 1, 2, \ldots, n$ are the scores of the first principal component.

For the second principal component, we want to find a direction that can maximize the variance not captured by the first principal component. The scores of the second principal component are defined as:

$$b_{i2} = \mathbf{v}_2^T\mathbf{x}_i = v_{21}x_{i1} + v_{22}x_{i2} + \cdots + v_{2p}x_{ip},$$

where $\mathbf{v}_2 = (v_{21}\ v_{22} \dots\ v_{2p})^T$. The loading vector $\mathbf{v}_2$ should be orthogonal to $\mathbf{v}_1$ so that b_{i2} are uncorrelated with b_{i1}. We can solve for $\mathbf{v}_2$ by maximizing $\mathbf{v}_2^T\mathbf{S}\mathbf{v}_2$ subject to constraints $\mathbf{v}_2^T\mathbf{v}_2 = 1$ and the orthogonality condition $\mathbf{v}_1^T\mathbf{v}_2 = 0$. Based on a similar argument as for the first principal component, the optimal solution of this optimization problem is given by the eigenvector of $\mathbf{S}$ with the second largest eigenvalue, λ_2. And $\mathbf{v}_2^T\mathbf{S}\mathbf{v}_2$, which is the sample variance of $\mathbf{v}_2^T\mathbf{x}_i$, is equal to λ_2. Continuing this procedure, for $k = 1, 2, \dots, p$, the kth principal component can be defined as:

$$b_{ik} = \mathbf{v}_k^T\mathbf{x}_i = v_{k1}x_{i1} + v_{k2}x_{i2} + \ldots + v_{kp}x_{ip},$$

where $\mathbf{v}_k = (v_{k1}\ v_{k2} \dots\ v_{kp})^T$. The loading vector of the kth principal component, $\mathbf{v}_k$, is exactly the eigenvector of $\mathbf{S}$ with eigenvalue λ_k, the kth largest eigenvalue of $\mathbf{S}$. The loading vectors of the p principal components are orthogonal to each other so that the corresponding principal components are uncorrelated. The sample variance of the kth principal component is equal to λ_k. It maximizes the variance that is not captured by the first $k-1$ principal components.

4.2.1 Proportion of Variance Explained

Because each eigenvalue λ_k, $k = 1, 2, \dots, p$, is the variance of the corresponding principal component, we can define the *proportion of variance explained* by the kth principal component as follows:

$$\begin{aligned}\text{Proportion of variance} &= \frac{\lambda_k}{\lambda_1 + \lambda_2 + \cdots + \lambda_p} \\ &= \frac{\lambda_k}{\text{tr}(\mathbf{S})}.\end{aligned} \tag{4.6}$$

The second equality is because the summation of all eigenvalues is equal to the trace of $\mathbf{S}$, i.e., $\sum_{k=1}^{p}\lambda_k = \text{tr}(\mathbf{S}) = \sum_{i=1}^{p}s_{ii}$, where s_{ii} is the sample variance of the variable x_i. So $\sum_{i=1}^{p}s_{ii}$ can be considered as the *total variance* present in a data set and (4.6) corresponds to the proportion of the total variance explained by the kth principal component. If most of the total variance of the data set can be explained by just a few of the principal components, the dimension of the data can be significantly reduced by using just these principal components without much loss of information. The scree plot introduced in Section 4.1 can be used to visualize the variance explained by each principal component to decide how many principal components should be retained.

Example 4.3 Consider the two variables, `displacement` and `horsepower`, of the `Auto` data set studied in Example 4.1. The sample covariance matrix of the data and its eigenvectors and eigenvalues are obtained using `R`:

```
> print(S <- cov(Auto.df[, 3:4]))
              displacement   horsepower
displacement     10950.368     3614.034
horsepower        3614.034     1481.569
> options(digits=3)
> eigen(S)
eigen() decomposition
$values
[1]  12172   260
$vectors
       [,1]     [,2]
[1,] -0.947   0.320
[2,] -0.320  -0.947
```

From the `R` outputs, the sample covariance matrix of these two variables are

$$\mathbf{S} = \begin{pmatrix} 10950 & 3614 \\ 3614 & 1482 \end{pmatrix}.$$

And the loading vectors of the two principal components are:

$$\mathbf{v}_1 = \begin{pmatrix} -0.947 \\ -0.320 \end{pmatrix} \text{ and } \mathbf{v}_2 = \begin{pmatrix} 0.320 \\ -0.947 \end{pmatrix}.$$

The loading vectors of principal components are consistent with those obtained in Example 4.1, except that the sign of $\mathbf{v}_1$ is flipped because each eigenvector and principal component is unique only up to a sign flip. It is also easy to verify that $\mathbf{v}_1^T\mathbf{v}_1 = 1$, $\mathbf{v}_2^T\mathbf{v}_2 = 1$, and $\mathbf{v}_1^T\mathbf{v}_2 = 0$. The two eigenvalues: $\lambda_1 = 12172$ and $\lambda_2 = 260$ are the variances explained by the first and the second principal components, respectively. The total variance of the data set is $\lambda_1 + \lambda_2 = s_{11} + s_{22} = 12432$. The first principal component explains $12172/12432 = 0.979 = 97.9\%$ of the total variance, which is the same as the output of PCA in Example 4.1.

Example 4.4 (Diagonal covariance matrix) Suppose the sample covariance matrix of a data set $\mathbf{S}$ has the form

$$\mathbf{S} = \begin{pmatrix} s_{11} & 0 & \dots & 0 \\ 0 & s_{22} & \dots & 0 \\ \vdots & \vdots & & \vdots \\ 0 & 0 & \dots & s_{pp} \end{pmatrix}.$$

The variables are uncorrelated because $\mathbf{S}$ is diagonal. The eigenvalues of $\mathbf{S}$ are $s_{ii}, i = 1, 2, \dots, p$ and the corresponding eigenvectors have a 1 in the ith position and 0 elsewhere:

$$\mathbf{v}_i = (0 \ \dots \ 0 \ 1 \ 0 \ \dots \ 0)^T.$$

It is easy to verify that $\mathbf{Sv}_i = \lambda_i \mathbf{v}_i$. Therefore $\mathbf{v}_i$ is a principal component loading vector of the diagonal covariance matrix $\mathbf{S}$ with variance λ_i, $i = 1, 2, \dots, p$. For any observation $\mathbf{x} = (x_1, \dots, x_p)^T$, its ith principal component score is

$$b_i = \mathbf{v}_i^T \mathbf{x} = x_i.$$

So when the sample covariance matrix is diagonal, the principal components are exactly the same as the original uncorrelated variables. In this case, the principal components do not provide a better summary of the data variation than the original variables.

4.2.2 Principal Components Obtained from the Correlation Matrix

When each (centered) variable is normalized by dividing its sample standard deviation, the sample covariance matrix of the normalized data is the same as the sample correlation matrix, $\mathbf{R}$, of the original data. Consequently, the kth principal component of the normalized data is the eigenvector of $\mathbf{R}$ with the kth largest eigenvalue. Because $\text{tr}(\mathbf{R}) = p$, the proportion of (normalized) sample variance explained by the kth principal component becomes:

$$\text{proportion of variance} = \frac{\lambda_k}{p}, \ k = 1, 2, \dots, p.$$

Example 4.5 For the five variables of `Auto` data set studied in Example 4.2, to perform PCA for the normalized data, we first find the sample correlation matrix and its eigenvectors and eigenvalues using:

```
R <- cor(Auto.df[, cont.var])
eigen(R)
```

```
> eigen(R)
eigen() decomposition
$values
[1] 3.9268 0.7120 0.2256 0.0829 0.0527

$vectors
        [,1]     [,2]     [,3]     [,4]     [,5]
[1,]  0.444   -0.304    0.839   0.0196   -0.075
[2,] -0.483    0.135    0.371  -0.4762    0.620
[3,] -0.484   -0.124    0.206   0.8256    0.160
[4,] -0.471    0.326    0.305  -0.1594   -0.744
[5,]  0.335    0.876    0.150   0.2565    0.178
```

From the R outputs, the loading vectors of the first two principal components obtained from the correlation matrix are:

$$\mathbf{v}_1 = \begin{pmatrix} 0.444 \\ -0.483 \\ -0.484 \\ -0.471 \\ 0.335 \end{pmatrix}, \text{ and } \mathbf{v}_2 = \begin{pmatrix} -0.304 \\ 0.135 \\ -0.124 \\ 0.326 \\ 0.876 \end{pmatrix}$$

The loading vectors of the first two principal components are consistent with those obtained in Example 4.2 with data normalization. The five eigenvalues: $\lambda_1 = 3.927$, $\lambda_2 = 0.712$, $\lambda_3 = 0.226$, $\lambda_4 = 0.083$, and $\lambda_5 = 0.053$ are the variances explained by each of the five principal components. The total variance is $\sum_{k=1}^{5} \lambda_k = p = 5$. The first principal component explains $3.927 / 5 = 0.785 = 78.5\%$ of the total variance of the normalized data, and the first two principal components together explain $(3.927 + 0.712) / 5 = 0.928 = 92.8\%$ of the total variance. These results are the same as the output of PCA in Example 4.2 with data normalization.

4.3 Geometric Interpretation of Principal Components

4.3.1 Interpretation Based on Rotation

Let $\mathbf{V}$ be the matrix whose columns are eigenvectors of $\mathbf{S}$:

$$\mathbf{V} = (\mathbf{v}_1 \quad \mathbf{v}_2 \quad \dots \quad \mathbf{v}_p).$$

For an observation $\mathbf{x}_i$ in the data set, $i = 1, 2, \ldots, n$, the vector of principal component scores, $\mathbf{b}_i = (b_{i1}, b_{i2}, \ldots, b_{ip})^T$ can be written as

$$\mathbf{b}_i = \mathbf{V}^T\mathbf{x}_i. \tag{4.7}$$

Since the columns of $\mathbf{V}$ are eigenvectors with unit norm, it is an orthogonal matrix with $\mathbf{V}^T\mathbf{V} = \mathbf{V}\mathbf{V}^T = \mathbf{I}$. And $\mathbf{V}^T$ is also an orthogonal matrix. From (4.7), $\mathbf{b}_i$ is the multiplication of an orthogonal matrix and $\mathbf{x}_i$, which can be considered as the coordinates of $\mathbf{x}_i$ under a rotation of the axes. It can be seen that after the rotation the distance of each observation to the origin is unchanged:

$$\mathbf{b}_i^T\mathbf{b}_i = (\mathbf{V}^T\mathbf{x}_i)^T(\mathbf{V}^T\mathbf{x}_i) = \mathbf{x}_i^T\mathbf{x}_i.$$

The rotated axes are given by $\mathbf{v}_k$, $k = 1, \ldots, p$. Since $\mathbf{v}_k$ has a unit length, geometrically the kth principal component score of $\mathbf{x}_i$, $b_{ik} = \mathbf{v}_k^T\mathbf{x}_i$, is the projected values of $\mathbf{x}_i$ onto the direction of $\mathbf{v}_k$. Equivalently, we can consider $\mathbf{v}_k^T\mathbf{x}_i$ as the kth coordinate value of $\mathbf{x}_i$ in the rotated coordinate system. The new variables $\mathbf{b}_i$ after the rotation are uncorrelated because based on (2.10) and eigendecomposition, the sample covariance matrix of $\mathbf{b}_i$ is equal to $\mathbf{V}^T\mathbf{S}\mathbf{V} = \mathbf{\Lambda}$, where $\mathbf{\Lambda}$ is a diagonal matrix.

Example 4.6 Suppose a data set is a random sample of a random vector $\mathbf{X} = \begin{pmatrix} X_1 \\ X_2 \end{pmatrix}$ following a bivariate normal distribution with zero mean. Assume the sample covariance matrix and population covariance matrix are equal and given as:

$$\mathbf{\Sigma} = \mathbf{S} = \begin{pmatrix} 2.5 & 1.5 \\ 1.5 & 2.5 \end{pmatrix}.$$

By solving the characteristic equation $|\mathbf{S} - \lambda\mathbf{I}| = 0$, the two eigenvalues of $\mathbf{S}$ are $\lambda_1 = 4$ and $\lambda_2 = 1$. Based on $\mathbf{S}\mathbf{v} = \lambda\mathbf{v}$, the corresponding eigenvectors can be obtained as

$$\mathbf{v}_1 = \begin{pmatrix} \frac{\sqrt{2}}{2} \\ \frac{\sqrt{2}}{2} \end{pmatrix}, \mathbf{v}_2 = \begin{pmatrix} \frac{-\sqrt{2}}{2} \\ \frac{\sqrt{2}}{2} \end{pmatrix}.$$

From Section 3.2, the contour of constant density for $\mathbf{X}$ is an ellipse with major axis in the direction of $\mathbf{v}_1$ and minor axis in the direction of $\mathbf{v}_2$. So the major and minor axes of the ellipse contour of constant density align exactly with the directions of the first and second principal components, as illustrated in the left panel of Figure 4.4. This is not surprising because intuitively the major axis of the density contour is the direction in which the data cloud has the maximum

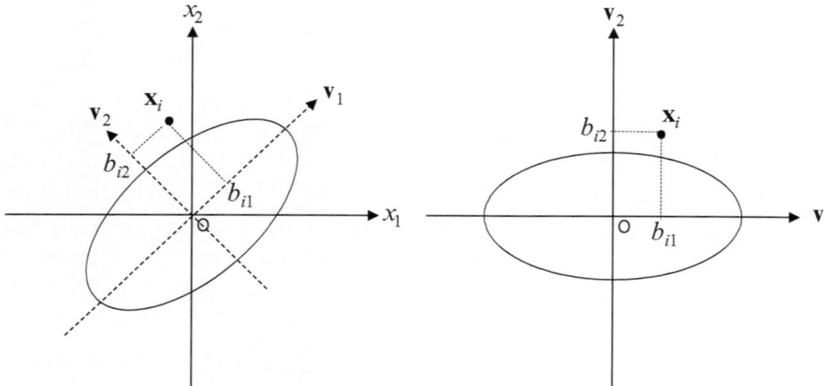

Figure 4.4 Principal components as rotation of axes for a bivariate distribution.

variance, which is the direction of the first principal component. In addition, $\mathbf{v}_1$ and $\mathbf{v}_2$ can be considered as rotation of the original x_1 and x_2 axes by an angle of $\frac{\pi}{4}$ counterclockwise, as illustrated in Figure 4.4. When the original x_1 and x_2 axes are rotated to $\mathbf{v}_1$ and $\mathbf{v}_2$, the major and minor axes of the ellipse contour of the data are aligned with the axes of the new coordinate system, as shown in the right panel of Figure 4.4. The new variables with the principal components as axes are uncorrelated, which can be seen from the shape of the rotated contour. For any observation $\mathbf{x}_i$ in the data set, its projection onto the directions of $\mathbf{v}_1$ and $\mathbf{v}_2$ are its principal component scores, b_{i1} and b_{i2}, respectively. They are also the coordinates under the rotated axes defined by the principal components.

4.3.2 Interpretation Based on Low-Dimensional Approximation

Another geometric interpretation of principal components is that they provide a low-dimensional approximation of the n observations in p-dimensional space with minimum errors. The first principal component can be considered as a line in p-dimensional space that approximate the observations with minimum error in terms of the average squared Euclidean distance. Figure 4.5 illustrates the approximation of the data with the line corresponding to the first principal component in a two-dimensional space. The dotted lines connected to each observation indicate the Euclidean distance to the first principal component line. This idea can be extended to spaces of higher dimensions. In general, the r ($r < p$) dimensional linear subspace defined by the first r principal components approximates the p dimensional observations with minimum error in terms of the average squared Euclidean distance.

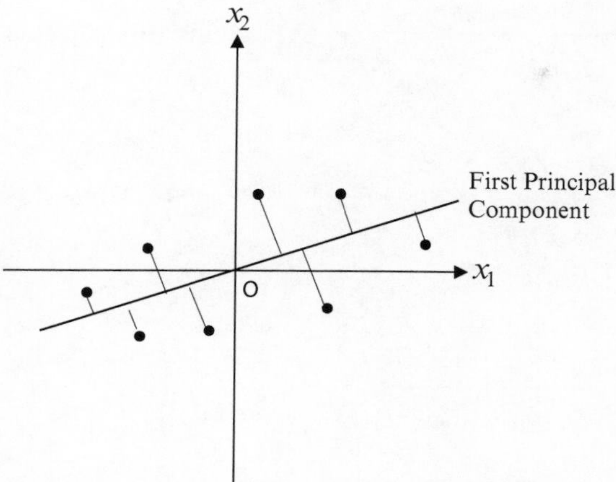

Figure 4.5 The first principal component approximation of the data.

Example 4.7 A data set contains $n = 4$ observations with $p = 2$ variables as given in the following table.

observation	x_1	x_2
1	−1.1	−1.1
2	−1.0	1.0
3	1.0	−1.0
4	1.1	1.1

The eigenvectors and eigenvalues of the sample covariance matrix can be obtained using the following R codes:

```
minerr.df <- data.frame(x1 = c(-1.1, -1, 1, 1.1),
                        x2 = c(-1.1, 1, -1, 1.1))
eigen(cov(minerr.df))
```

From the outputs of the above R codes, the loading vector of the first principal component is $\begin{pmatrix} 0.707 \\ 0.707 \end{pmatrix}$ with variance equal to $\lambda = 1.61$. In Figure 4.6, the line corresponding to the first principal component is drawn, which passes the two data points $(-1.1, -1.1)$ and $(1.1, 1.1)$. Among all straight lines, the first principal component line minimizes the squared distances to the four data points. The distances to the other two points, $(-1, 1)$ and $(1, -1)$ are marked by

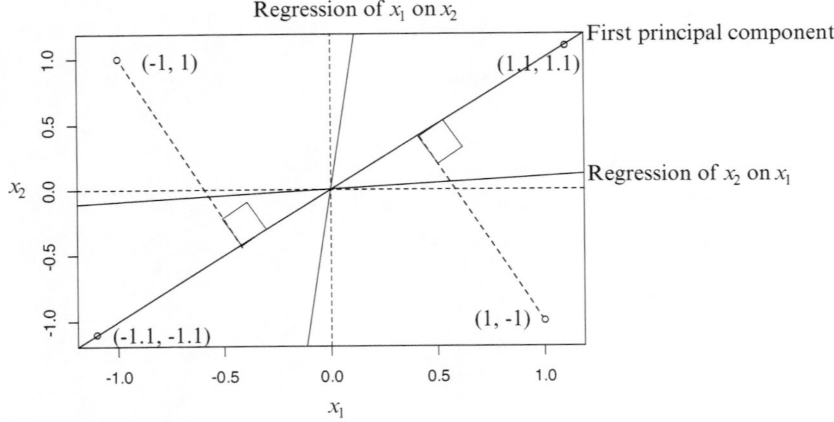

Figure 4.6 First principal component line compared with regression lines.

the dashed lines in Figure 4.6. Note that in PCA, the distances from an observation to the principal component line are measured by the "perpendicular" distance to the line. A linear regression model that will be introduced in Chapter 5 also uses a straight line to approximate the data. However, in linear regression, the approximation error is not measured by "perpendicular" distance. Therefore, the first principal component line is different from the linear regression lines in general. For this example, the lines for regression of x_2 on x_1 and regression of x_1 on x_2 are both shown in Figure 4.6. It is clear that the two regression lines are quite different from the first principal component line for this example.

Bibliographic Notes

Principal component analysis was first invented in 1901 by Karl Pearson [1901]. It is also named the Karhunen–Loéve transform (KLT) in the signal processing field. It is a popular method in both statistics and machine learning and discussed in books from both fields such as Johnson et al. [2002], Rencher [2003], Anderson [2003], Hastie et al. [2009], and Bishop [2006].

Exercises

1. A data set with two variables has sample covariance matrix given by

$$\mathbf{S} = \begin{pmatrix} s_{11} & s_{12} \\ s_{12} & s_{22} \end{pmatrix}.$$

If $s_{11} = s_{22}$, show that the loading vector of the first principal component is $\mathbf{v}_1 = \begin{pmatrix} \frac{\sqrt{2}}{2} \\ \frac{\sqrt{2}}{2} \end{pmatrix}$.

2. A data set with three variables has sample covariance matrix given by

$$\mathbf{S} = \begin{pmatrix} 1 & -2 & 0 \\ -2 & 5 & 0 \\ 0 & 0 & 3 \end{pmatrix}.$$

Find all the principal components based on $\mathbf{S}$. What is the proportion of the total variance explained by the first principal component and the first two principal components, respectively?

3. A data set with two variables has sample covariance matrix given by

$$\mathbf{S} = \begin{pmatrix} 1 & 4 \\ 4 & 100 \end{pmatrix}.$$

   (a) Find all the principal components based on $\mathbf{S}$. What is the proportion of the total variance explained by the first principal component?
   (b) Find all the principal components for the normalized data. What is the proportion of the total (normalized) variance explained by the first principal component? Compare the results with part (a).

4. For the `auto_spec` data set, use `R` to perform the principal component analysis for the natural logarithm of the following three variables: `length`, `width`, and `height`.
   (a) Find the loadings of all three principal components and the proportion of total variance explained by each of the principal components.
   (b) Write the equations for the first two principal components and provide a subject-matter interpretation of these two components.
   (c) Draw the scree plot and comment on how many principal components should be retained to summarize most of the data variation.

5. Consider the three variables in the `auto_spec` data set that are considered in Exercise 4.
   (a) Find the sample covariance matrix of these three variables.

(b) Obtain the eigenvalues and eigenvectors of the sample covariance matrix and use them to find the loading vectors and the proportion of total variance explained of each principal component. Compare the results with the `R` outputs from Exercise 4.

6. For the `auto_spec` data set, use `R` to perform the principal component analysis for the following four variables: `curb.weight`, `horsepower`, `city.mpg`. and `highway.mpg`.
   (a) Perform the PCA for the non-normalized data. Find the loadings of all principal components and the proportion of total variance explained by each of the principal components.
   (b) Perform the PCA for the normalized data. Find the loadings of all principal components and the proportion of total variance explained by each of the principal components.
   (c) Draw the biplot of the first two principal components for the normalized data.
   (d) Comment on which provides more useful information: PCA on non-normalized data or PCA on normalized data.
7. Use `R` to perform PCA for the Stand 5 side temperature data in `stand_5_side_temp.csv` for a hot rolling process. Find the proportion of total variance explained by the first 10 principal components. Use the scree plot to find a proper number of principal components to retain for dimension reduction.

5

Linear Model for Numerical and Categorical Response Variables

In many applications with data collected on multiple variables, there is a particular variable of interest called the *response variable*. In the literature, the response variable is also called the dependent variable, output, outcome, target, or y-variable. One important goal of data analytics is to build a model to describe the relationship between the response variable and other variables. In other words, we want to know how the values of other variables affect the response variable. Such a model can be used to gain useful insights to the factors affecting the response variable and support decisions such as process adjustment and control, system design, and maintenance. This section introduces models for both numerical and categorical response variables that can be represented by a linear function.

5.1 Numerical Response – Linear Regression Models

A linear regression model is used for a numerical response variable. The following example is used to illustrate the application of a linear regression model.

Example 5.1 Combined Cycle Power Plant Data. Consider the combined cycle power plant data set [Tüfekci, 2014; Kaya et al., 2012], where we are interested in predicting the electrical power output (`PE`), which is considered as the response variable, using other variables, including Ambient Temperature (`AT`), Exhaust Vacuum (`V`), Ambient Pressure (`AP`), and Relative Humidity (`RH`), which are called the *predictor* variables. In statistics and data mining literature, the predictor variable is also called the input, feature, regressor, or independent variable. A mathematical model describing the relationship

Industrial Data Analytics for Diagnosis and Prognosis: A Random Effects Modelling Approach, First Edition. Shiyu Zhou and Yong Chen.

between the response variable and the predictor variables is helpful to make such predictions. Perhaps the simplest such model is represented by a linear function as:

$$\mathrm{PE} = \beta_0 + \beta_1 \times \mathrm{AT} + \beta_2 \times \mathrm{V} + \beta_3 \times \mathrm{AP} + \beta_4 \times \mathrm{RH} + \epsilon, \tag{5.1}$$

where ϵ represents the zero-mean random error term that captures the information that cannot be explained by the predictor variables. The intercept β_0 and coefficients $\beta_i, i = 1, \ldots, 4$ of the linear function in (5.1) are model parameters that should be estimated from the data. A *linear regression* model is a model that uses a linear function with respect to the model parameters to predict the response variable, as in (5.1). The `lm()` function in `R` can be used to fit a linear regression model to the data:

```
CCPP.df <- read.csv("CCPP.csv", header = T)
CCPP.lm <- lm(PE ~ ., data = CCPP.df)
```

The model parameter estimates that best fit the data can be displayed using `summary()`:

```
> summary(CCPP.lm)

Call:
lm(formula = PE ~ ., data = CCPP.df)

Residuals:

    Min       1Q   Median       3Q      Max
-43.435   -3.166   -0.118    3.201   17.778

Coefficients:

              Estimate Std. Error  t value   Pr(>t)
(Intercept) 454.609274   9.748512   46.634  < 2e-16 ***
AT           -1.977513   0.015289 -129.342  < 2e-16 ***
V            -0.233916   0.007282  -32.122  < 2e-16 ***
AP            0.062083   0.009458    6.564 5.51e-11 ***
RH           -0.158054   0.004168  -37.918  < 2e-16 ***
---
Signif. codes: 0 '***' 0.001 '**' 0.01 '*' 0.05 '.' 0.1 ' ' 1

Residual standard error: 4.558 on 9563 degrees of freedom
Multiple R-squared: 0.9287, Adjusted R-squared: 0.9287
F-statistic: 3.114e+04 on 4 and 9563 DF, p-value: < 2.2e-16
```

From the R output, the estimated value of the intercept is 454.6 and the estimated coefficients for AT, V, AP, and RH are -1.9775, -0.234, 0.062, and -0.158, respectively. Substituting these estimated model parameters into (5.1), we have

$$\text{PE} = 454.6 - 1.9775 \times \text{AT} - 0.234 \times \text{V} + 0.062 \times \text{AP} - 0.158 \times \text{RH} + \epsilon.$$

Since ϵ has a zero mean, we can predict PE using the equation

$$\widehat{\text{PE}} = 454.6 - 1.9775 \times \text{AT} - 0.234 \times \text{V} + 0.062 \times \text{AP} - 0.158 \times \text{RH}, \quad (5.2)$$

where $\widehat{\text{PE}}$ denotes the prediction of the electrical power output PE. For example, if the ambient temperature is 20°C, the exhaustive vacuum is 40 cm Hg, the ambient pressure is 1000 millibar, and the relative humidity is 50%, the predicted electrical power output can be calculated as $454.6 - 1.9775 \times 20 - 0.234 \times 40 + 0.062 \times 1000 - 0.158 \times 50 = 459.76$ MW. We plot the predicted power output for the entire data set, which can be obtained from the `fitted.values` value of `lm()`, versus the actual output in Figure 5.1 based on the following R codes.

```
plot(CCPP.lm$fitted.values~CCPP.df$PE,
   xlab = "Actual PE", ylab = "Predicted PE")
abline(0, 1, col = "green", lwd = 2)
```

We draw a line of slope equal to one on the plot. If the predicted power output matches exactly the actual output, the corresponding data point in Figure 5.1

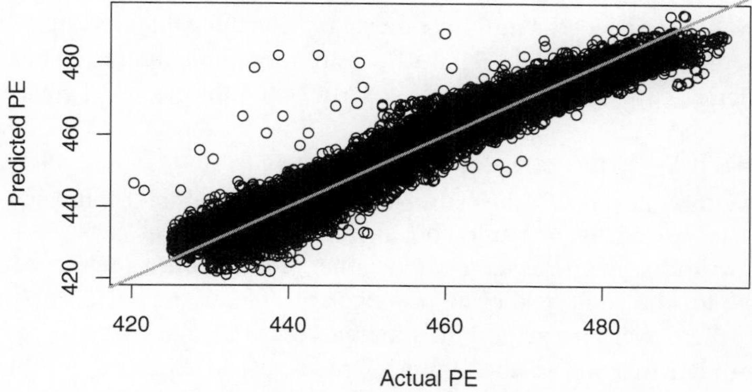

Figure 5.1 Predicted versus actual power energy outputs.

will be right on the line. A point above the line corresponds to a case of overprediction while a point below the line a case of underprediction. From Figure 5.1, most of the points are in a band close to the slope line. The largest prediction errors are more likely due to overprediction.

5.1.1 General Formulation of Linear Regression Model

In general, let Y denote the response variable and $X_1, X_2, \dots, X_p$ denote p predictor variables. A linear regression model takes the form

$$Y = \beta_0 + \beta_1 X_1 + \beta_2 X_2 + \cdots + \beta_p X_p + \epsilon, \tag{5.3}$$

where ϵ is the random error term with zero mean. The model parameters $\beta_i, i = 0, 1, \dots, p$ are unknown and should be estimated from the data. Estimation of model parameters in a linear regression model is discussed in detail in Section 5.2. Given parameter estimates $\widehat{\beta}_i, i = 0, 1, \dots, p$, the response variable can be predicted by

$$\widehat{Y} = \widehat{\beta}_0 + \widehat{\beta}_1 X_1 + \widehat{\beta}_2 X_2 + \cdots + \widehat{\beta}_p X_p. \tag{5.4}$$

5.1.2 Significance and Interpretation of Regression Coefficients

After we fit a linear regression model for a data set, we would like to know which predictors have a statistically significant non-zero effect on the response variable. This problem is answered by testing the null hypothesis $H_0 : \beta_i = 0$ for the ith regression coefficient, $i = 1, \dots, p$. In the `R` output from the regression model for Example 5.1, the last column of the coefficients table gives the p-value of this hypothesis test. If the p-value is smaller than a significance level α, we reject the null-hypothesis and conclude that β_i is significantly non-zero. For Example 5.1, all p-values are very small, indicating that all the predictors have a significantly non-zero effect on the electrical power output.

From (5.4), if X_i is increased by 1, $\widehat{Y}$ is increased by $\widehat{\beta}_i, i = 1, 2, \dots, p$. Therefore we can interpret $\widehat{\beta}_i$, the estimate of the ith regression coefficient, as: if X_i is increased by one unit, on average the response variable is increased by $\widehat{\beta}_i$ units, given the values of the other predictors are unchanged. For example, for the combined cycle power plant data, based on (5.2), we can say that the power energy output is increased by 0.062 MV on average for every millibar increase of the ambient pressure (`AP`). Similarly, if the ambient temperature (`AT`) is increased by 1°C, an average decrease of 1.9775 MW of power output is expected since the regression coefficient of `AT` is negative.

5.1.3 Other Types of Predictors in Linear Models

In the combined cycle power plant (CCPP) data example, all the predictors are numerical variables in the data set. In general, predictors in a linear regression model can also be (coding of) categorical variables, interactions between variables, and nonlinear transformations of numerical variables.

Categorical Predictors

We use the following example to illustrate how other types of predictors are handled in linear regression.

Example 5.2 We consider the auto_spec data set from Chapter 2, which is similar to the Auto data set but contains more vehicle specification variables. We first fit a linear regression model to predict the highway miles per gallon (highway.mpg) using five variables as predictors: fuel.type, body.style, height, length, and horsepower. Among the five predictor variables, height, length, and horsepower are all numerical variables and can be directly used as predictor X_j in the linear regression model in (5.3). The other two variables, fuel.type and body.style, are both categorical variables and cannot be used in the linear regression model directly.

The variable fuel.type has two possible values, diesel and gas. For a predictor variable with only two levels such as fuel.type, we can simply create a numerical indicator or dummy variable taking the form

$$X_{\text{gas}} = \begin{cases} 1 & \text{if fuel type is gas} \\ 0 & \text{if fuel type is diesel} \end{cases}.$$

Since the indicator variable X_{gas} has a numerical coding, it can be used as a predictor in the linear regression model.

The variable body.style has five possible values (levels): hardtop, wagon, sedan, hatchback, and convertible. For a categorical variable with more than two levels, a single dummy variable cannot represent all its possible values. In general, the number of dummy variables is one fewer than the number of levels. For example, we need the following $5-1=4$ dummy variables to represent body.style:

$$X_j = \begin{cases} 1 & \text{if body style is equal to } j \\ 0 & \text{if body style is not equal to } j \end{cases}, \; j = \text{hardtop, hatchback, sedan, and wagon.}$$

A dummy variable is created for each level of body.style except for convertible because we (arbitrarily) chose convertible as the *baseline* level. If a car is a convertible, all four dummy variables of the car will be equal to zero. For any other body style, one of the four dummy variables will be equal to one. The dummy variables are numerical coding of a categorical variable, and can be used in the linear regression model. The `lm()` function in `R` automatically creates the dummy variables for the categorical variables in the data set when fitting a linear regression model.

```
auto.spec.df<-read.csv("auto_spec.csv", header = T)
auto.spec.lm1<-lm(highway.mpg~fuel.type+body.style
                  +height+width+horsepower,
                  data = auto.spec.df)
summary(auto.spec.lm1)
```

```
> summary(auto.spec.lm1)

Call:
   lm(formula = highway.mpg ~ fuel.type + body.style +
    height + width + horsepower, data = auto.spec.df)

Residuals:

   Min      1Q    Median      3Q       Max
-9.155  -1.763    0.023    1.636    20.228

Coefficients:

                     Estimate Std. Error t value Pr(>t)
(Intercept)           136.1206  12.6512  10.76  < 2e-16 ***
fuel.typegas           -5.2315   1.0277  -5.09  8.4e-07 ***
body.stylehardtop       3.1142   2.0565   1.51    0.132
body.stylehatchback     3.2453   1.6314   1.99    0.048 *
body.stylesedan         3.1058   1.6587   1.87    0.063 .
body.stylewagon         1.1677   1.8944   0.62    0.538
height                 -0.2372   0.1523  -1.56    0.121
width                  -1.2418   0.1991  -6.24  2.8e-09 ***
horsepower             -0.0854   0.0106  -8.07  7.4e-14 ***
---

Signif. codes: 0 '***' 0.001 '**' 0.01 '*' 0.05 '.' 0.1 ' ' 1

Residual standard error: 3.78 on 194 degrees of freedom
(2 observations deleted due to missingness)
Multiple R-squared: 0.713, Adjusted R-squared: 0.701
F-statistic: 60.2 on 8 and 194 DF, p-value: <2e-16
```

In the above R outputs, regression coefficients are estimated for the single dummy variable for `fuel.type` and the four dummy variables for `body.style`. These estimated coefficients can be interpreted similarly to the numerical predictors. For example, the $\widehat{\beta}_j$ for the dummy variable of `fuel.type` is -5.2315, meaning that the average highway MPG of a car with gas fuel is 5.2315 lower than that of a car with diesel fuel, given that the other predictors have the same values. The $\widehat{\beta}_j$ for the dummy variable created for the `hatchback` value of `body.style` is 3.2453, meaning that, given that all other predictors are the same, the average highway MPG of a car of hatchback style is 3.2353 higher than a car of convertible style, the baseline level. In general, for a categorical variable with more than two levels, the estimated coefficient of a dummy variable can be interpreted as the difference in the average response from the baseline level.

Interactions and Nonlinear Transformation of Variables

In addition to categorical variables, sometimes we want to include nonlinear transformations of variables, such as the product of two or more variables, and square, square-root, or log of some numerical variables. All these nonlinear terms can be easily incorporated into the linear regression model because the linear regression model is *linear* in the model parameters $\boldsymbol{\beta}$, but not necessarily linear in the predictors. For example, in the `auto_spec` data set, it is interesting to study how `width×height` affects the response because the product of `width` and `height` can be considered as an approximation of the cross-sectional area of a vehicle, which can be a meaningful measure of the vehicle size. In statistics, the product of two or more variables is referred to as an *interaction* effect. An interaction effect is needed if the effect of a variable may depend on the value of another variable, which could be the case for `width` and `height`.

Now consider the relationship between `highway.mpg`, the response variable, and the predictor `horsepower`. The scatter plot of these two variables is shown in Figure 5.2. The regression fit of `highway.mpg` on `horsepower` is shown by the straight line. It can be seen that the observations with low and high horsepower significantly deviated from the linear regression line, indicating the relationship between the two variables may be nonlinear. We can introduce a new predictor corresponding to the square of `horsepower` in the linear regression model to capture the nonlinear relationship. The linear regression model that considers both the interaction between `width` and `height` and the quadratic term of `horsepower` can be written as

$$\begin{aligned}\text{highway.mpg} &= \beta_0 + \beta_1 \times \text{width} + \beta_2 \times \text{height} + \beta_3 \times (\text{width} \times \text{height}) \\ &\quad + \beta_4 \times \text{horsepower} + \beta_5 \times \text{horsepower}^2 + \epsilon \end{aligned} \tag{5.5}$$

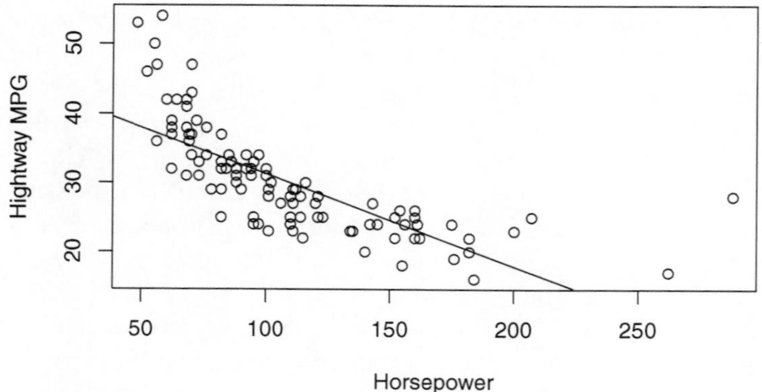

Figure 5.2 Scatter plot of Highway MPG versus Horsepower, with linear regression fit shown

For illustration purpose, we do not include other predictors in this model. The model in (5.5) is still a linear regression model because the right-hand side is a linear function of the model parameters β, although the response variable has a nonlinear relationship with the original predictor variables. The R codes used to fit the model in (5.5) are

```
auto.spec.lm2 <-lm(highway.mpg ~ height*width
                   + horsepower + I(horsepower^2),
                   data = auto.spec.df)
summary(auto.spec.lm2)
```

The syntax `height*width` used in the `formula` argument of `lm()` simultaneously includes `height`, `width`, and the interaction term `height×width` as predictors. It can be also written as `height+width+height:width`, where `height:width` represents the interaction term. The quadratic term of `horsepower` is represented by `I(horsepower^2)` instead of `horsepower^2` because the ^ has a special meaning in a formula. The outputs of the above R codes are

```
> summary(auto.spec.lm2)

Call:
lm(formula = highway.mpg ~ height * width + horsepower +
    I(horsepower^2), data = auto.spec.df)
```

```
Residuals:
    Min        1Q    Median        3Q       Max
-8.3923   -1.2400   -0.0469    1.6558   12.4395

Coefficients:
                   Estimate  Std. Error  t value  Pr(>t)
(Intercept)       5.000e+02   1.747e+02    2.862   0.00467 **
height           -7.388e+00   3.176e+00   -2.326   0.02101 *
width            -6.554e+00   2.656e+00   -2.468   0.01446 *
horsepower       -3.955e-01   2.665e-02  -14.840   < 2e-16 ***
I(horsepower^2)   1.086e-03   9.794e-05   11.085   < 2e-16 ***
height:width      1.094e-01   4.829e-02    2.265   0.02461 *
---
Signif. codes: 0 '***' 0.001 '**' 0.01 '*' 0.05 '.' 0.1 ' ' 1

Residual standard error: 3.182 on 197 degrees of freedom
    (2 observations deleted due to missingness)
Multiple R-squared: 0.7939, Adjusted R-squared: 0.7886
F-statistic: 151.7 on 5 and 197 DF, p-value: < 2.2e-16
```

From the estimated regression coefficients, we can see that the p-values of `height`, `width`, and their interaction are all statistically significant. So it is clear that the interaction term should be included in the model. In addition, the p-value of the quadratic term of `horsepower` is very small. So the effect of `horsepower`^{2} is highly significant, which further verifies that the relationship between `highway.mpg` and `horsepower` is indeed nonlinear.

5.2 Estimation and Inferences of Model Parameters for Linear Regression

Suppose we have a data set of size n with observations of response variable given by $y_1, y_2, \ldots, y_n$. And the corresponding predictors of y_i are $\{x_{i1}, x_{i2}, \ldots, x_{ip}\}$ with $n > p+1$. The relationship between the response variables and the corresponding predictors can be described by the linear regression model:

$$y_i = \beta_0 + \beta_1 x_{i1} + \beta_2 x_{i2} + \cdots + \beta_p x_{ip} + \epsilon_i,\ i = 1, \ldots, n. \tag{5.6}$$

The goal of model parameter estimation and inference is to estimate β_i, $i = 0, 1, \ldots, p$ and test hypothesis related to these model parameters based on a given data set.

5.2.1 Least Squares Estimation

The most popular parameter estimation method for linear regression models is the method of *least squares*. The idea is to find $\hat{\beta} = (\hat{\beta}_0\ \hat{\beta}_1\ \ldots\ \hat{\beta}_p)^T$ to minimize the sum of the squared prediction errors:

$$\begin{aligned} \text{SSE} &= \sum_{i=1}^{n} (y_i - \hat{y}_i)^2 \\ &= \sum_{i=1}^{n} (y_i - \hat{\beta}_0 - \hat{\beta}_1 x_{i1} - \cdots - \hat{\beta}_p x_{ip})^2. \end{aligned}$$

Let $\mathbf{y} = (y_1\ \ldots\ y_n)^T$, and

$$\mathbf{X} = \begin{pmatrix} 1 & x_{11} & x_{12} & \cdots & x_{1p} \\ 1 & x_{21} & x_{22} & \cdots & x_{2p} \\ \vdots & \vdots & \vdots & & \vdots \\ 1 & x_{n1} & x_{n2} & \cdots & x_{np} \end{pmatrix}.$$

The matrix $\mathbf{X}$ is also referred to as the *design matrix* of the linear regression model. Then the sum of squared errors can be written in matrix notation as

$$\text{SSE} = (\mathbf{y} - \mathbf{X}\hat{\beta})^T (\mathbf{y} - \mathbf{X}\hat{\beta}). \tag{5.7}$$

To minimize SSE, we set the first derivative to zero as

$$\frac{\partial \text{SSE}}{\partial \hat{\beta}} = -2\mathbf{X}^T (\mathbf{y} - \mathbf{X}\hat{\beta}) = \mathbf{0} \Rightarrow (\mathbf{X}^T\mathbf{X})\hat{\beta} = \mathbf{X}^T\mathbf{y}. \tag{5.8}$$

Assuming $\mathbf{X}$ has full column rank so that $\mathbf{X}^T\mathbf{X}$ is invertible, the unique solution of (5.8) is given by

$$\hat{\beta} = (\mathbf{X}^T\mathbf{X})^{-1}\mathbf{X}^T\mathbf{y}. \tag{5.9}$$

Because the Hessian of SSE $\frac{\partial^2 \text{SSE}}{\partial \hat{\beta}\, \partial \hat{\beta}^T} = 2\mathbf{X}^T\mathbf{X}$ is a positive definite matrix when $\mathbf{X}$ has full column rank, SSE is a strictly convex function of $\hat{\beta}$ and (5.9) gives the unique optimal solution that minimizes SSE.

Based on the least squares estimate $\hat{\beta}$, the predicted values, also called *fitted values*, of the response variable can be obtained as

$$\hat{\mathbf{y}} = \mathbf{X}\hat{\beta} = \mathbf{X}(\mathbf{X}^T\mathbf{X})^{-1}\mathbf{X}^T\mathbf{y} = \mathbf{H}\mathbf{y},$$

where $\hat{\mathbf{y}} = (\hat{y}_1 \cdots \hat{y}_n)^T$ and the matrix $\mathbf{H} = \mathbf{X}(\mathbf{X}^T\mathbf{X})^{-1}\mathbf{X}^T$ is called the *hat* matrix because it puts the "hat" on the response $\mathbf{y}$. Figure 5.3 illustrates an geometrical interpretation of the least squares estimate. Consider the vector of responses $\mathbf{y}$

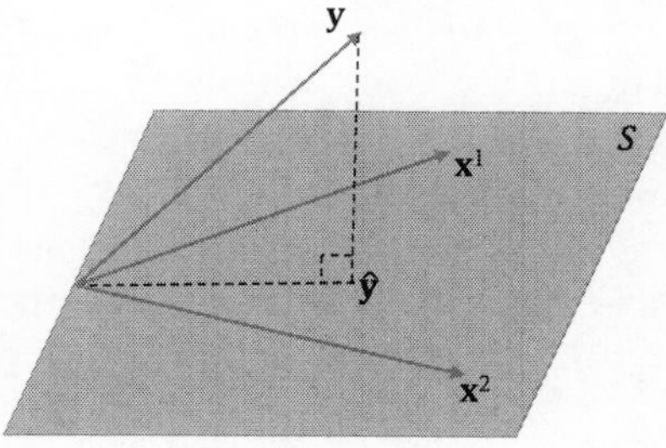

Figure 5.3 Geometrical illustration of least squares estimation.

in $\mathcal{R}^n$. Let $\mathbf{x}^j$ denote the jth column of the matrix $\mathbf{X}$, which is also a vector in $\mathcal{R}^n$, $j = 1, \ldots, p$. For any parameter estimate $\hat{\beta}$, the fitted value $\hat{\mathbf{y}} = \mathbf{X}\hat{\beta}$ is in the subspace in $\mathcal{R}^n$ spanned by $\mathbf{x}^1, \mathbf{x}^2, \ldots, \mathbf{x}^p$, which is denoted by $\mathcal{S}$. Figure 5.3 illustrates the case when $n = 3$, $p = 2$ and the two-dimensional plane is the subspace spanned by $\mathbf{x}^1$ and $\mathbf{x}^2$. In the least squares estimation, we look for $\hat{\beta}$ that minimizes $\text{SSE} = \| \mathbf{y} - \mathbf{X}\hat{\beta} \|^2$, which is equivalent to minimizing the Euclidean distance between $\mathbf{y}$ and $\hat{\mathbf{y}}$. Intuitively, as illustrated in Figure 5.3, the $\hat{\mathbf{y}}$ that minimizes the Euclidean distance between $\mathbf{y}$ and $\hat{\mathbf{y}}$ should be the orthogonal projection of $\mathbf{y}$ onto $\mathcal{S}$. The least squares estimate exactly corresponds to the orthogonal projection of $\mathbf{y}$ onto $\mathcal{S}$, where the orthogonality between $\mathbf{y} - \hat{\mathbf{y}}$ and $\mathcal{S}$ can be seen from (5.8). Since multiplication of the hat matrix $\mathbf{H} = \mathbf{X}(\mathbf{X}^T\mathbf{X})^{-1}\mathbf{X}^T$ projects $\mathbf{y}$ onto the space spanned by the columns of $\mathbf{X}$, $\mathbf{H}$ is a matrix known as the *projection matrix*. Since the projection of a projected vector onto the same subspace is the same as the projected vector, all projection matrix $\mathbf{H}$ satisfy $\mathbf{H}^2 = \mathbf{H}$. This can be easily verified for the hat matrix based on its definition.

Now we assume that $\mathbf{X}$ is fixed (non-random) and the random errors ϵ_i in (5.6) are uncorrelated with zero mean and constant variance σ^2, which is also equal to the variance of y_i. It is easy to see that the mean of $\hat{\beta}$ is equal to β, indicating the least squares estimate is unbiased. The covariance matrix of $\hat{\beta}$ can be obtained as

$$\text{cov}(\hat{\beta}) = (\mathbf{X}^T\mathbf{X})^{-1}\mathbf{X}^T(\sigma^2\mathbf{I})\mathbf{X}(\mathbf{X}^T\mathbf{X})^{-1} = \sigma^2(\mathbf{X}^T\mathbf{X})^{-1}. \tag{5.10}$$

The variance σ^2 can be estimated by

$$\hat{\sigma}^2 = \frac{1}{n-p-1}\sum_{i=1}^{n}(y_i - \hat{y}_i)^2 = \frac{1}{n-p-1}(\mathbf{y} - \mathbf{X}\hat{\beta})^T(\mathbf{y} - \mathbf{X}\hat{\beta}). \tag{5.11}$$

The estimate in (5.11) is similar to the sample variance of the model residuals $\mathbf{y} - \mathbf{X}\hat{\beta}$. The $n-p-1$ used in the denominator is to make sure $\hat{\sigma}^2$ is an unbiased estimate of σ^2. When σ^2 is unknown, the covariance matrix of $\hat{\beta}$ can be estimated by substituting $\hat{\sigma}^2$ in (5.11) for σ^2 in (5.10).

Example 5.3 Revisit Example 4.7, where we considered a data set containing $n = 4$ observations with two variables, x_1 and x_2, as given in the following table.

observation	x_1	x_2
1	−1.1	−1.1
2	−1.0	1.0
3	1.0	−1.0
4	1.1	1.1

Instead of performing PCA on the data set, we fit a linear regression model with x_2 being the response variable and x_1 the predictor. The linear regression model has the form

$$x_2 = \beta_0 + \beta_1 x_1 + \epsilon. \tag{5.12}$$

A linear regression model with a single predictor variable, such as the model in (5.12), is often called *simple linear regression*. For this data set, we have

$$\mathbf{X} = \begin{pmatrix} 1 & -1.1 \\ 1 & -1.0 \\ 1 & 1.0 \\ 1 & 1.1 \end{pmatrix} \quad \text{and} \quad \mathbf{y} = \begin{pmatrix} -1.1 \\ 1.0 \\ -1.0 \\ 1.1 \end{pmatrix}.$$

The least squares estimate of the regression coefficients can be calculated as

$$\hat{\beta} = \begin{pmatrix} \hat{\beta}_0 \\ \hat{\beta}_1 \end{pmatrix} = (\mathbf{X}^T\mathbf{X})^{-1}\mathbf{X}^T\mathbf{y} = \begin{pmatrix} 0.000 \\ 0.095 \end{pmatrix}.$$

Based on the least squares estimates, the fitted value of x_2 is given by $\hat{x}_2 = 0.095x_1$, which is a line with zero intercept and slope equal to 0.095. This regression line is shown in Figure 4.6, together with the line for the first principal component, which has a slope equal to 1. So the principal component line is very different from the regression line for this example. Although both the principal components and the least squares method can be interpreted as approximation of the observations by minimizing some kind of squared errors, there is an essential difference in how the errors are defined in these two methods, which is illustrated in Figure 5.4. The approximation errors of the

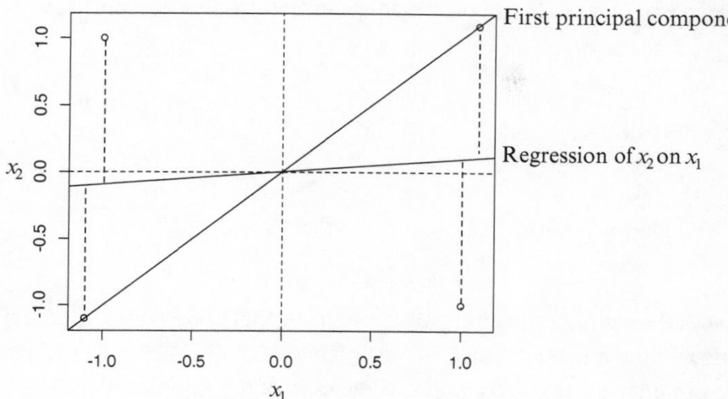

Figure 5.4 Errors are measured by vertical distance in regression.

line corresponding to the first principal component is measured by the distance perpendicular to the line, as marked by the dashed lines in Figure 4.6. In linear regression, however, we care about the approximation of the response variable only, not all the variables. Thus the approximation errors for a regression line are measured by the vertical distances from the observations to the regression line, as marked by the dashed lines in Figure 5.4.

Example 5.4 Consider the linear regression model fitting for the combined cycle power plant (CCPP) data set in Example 5.1. We will use the following R codes to obtain the least squares estimate of β and its covariance matrix.

```
CCPP.n <- nrow(CCPP.df)
CCPP.X <- as.matrix(cbind(intercept = rep(1, CCPP.n),
                    CCPP.df[, 1:4]))
CCPP.y <- CCPP.df[, 5]
# Least squares estimate of beta
CCPP.beta <- solve(t(CCPP.X)%*%CCPP.X)%*%t(CCPP.X)%*%
                    CCPP.y
# estimate of variance
sigma2.hat <- 1/(CCPP.n-4-1)*t(CCPP.y-CCPP.X%*%
               CCPP.beta)%*%(CCPP.y-CCPP.X%*%CCPP.beta)
# covariance matrix of estimate of beta
cov.beta <- drop(sigma2.hat)*solve(t(CCPP.X)%*%CCPP.X)
```

The R codes above use (5.9) to estimate model parameters. The results are:

```
> CCPP.beta
                [,1]
intercept  454.6093
AT          -1.9775
V           -0.2339
AP           0.0621
RH          -0.1581
```

These values match the estimated coefficients from the output of the `lm()` function for the CCPP data as shown in Section 5.1. We estimate σ^2 using (5.11) and evaluate its square root to obtain the estimated standard deviation of the error term:

```
> sqrt(sigma2.hat)
        [,1]
[1,]    4.56
```

The result matches the `Residual standard error` in the output of the `lm()` function for the CCPP data. Substituting $\hat{\sigma}^2$ for σ^2 in (5.10), the estimated covariance matrix of $\hat{\boldsymbol{\beta}}$ is obtained as

```
> cov.beta
          intercept    AT            V          AP           RH
intercept 95.0335  -6.26e-02   8.10e-03 -9.21e-02  -1.22e-02
AT        -0.0626   2.34e-04 -9.07e-05  5.94e-05   3.77e-05
V          0.0081  -9.07e-05  5.30e-05 -8.32e-06  -1.04e-05
AP        -0.0921   5.94e-05 -8.32e-06  8.95e-05   1.06e-05
RH        -0.0122   3.77e-05 -1.04e-05  1.06e-05   1.74e-05
```

The standard deviation of $\hat{\beta}_i$, $i = 1, \ldots, p$, can be obtained by taking the square root of the diagonal elements of $\text{cov}(\hat{\boldsymbol{\beta}})$:

```
> sqrt(diag(cov.beta))
intercept        AT          V         AP         RH
9.74851    0.01529    0.00728    0.00946    0.00417
```

These values match the `Std.Error` column of the coefficient table in the output of `lm()` function for the CCPP data.

5.2.2 Maximum Likelihood Estimation

In this subsection, we study the maximum likelihood estimation (MLE) of model parameters for linear regression. For this purpose, we make an additional assumption for the model as given in (5.6) that the random errors are i.i.d. and ϵ_i follow a normal distribution, that is, $\epsilon_i \sim \mathcal{N}(0,\sigma^2)$. Under this assumption, the likelihood function can be written as

$$L(\boldsymbol{\beta}, \sigma) = (2\pi\sigma^2)^{-\frac{n}{2}} \exp\left(-\frac{1}{2\sigma^2}\sum_{i=1}^{n}(y_i - \mathbf{x}_i^T\boldsymbol{\beta})^2\right),$$

where $\mathbf{x}_i$ is the ith row of $\mathbf{X}$. By taking the log of the likelihood function, we obtain the negative log-likelihood

$$-l(\boldsymbol{\beta}, \sigma^2) = \frac{n}{2}\ln(2\pi) + \frac{n}{2}\ln\sigma^2 + \frac{1}{2\sigma^2}\sum_{i=1}^{n}(y_i - \mathbf{x}_i^T\boldsymbol{\beta})^2. \tag{5.13}$$

Based on the method of maximum likelihood, we would like to find $\widehat{\boldsymbol{\beta}}$ and $\widehat{\sigma}^2$ that minimize the negative log-likelihood. It is easy to see from (5.13) that to minimize $-l(\boldsymbol{\beta}, \sigma)$ we need to choose $\boldsymbol{\beta}$ to minimize $\sum_{i=1}^{n}(y_i - \mathbf{x}_i^T\boldsymbol{\beta})^2$, which is exactly the same as the SSE defined in (5.7). Consequently, the MLE of the regression coefficients $\boldsymbol{\beta}$ is the same as its least squares estimate. That is, the MLE of $\boldsymbol{\beta}$ is

$$\widehat{\boldsymbol{\beta}} = (\mathbf{X}^T\mathbf{X})^{-1}\mathbf{X}^T\mathbf{y}. \tag{5.14}$$

To find the MLE of σ^2, we substitute $\boldsymbol{\beta}$ with $\widehat{\boldsymbol{\beta}}$ in the negative log-likelihood and take the partial derivative with respect to σ^2:

$$\begin{aligned}\frac{\partial}{\partial\sigma^2}(-l(\widehat{\boldsymbol{\beta}}, \sigma^2)) &= \frac{\partial}{\partial\sigma^2}\left(\frac{n}{2}\ln(2\pi) + \frac{n}{2}\ln\sigma^2 + \frac{1}{2\sigma^2}\sum_{i=1}^{n}(y_i - \mathbf{x}_i^T\widehat{\boldsymbol{\beta}})^2\right)\\ &= \frac{1}{2\sigma^2}\left(n - \frac{1}{\sigma^2}\sum_{i=1}^{n}(y_i - \mathbf{x}_i^T\widehat{\boldsymbol{\beta}})^2\right).\end{aligned} \tag{5.15}$$

Setting the partial derivative in (5.15) to zero, we obtain the MLE of σ^2 as

$$\widehat{\sigma}^2 = \frac{1}{n}\sum_{i=1}^{n}(y_i - \mathbf{x}_i^T\widehat{\boldsymbol{\beta}})^2 = \frac{1}{n}(\mathbf{y} - \mathbf{X}\widehat{\boldsymbol{\beta}})^T(\mathbf{y} - \mathbf{X}\widehat{\boldsymbol{\beta}}).$$

Comparing with (5.11), the difference between the MLE of σ^2 and its least squares estimate is in the denominator, where MLE uses n and the least squares estimate uses $n-p-1$. So the MLE of σ^2 is biased. For fixed p and large sample size n, MLE and the least squares estimate of σ^2 become very close and both converge to the true σ^2.

So far we have assumed that the random error $\boldsymbol{\epsilon} = \left(\epsilon_1\ \epsilon_2 \dots \epsilon_n\right)^T$ is distributed as $\boldsymbol{\epsilon} \sim \mathcal{N}(0,\sigma^2\mathbf{I})$. If we allow a more general variance-covariance structure for $\boldsymbol{\epsilon}$ such that

$$\boldsymbol{\epsilon} \sim \mathcal{N}(\mathbf{0}, \mathbf{V}),$$

where $\mathbf{V}$ is a general covariance matrix of an arbitrary structure, the MLE of $\boldsymbol{\beta}$ can be obtained as

$$\hat{\beta} = (\mathbf{X}\mathbf{V}^{-1}\mathbf{X})^{-1}\mathbf{X}\mathbf{V}^{-1}\mathbf{y}. \tag{5.16}$$

The estimator in (5.16) is called the *generalized least squares* estimator, while (5.14) is also referred to as the ordinary least squares estimate.

5.2.3 Variable Selection in Linear Regression

In practice, one often has more predictors Xs than needed for predicting the response variable Y. Some of the predictors may be redundant and could be discarded. Removing the unnecessary predictor variables can save the costs in collecting a full range of predictors for future predictions. With more resources focusing on the fewer predictors, the measurement accuracy can be improved and the chances of missing values could be reduced. The more parsimonious model resulted from removing some predictors and focusing on the most important ones can lead to easier interpretation of the model. Statistically, removing a predictor from the model can reduce the variances of the regression coefficient estimates and the predictions. On the other hand, dropping predictors that are actually correlated to the response variable can increase the bias of predictions. Sometimes sacrificing a little bit of bias by dropping some predictors to reduce the variance of the predictions may improve the overall prediction accuracy. This is called the *bias-variance trade-off*, which plays a central role in determining the appropriate level of model complexity to improve prediction accuracy.

Two of the most popular approaches for variable selection are (1) best subset selection and (2) stepwise selection. In the best subset selection, all possible subsets of predictors are examined and compared. To select the best subset, we need a criterion to compare the models with different subsets of predictors. One idea is to use the maximum value of the likelihood function, denoted by L^*, as a criterion. However, the issue of this idea is that adding more predictors to a model almost always increases L^*, making it very difficult to use L^* to select the number of predictors. Two of the most popular criteria for model selection are the Akaike Information Criterion (AIC) and the Schwartz's Bayesian Information Criterion (BIC), which are defined as

$$\text{AIC} = -2\ln L^* + 2k$$
$$\text{BIC} = -2\ln L^* + k\ln n,$$

where k is the number of estimated parameters in the model, including the intercept and the variance of random errors. The models with smaller AIC and BIC values are considered better. Both AIC and BIC include a second term that penalizes large models. When additional predictors are added to the model, it will reduce the negative log-likelihood in the first term, while increasing the penalty in the second term. The best model is a result of the trade-off between the goodness-of-fit measured by the likelihood and the model complexity measured by the number of parameters.

The best subset selection becomes computationally infeasible if the number of predictors is large (say, much larger than 40). In a stepwise selection method, rather than search through all possible subsets, we seek a good path through them. Popular stepwise selection algorithms are *forward selection* and *backward elimination*. Forward selection starts with a model with only the intercept, and then sequentially adds into the model the predictor that most improves the model fit. Backward elimination starts with the full model containing all the predictors, and sequentially deletes the predictor that has the least impact on the model fit.

Another alternative of stepwise selection is a hybrid strategy that considers both forward and backward moves at each iteration, and selects the "best" of the two. For example, the `R` function `step` uses the AIC criterion for comparing the choices, and at each iteration performs an add or drop that minimizes the AIC value.

5.2.4 Hypothesis Testing

In this section, we study some basic tests on the regression coefficients $\boldsymbol{\beta}$. We adopt the same model assumptions as in Section 5.2.2. Because the random error ϵ_i follows normal distribution, so is the observation y_i. Being a linear transformation of $\mathbf{y}$, which is a normally distributed random vector, the estimate of regression coefficients, $\widehat{\boldsymbol{\beta}}$, follows a multivariate normal distribution. We have

$$\widehat{\boldsymbol{\beta}} \sim \mathcal{N}(\boldsymbol{\beta}, \sigma^2(\mathbf{X}^T\mathbf{X})^{-1}) \tag{5.17}$$

The distribution of the unbiased estimate of σ^2, as given in (5.11) can be obtained as

$$(n-p-1)\frac{\widehat{\sigma}^2}{\sigma^2} \sim \chi^2_{n-p-1}, \tag{5.18}$$

which is a chi-squared distribution with $n-p-1$ degrees of freedom. And $\widehat{\sigma}^2$ and $\widehat{\boldsymbol{\beta}}$ are statistically independent. To test for a particular regression coefficient, we test the null hypothesis

$$H_0: \beta_j = 0$$

versus the alternative

$$H_1 : \beta_j \neq 0.$$

We form the following test statistic

$$t_j = \frac{\hat{\beta}_j}{\sqrt{\widehat{\text{var}}(\hat{\beta}_j)}}, \tag{5.19}$$

where $\widehat{\text{var}}(\hat{\beta}_j)$ is the jth diagonal element of $\hat{\sigma}^2(\mathbf{X}^T\mathbf{X})^{-1}$, the estimated covariance matrix of $\hat{\boldsymbol{\beta}}$. Intuitively, if $|t_j|$ is large, we can say that the magnitude of $\hat{\beta}_j$ is significantly larger than its estimated standard deviation, indicating there is strong evidence that β_j is non-zero. Based on the distributional properties in (5.17) and (5.18), under the null hypothesis the test statistic t_j follows a Student's t-distribution with $n-p-1$ degrees of freedom. Therefore we reject H_0 at significance level α if $|t_j| > t_{\alpha/2,n-p-1}$. The p-value of the test can be calculated as

$$P = 2\ \Pr(T(n-p-1) > |t_j|), \tag{5.20}$$

where $T(n-p-1)$ denotes a random variable distributed as t_{n-p-1}. The null hypothesis is rejected if $P < \alpha$.

Besides testing a regression coefficients individually, sometimes we want to test all the regression coefficients simultaneously to see if there is at least one non-zero coefficient in the model. The null hypothesis for this test is formulated as

$$H_0 : \beta_1 = \beta_2 = \cdots = \beta_p = 0 \tag{5.21}$$

versus the alternative

$$H_1 : \text{ at least one } \beta_j \text{ is non-zero, } j = 1, \dots, p.$$

Note that the intercept β_0 is not included in the hypothesis so that the data can have a non-zero mean. We call the model with all p predictors as the *full* model and the one under the null hypothesis in (5.21) the *reduced* model. To design a test statistic, we consider the sum of squared errors (SSE) defined in (5.7). For the reduced model, the only non-zero regression coefficient is β_0 with $\hat{\beta}_0 = \overline{y}$, where $\overline{y}$ is the sample mean of $\mathbf{y}$. The SSE of the reduced model can be written as

$$\text{SSE}_R = (\mathbf{y} - \overline{y})^T(\mathbf{y} - \overline{y}),$$

which is the total sum of squares of the data adjusted for the mean. We form the test statistic for the hypothesis in (5.21) as

$$F = \frac{(\text{SSE}_R - \text{SSE}_F)/(df_R - df_F)}{\text{SSE}_F / df_F} = \frac{(\text{SSE}_R - \text{SSE}_F)/p}{\text{SSE}_F/(n-p-1)}, \tag{5.22}$$

where SSE_F is the SSE of the full model, and $df_R = n-1$ and $df_F = n-p-1$ are the degrees of freedom of the reduced model and the full model, respectively. The difference between SSE_R and SSE_F in the numerator of the F statistic is the sum of squares explained by the full regression model. Intuitively, if the F statistic in (5.22) has a large value, we can say that the sum of squares explained by the full regression model is significantly larger than the residual sum of squares, indicating there is strong evidence that some of the p regression coefficients are non-zero. It can be shown that the F-statistic in (5.22) is distributed as $F_{p,n-p-1}$ under the null hypothesis. We reject H_0 if $F > F_{\alpha,p,n-p-1}$. And the p-value of the test can be found as

$$P = \Pr(F(p, n-p-1) > F),$$

where $F(p, n-p-1)$ denotes a random variable distributed as $F_{p,n-p-1}$.

Another commonly used statistic similar to the F statistic is the so-called *coefficient of determination*, which is denoted by R^2:

$$R^2 = \frac{\text{regression sum of squares}}{\text{total sum of squares}} = \frac{\text{SSE}_R - \text{SSE}_F}{\text{SSE}_R}. \tag{5.23}$$

The R^2 statistic measures the proportion of the total variation in the data that can be explained by the predictors in a linear regression model. It is a commonly used measure of model fit – a large R^2 value (close to one) indicates the model fits the data well. It is easy to see that the F statistic in (5.22) and the R^2 statistic in (5.23) are related as:

$$F = \frac{n-p-1}{p} \frac{R^2}{1-R^2}.$$

Example 5.5 Consider again the combined cycle power plant (`CCPP`) data. We continue from the `R` codes used for least squares in Section 5.2.1. The following `R` codes calculate the test statistic and p-value based on (5.19) and (5.20) to test if the regression coefficient of the Ambient Temperature (`AT`) is zero.

```
AT.sigma <- sqrt(cov.beta[1,1])
t1 <- CCPP.beta[1]/AT.sigma #t statistic for AT
AT.p <- 2*(1-pnorm(abs(t1))) #p-value
```

```
> AT.sigma
[1] 9.749
> t1
[1] 46.63
> AT.p
[1] 0
```

These three outputs match the "Std.Error", "t value", and "Pr(>|t|)" columns, respectively, of the coefficient table obtained using the `lm()` function in Section 5.1.

The following R codes are used to find the F statistic and R^2 of the linear regression model for the CCPP data based on (5.22) and (5.23).

```
SSE.F <- t(CCPP.y-CCPP.X%*%CCPP.beta)%*%
         (CCPP.y-CCPP.X%*%CCPP.beta)
SSE.R <- t(CCPP.y-mean(CCPP.y))%*%(CCPP.y-mean(CCPP.y))
#F statistic
CCPP.F <- ((SSE.R-SSE.F)/4)/(SSE.F/(CCPP.n-4-1))
#R^2
CCPP.R2 <- (SSE.R-SSE.F)/SSE.R
```

```
> CCPP.F
      [,1]
[1,] 31138
> CCPP.R2
       [,1]
[1,] 0.9287
```

The obtained F statistic and R^2 are the same as the `lm()` output for the CCPP data.

5.3 Categorical Response – Logistic Regression Model

In many applications, the response variable is a categorical variable with just a few possible values. A logistic regression model can be used to study how other variables affect a categorical response variable. We will focus on problems with binary response variables, where the response variable can only have two possible values.

Example 5.6 Failure of silver–zinc cells. Consider the data that record the failures of silver–zinc cells used in satellite applications. The data set is based on a subset of observations from an experiment reported by Sidik et al. [1980].

The response variable of interest in this application is the `failure` variable, which indicates if the cell fails within 200 charge–discharge cycles. We are interested in characterizing how the five use condition variables: charge rate (`CR`), discharge rate (`DR`), depth of discharge (`DOD`), ambient temperature (`T`), and end of charge voltage (`ECV`), affect the probability of failure. As in linear regression, we consider a model of simple form using a linear function such as

$$\beta_0 + \beta_1 \times \mathrm{CR} + \beta_2 \times \mathrm{DR} + \beta_3 \times \mathrm{DOD} + \beta_4 \times \mathrm{T} + \beta_5 \times \mathrm{ECV}. \qquad (5.24)$$

Since the output of the function in (5.24) is continuous, it cannot be directly used to predict a categorical variable such as the binary `failure` variable which can only be equal to "Yes" or "No". Instead, we can consider prediction of the probability of failure, P, which has continuous numerical values. However, any probability is bounded in the interval [0, 1]. But it is difficult to ensure the value of the linear function in (5.24) be always bounded in [0, 1]. To address this issue, we can consider some monotone transformation of P as the response variable. One popular transformation of a probability, which is used in areas such as horse-racing, gambling, and epidemiology, is the *odds* of an event occurring. The odds is defined as the ratio of the probability an event occurs to the probability it does not occur. Mathematically, the odds is defined as

$$\text{Odds}\,(\text{failure} = \text{“Yes”}) = \frac{P}{1-P}.$$

Table 5.1 shows the odds corresponding to different values of probabilities. Obviously knowing the odds is equivalent to knowing the corresponding probability. From Table 5.1, it can be seen that there is no upper bound for the possible values of odds. However, an odds has to be a positive number and it is still not easy to ensure the linear function in (5.24) be always positive. This issue can be easily addressed by taking the natural logarithm of the odds. Table 5.1 also shows the *log-odds* values of corresponding probabilities. The log-odds can

Table 5.1 Odds and log odds of different probabilities

Probability	Odds	Log-odds
0.001	1:999 or 0.001001	−6.9
0.01	1:99 or 0.0101	−4.6
0.5	50:50 or 1	0
0.9	90:10 or 9	2.19
0.999	999:1 or 999	6.9

be any real numbers, positive or negative, and is also called the *logit*. A logit of a very small negative value corresponds to very low probability, while a logit of a large positive value corresponds to a very high probability. A logit equal to zero corresponds to the even odds, or probability of 0.5. In general, a larger logit corresponds to a larger odds and larger probability. For the cell failure data set, by writing the logit of cell failure as a linear function of the predictor variables, we have

$$\ln\left(\frac{P}{1-P}\right) = \beta_0 + \beta_1 \times \mathrm{CR} + \beta_2 \times \mathrm{DR} + \beta_3 \times \mathrm{DOD} + \beta_4 \times \mathrm{T} + \beta_5 \times \mathrm{ECV}. \tag{5.25}$$

The model that uses a linear function of predictors to model the logit, as in (5.25), is a *logistic regression* model.

We can fit a logistic regression model to the cell failure data using `glm()` function in `R` as:

```
cell.df <- read.csv("cell_failures.csv", header = T)
cell.logit <- glm(failure ~ .-ID, data = cell.df,
     family = "binomial")
```

The estimated model parameters can be displayed by `summary()` as

```
> summary(cell.logit)
Call:
glm(formula = failure ~ .-ID, family = "binomial",
data = cell.df)

Deviance Residuals:
    Min         1Q     Median       3Q       Max
-2.0929   -0.3165    0.2481   0.4897    2.0804

Coefficients:
           Estimate Std.   Error z  value    Pr(>|z|)
(Intercept)   -40.52977   70.97517 -0.571 0.567972
CR              1.76165    0.83665  2.106 0.035240 *
DR             -0.03692    0.36863 -0.100 0.920232
DOD             0.10346    0.02915  3.549 0.000387 ***
T              -0.15062    0.04491 -3.354 0.000796 ***
ECV            18.88006   35.46369  0.532 0.594465
```

```
---
Signif. codes: 0 '***' 0.001 '**' 0.01 '*' 0.05 '.' 0.1 ' ' 1
(Dispersion parameter for binomial family taken to be 1)
Null deviance: 92.105 on 79 degrees of freedom
Residual deviance: 54.073 on 74 degrees of freedom
AIC: 66.073
```

From the coefficients table in the above `R` output, the logistic regression model for the cell failure data can be written as

$$\ln\left(\frac{P}{1-P}\right) = -40.53 + 1.762 \times \text{CR} - 0.037 \times \text{DR} + 0.1035 \times \text{DOD} - 0.151 \times \text{T} + 18.88 \times \text{ECV}$$

The fitted logistic regression model can be used to predict the failure probability for given conditions. For example, if for a cell the charge rate (`CR`) is 0.375 amperes, discharge rate (`DR`) is 1.25 amperes, depth of discharge (`DOD`) is 43.2%, ambient temperature (`T`) is 10°C, and end of charge voltage (`ECV`) is 1.99 volts, the predicted logit can be calculated as

$$\ln\left(\frac{\hat{P}}{1-\hat{P}}\right) = -40.53 + 1.762 \times 0.375 - 0.037 \times 1.25 + 0.1035 \times 43.2 - 0.151 \times 10 + 18.88 \times 1.99 = 0.62$$

Then we can find the predicted probability of failure as $\hat{P} = \frac{1}{1+e^{-0.62}} = 0.65$.

5.3.1 General Formulation of Logistic Regression Model

Without loss of generality, we assume that binary response variable Y has two possible values: 0 and 1. In general, the logistic regression model with response variable Y and p predictors $X_1, X_2, \ldots, X_p$ has the form

$$\text{logit} = \ln\left(\frac{P}{1-P}\right) = \beta_0 + \beta_1 X_1 + \beta_2 X_2 + \cdots + \beta_p X_p,$$

where $P = \Pr(Y = 1)$. The logistic regression model can also be written for the odds and the probability as

$$\text{odds}(Y = 1) = \frac{P}{1-P} = e^{\beta_0 + \beta_1 X_1 + \ldots + \beta_p X_p},$$

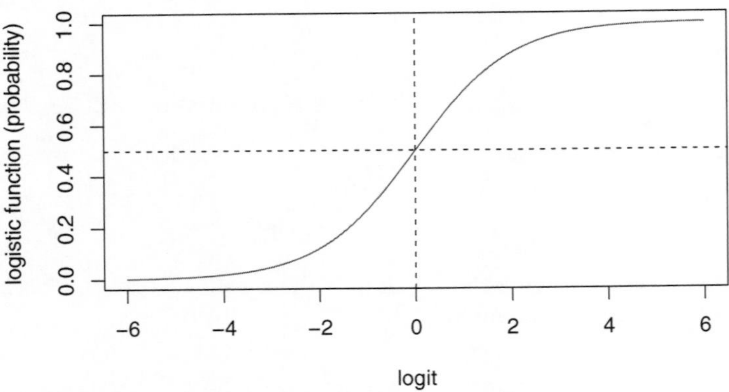

Figure 5.5 Logistic function (probability versus logit)

and

$$\Pr(Y=1)=P=\frac{e^{\beta_0+\beta_1X_1+\cdots+\beta_pX_p}}{1+e^{\beta_0+\beta_1X_1+\cdots+\beta_pX_p}}=\frac{e^{\text{logit}}}{1+e^{\text{logit}}}. \tag{5.26}$$

The function in (5.26) is called the *logistic function*. It is easy to see that the output of a logistic function is always between 0 and 1, giving a valid probability value. The plot of the logistic function versus the logit is shown in Figure 5.5. From Figure 5.5, it can be seen that the logistic function is an S-shaped curve. While the logit can be equal to any real number, the logistic function is bounded between 0 and 1. When the value of the logit is very small or large, the probability value of the logistic function approaches to 0 or 1, respectively. When the logit is equal to 0, the probability is 0.5.

5.3.2 Significance and Interpretation of Model Coefficients

After fitting a logistic regression model, one question of interest is whether the effect of a predictor is statistically significant, which is to test the null hypothesis $H_0: \beta_i = 0$, for a given $i = 1, \ldots, p$. For example, in the `R` outputs of the logistic regression model for the cell failure data, the predictors `CR`, `DOD`, and `T` have p-values smaller than 0.05, indicating that the coefficients of these three variables are significantly non-zero and they have statistically significant effects on failure probability at the significance level $\alpha = 0.05$.

In a logistic regression model, the logit, or log odds, has a linear relationship with the predictors. So the interpretation of model coefficients in terms of effects on the log odds is similar to that of the linear regression model. If the predictor X_i is increased by one unit, the log odds will increase by β_i. And increasing the log odds by β_i is equivalent to multiplying the odds by a factor of e^{β_i}. However, such a simple interpretation does not exist for the probability, which has a nonlinear relationship with the predictors. Since an increase in the logit corresponds to an increase in the probability, if β_i is positive (negative), increasing X_i will increase (decrease) both the logit and the probability. As an example, for the cell failure data set, the estimated coefficient of the depth of discharge (DOD) is 0.1035. We can say that an increase in the depth of discharge will increase the failure probability. If the depth of discharge is increased by 1 unit, the log odds of failure will increase by 0.1035, or the odds will increase by a (multiplicative) factor of $e^{0.1035} = 1.11$. Similarly, the coefficient of the ambient temperature (T) is -0.151. So increasing the ambient temperature will decrease the failure probability. An increase of ambient temperature by 1°C will decrease the log odds of failure by 0.151, or decrease the odds by a (multiplicative) factor of $e^{-0.151} = 0.86$.

5.3.3 Maximum Likelihood Estimation for Logistic Regression

Logistic regression models are usually fited by the maximum likelihood method. Suppose there are n observations, y_i, $i = 1, 2, \ldots, n$ in a data set. Each observation has two possible values, which are coded by 0 and 1. The predictors corresponding to y_i are $\{x_{i1}, x_{i2}, \ldots, x_{ip}\}$. Following the logistic regression model in (5.26), we define $P(\mathbf{x}_i; \boldsymbol{\beta})$ as

$$P(\mathbf{x}_i; \boldsymbol{\beta}) = \frac{e^{\boldsymbol{\beta}^T \mathbf{x}_i}}{1 + e^{\boldsymbol{\beta}^T \mathbf{x}_i}},$$

where $\boldsymbol{\beta} = (\beta_0\ \beta_1\ \cdots\ \beta_p)$ and $\mathbf{x}_i = (1\ x_{i1}\ \cdots\ x_{ip})$. The likelihood function can be written as

$$L(\boldsymbol{\beta}) = \prod_{i=1}^{n} P(\mathbf{x}_i; \boldsymbol{\beta})^{y_i} (1 - P(\mathbf{x}_i; \boldsymbol{\beta}))^{1-y_i}.$$

And the negative log-likelihood is

$$\begin{aligned} -l(\boldsymbol{\beta}) &= -\sum_{i=1}^{n} \{y_i \ln P(\mathbf{x}_i; \boldsymbol{\beta}) + (1 - y_i) \ln(1 - P(\mathbf{x}_i; \boldsymbol{\beta})\} \\ &= -\sum_{i=1}^{n} \{y_i \boldsymbol{\beta}^T \mathbf{x}_i - \ln(1 + e^{\boldsymbol{\beta}^T \mathbf{x}_i})\}. \end{aligned} \tag{5.27}$$

The maximum likelihood method finds $\hat{\boldsymbol{\beta}}$ that minimizes the negative log-likelihood in (5.27). Unlike MLE for linear regression models, for logistic regression there is no longer a closed-form solution, due to the nonlinearity of the logistic function. Iterative algorithms such as the Newton–Raphson algorithm can be used to find MLE for logistic regression. The details of such optimization algorithms are beyond the scope of this book.

Bibliographic Notes

Both linear regression and logistic regression models are discussed in many statistics books such as Agresti [2015] and Casella and Berger [2002]. They are also basic tools in machine learning and introduced in machine learning texts such as Hastie et al. [2009] and Bishop [2006]. Subset selection methods including forward selection and backward elimination are reviewed by Miller [2002]. Another approach that has become popular in recent years for variable selection is based on l_1-regularized regression, or *lasso*, which was proposed by Tibshirani [1996] and thoroughly reviewed in Hastie et al. [2015].

Exercises

1. Consider the following data set

x	y
10	15
5	9
6	3
18	25
11	8
8	12

   (a) Fit the linear regression model $Y = \beta_0 + \beta_1 X + \epsilon$ with $\text{var}(\epsilon) = \sigma^2$ by calculating the least squares estimates of $\boldsymbol{\beta} = (\beta_0\ \beta_1)^T$.

   (b) Find an unbiased estimate of σ^2. Based on $\hat{\sigma}^2$, estimate the covariance matrix of $\hat{\boldsymbol{\beta}}$.

2. Consider the following data set

x_1	x_2	y
2	10	15
3	5	9
3	6	3
6	18	25
7	11	8
9	8	12

   (a) Fit the linear regression model $Y = \beta_0 + \beta_1 X_1 + \beta_2 X_2 + \epsilon$ with $\text{var}(\epsilon) = \sigma^2$ by calculating the least squares estimates of $\boldsymbol{\beta} = (\beta_0\ \beta_1\ \beta_2)^T$.

   (b) Find an unbiased estimate of σ^2. Based on $\hat{\sigma}^2$, estimate the covariance matrix of $\hat{\boldsymbol{\beta}}$.

3. Consider the `Auto` data set in `Auto.csv`. Use `R` to fit a linear regression model with `mpg` as the response variable and all other variables except `name` as the predictors.
   (a) Write the equation for the fitted linear regression model.
   (b) Which predictors have a statistically significant relationship with the response variable?
   (c) Draw a scatter plot and compare the predicted values based on the linear regression model versus the actual values of `mpg`.
   (d) Using the estimated linear regression model, what `mpg` value is predicted for a 4-cylinder car that has the following specification: `displacement`=97, `horsepower`=46, `weight`=1835, `acceleration`=20.5, `year`=70, and `origin`=2.
   (e) Provide an interpretation for the estimated coefficient of each statistically significant predictors.
   (f) Based on the estimated linear regression model, what type of cars is most likely to have a high `mpg`?
   (g) Calculate the least squares estimate of β and σ^2 based on (5.9) and (5.11). Compare with the `R` outputs from the `lm()` function.
   (h) Use (5.19) and (5.20) to find the t-statistic and the corresponding p-value to test the significance of the predictors `cylinders` and `displacement`, respectively. State the conclusion of the hypothesis testing at significance level of $\alpha = 0.01$. Compare the results with the `R` outputs from the `lm()` function.
   (i) Use (5.22) and (5.23) to find the F-statistic and the R^2 values of this linear regression model. Compare with the `R` outputs from the `lm()` function.
4. Consider `Auto` data set to predict the variable `mpg`.
   (a) In addition to the predictors considered in Exercise 3, fit a linear regression model that also includes the two-way interaction effects as the predictors. Do any interactions appear to be statistically significant?
   (b) Try a few different nonlinear transformations of the variables, such as $\ln X$, $\sqrt{X}$, and X^2. Comment on the statistical significance of the transformed predictors.
   (c) Compare the linear regression models in part (a) and part (b) with the model obtained in Exercise 3 in terms of AIC and BIC. You can use the `R` function `AIC()` and `BIC()` to find AIC or BIC of a model.
5. Consider the `auto_spec` data set in `auto_spec.csv` and predict `city.mpg` using the following variables as predictors: `fuel.type`, `body.style`, `engine.type`, `compression.ratio`, `length`, and `curb.weight`.
   (a) Use `R` to fit a linear regression model and write the equation for the fitted model.

(b) Which predictors have a statistically significant relationship with the response variable?

(c) Draw a scatter plot and compare the predicted values based on the linear regression model versus the actual values of `city.mpg`.

(d) Provide an interpretation for the estimated coefficient of each statistically significant predictor.

(e) Based on the estimated linear regression model, what type of car is most likely to have a high `city.mpg`?

6. Consider the `auto_spec` data set. We are interested in a model that describes how the variable `fuel.type` is related to the variables `horsepower` and `peak.rpm`.

   (a) Use `R` to fit a logistic regression model with `fuel.type` as the response variable and `horsepower` and `peak.rpm` as predictors. Write the estimated model equation in three formats:

      i. The logit as a function of the predictors
      ii. The odds as a function of the predictors
      iii. The probability as a function of the predictors

   (b) Consider a new car with horsepower $=100$ and peak RPM $=4500$. From your logistic regression model, estimate the logit, the odds, and the probability of car being gas powered.

   (c) For cars with horsepower $=75$, find the range of values of the peak RPM so that the fuel type would be predicted to be more likely to be `gas` instead of `diesel` based on the logistic regression model.

   (d) Which regression coefficients in the logistic regression model are significantly non-zero?

   (e) Interpret the estimated coefficients for the two predictors in terms of the odds of a car being gas powered.

7. Combine the side temperature data at Stand 5 for defective billets from a hot rolling process (`side_temp_defect.csv`) with the side temperature data at Stand 5 for normal billets (`side_temp_normal.csv`). Based on the combined data, predict if a billet is defective based on the side temperature measurements at locations 2, 40, and 78 at Stand 5.

   (a) Use `R` to fit a logistic regression model to predict whether a billet is defective with side temperature measurements at locations 2, 40, and 78 as predictors. Write the estimated model equation for the log odds of being defective.

   (b) Which side temperature measurements have statistically significant effect on the log odds of a defect?

   (c) If the side temperature measurements for a billet at locations 2, 40, and 78 are 1930, 1848, and 1864, respectively, what is the estimated probability that the billet is defective?

6

Linear Mixed Effects Model

In this chapter, we will introduce the basic statistical theory regarding the linear mixed effects (LME) model including model structure, model inference, and hypothesis testing on model parameters. This chapter provides the theoretical foundation for the diagnosis and prognosis methods introduced in the following chapters.

6.1 Model Structure

In Chapter 5, we presented the approaches for setting up a linear regression model and its inference. One important assumption of linear regression models is that the data are drawn from the same homogeneous population and the data are independent and identically distributed (i.i.d.). However, in practice, the data could be more complex. Let's look at the linear regression model again

$$y_j = \mathbf{x}_j^T \boldsymbol{\beta} + \epsilon_j, \tag{6.1}$$

where j is the index of observations and the error term ϵ_j is an i.i.d. random variable with zero mean and constant variance σ_ϵ^2. The model parameter $\boldsymbol{\beta}$ is constant for all the observations. Thus it is called *fixed effects* or *population parameters*. In this model, only the observation y_j and the error term ϵ_j are random variables. However, in industrial practice, many scenarios call for more complex model structure or more random factors to be modeled.

The first common scenario is the modeling for clustered data. In practice, we often collect repeated observations from the same equipment over time. The

Industrial Data Analytics for Diagnosis and Prognosis: A Random Effects Modelling Approach, First Edition. Shiyu Zhou and Yong Chen.

battery resistance data shown in Figure 1.3 is a typical example of clustered data. In that data set, multiple battery resistance observations are collected over time from each of the batteries. If we use i as the index for battery unit, then the data collected from N batteries can be expressed as $\{(y_{i1}, t_{i1})..., (y_{ij}, t_{ij}), ..., (y_{in_i}, t_{in_i})\}$, $i = 1, ..., N$, where y_{ij} is the battery internal resistance observation, t_{ij} is the time when the jth observation is collected from the ith battery, and n_i is the total number of observations from the ith battery. The data collected from the same battery unit can be viewed as being in the same cluster. As shown in Figure 1.3, the relationship between y_{ij} and t_{ij} is obviously different for different batteries. In other words, each data cluster has its unique characteristic. It is unreasonable to use a single population model such as that in (6.1) to describe the evolving paths of the battery internal resistance for multiple batteries. One simple way to address this issue is to fit a model individually for each unit i only using the data collected from that unit. However, the data from a single unit is often limited. Furthermore, by observing the data shown in Figure 1.3, we can clearly see that although each individual path is distinct, the basic trends and shapes of these paths bear some similarities. It will be beneficial to model these data together and share the information from all the battery units. Is there a way to make the structure of a regular regression model more sophisticated to cover multiple clusters of data? The answer to this question is to introduce random effects in the population model. For example, a direct extension of the population regression model (6.1) is

$$y_{ij} = \mathbf{x}_{ij}^T \boldsymbol{\beta}_i + \epsilon_{ij}, \tag{6.2}$$

where for each battery unit i, $\boldsymbol{\beta}_i$ can pick a specific value, while across multiple units, $\{\boldsymbol{\beta}_1, ..., \boldsymbol{\beta}_i, ..., \boldsymbol{\beta}_N\}$ are independent samples from a common probability distribution. Thus, $\boldsymbol{\beta}_i$ is viewed as a random vector, also called *random effects*. The model in (6.2) is called a regression model with random coefficients, which is a special form of the general *linear mixed effects model*. The linear mixed effects model, as the name implies, includes both fixed effects and random effects in the model, which is a very powerful modeling tool for clustered data.

The second scenario is the need for modeling the uncertainties in the physical system. A typical example is the automotive assembly process as shown in Section 1.1. In an automotive assembly process, the variation in the final quality observation is not only due to the measurement noise, but also due to variations in the parts to be assembled and the variations in the fixture locators. As a result, additional random effects representing the variation of the locators should be included in the model as predictors to predict the final quality. If we denote y_{ij} as the jth key quality characteristic (KQC) on the ith unit, $\boldsymbol{\beta}_i$ as the key control characteristics (KCC) for the assembly process (e.g., fixture locator positions), and ϵ_{ij} as the error term due to measurement noise, then (6.2) can be used to represent the linearized KQC–KCC relationships. One special feature of this model for the assembly process is that the number of observations from different units are typically the same and the design vectors $\mathbf{x}_{ij}^T$ are also the same for different unit i because the physical principles of the assembly process are the same for different units. In other words, we can simply drop the index i from the notation of $\mathbf{x}_{ij}^T$.

A general linear mixed effects model is written as

$$y_{ij} = \mathbf{x}_{ij}^T \boldsymbol{\beta} + \mathbf{z}_{ij}^T \mathbf{b}_i + \epsilon_{ij}.$$

In this model, $\boldsymbol{\beta} \in \mathcal{R}^{p\times 1}$ and $\mathbf{b}_i \in \mathcal{R}^{q\times 1}$ represent the fixed effects and the random effects, respectively. The index i is the unit (or cluster) index from 1 to N and the index j is the observation index from 1 to n_i. Please note that the number of observations from different units may be different. We can stack the observations y_{ij} and the error term ϵ_{ij} together to form vectors $\mathbf{y}_i \in \mathcal{R}^{n_i\times 1}$ and ϵ_i. Similarly, we can stack the row vectors $\mathbf{x}_{ij}^T$ and $\mathbf{z}_{ij}^T$ to form matrices $\mathbf{X}_i \in \mathcal{R}^{n_i\times p}$ and $\mathbf{Z}_i \in \mathcal{R}^{n_i\times q}$, respectively. Then we can get the matrix form of the model as

$$\mathbf{y}_i = \mathbf{X}_i \boldsymbol{\beta} + \mathbf{Z}_i \mathbf{b}_i + \epsilon_i. \tag{6.3}$$

A multivariate normal distribution is often assumed for the random variables in the mixed effects model as $\mathbf{b}_i \sim \mathcal{N}(\mathbf{0}, \mathbf{G})$ and $\epsilon_i \sim \mathcal{N}(\mathbf{0}, \mathbf{R}_i)$. For simplicity, a common choice for $\mathbf{R}_i$ is $\mathbf{R}_i = \sigma_\epsilon^2 \mathbf{I}_{n_i}$, where $\mathbf{I}_{n_i}$ is the identity matrix with dimension of $n_i \times n_i$. In this book, we will mainly focus on this simplified case.

The model in (6.3) is very general and has several interesting special forms.

- If $n_i = n$ and $\mathbf{Z}_i = \mathbf{Z}$, $\forall i \in \{1, \ldots, N\}$, then the model (6.3) is called *balanced* LME model. It is relatively simpler to analyze a balanced model comparing to the general model.
- When $\mathbf{X}_i = \mathbf{Z}_i$, $\forall i \in \{1, \ldots, N\}$, the model in (6.3) becomes the regression model with random coefficients (6.2). We will use this model for degradation prognosis in Chapter 10. Further, if we additionally have $\mathbf{X}_i = \mathbf{X}$, $\forall i \in \{1, \ldots, N\}$, the model is called a balanced regression model with random coefficients.
- The variance component model that is often used to model the within cluster variance, $y_{ij} = \beta_0 + b_i + \epsilon_{ij}$, where $b_i \sim \mathcal{N}(0, \sigma_b^2)$ and $\epsilon_{ij} \sim \mathcal{N}(0, \sigma_\epsilon^2)$, is obviously a special case of the mixed effects model.
- Another special form of (6.3) is the case when $n_i = 1$, $\forall i \in \{1, \ldots, N\}$. This will lead to a linear regression model with heteroscedastic errors because in this case $\mathbf{y}_i$ becomes a scalar and $\mathbf{Z}_i \mathbf{b}_i + \epsilon_i$ can be represented by a univariate random variable with different variance for different i. Methods addressing this model can be found in Carroll [2017].

It is also worth noting that in (6.3), instead of letting the random effects $\mathbf{b}_i$ follow a fixed parametric distribution, we can treat $\mathbf{b}_i$ as the response of another model. Then the model in (6.3) becomes a *multilevel model*. The unknown parameters in the LME model (6.3) include the fixed parameter β, the covariance parameters for random effects and error term, $\mathbf{G}$ and $\mathbf{R}_i$, respectively. The random effects $\mathbf{b}_i$ is not observable and often treated as a nuisance parameter. However, in some scenarios, we may also want to estimate the value of $\mathbf{b}_i$ for a specific unit i. We will discuss such case in Chapter 10. In this chapter, we will focus on the estimation of β, $\mathbf{G}$, and $\mathbf{R}_i$.

The model in (6.3) is in a linear form. However, because the covariance parameters are unknown, the model estimation is nonlinear. Denoting the covariance matrix of $\mathbf{y}_i$ as $\boldsymbol{\Sigma}_i$ and noting that the random effects $\mathbf{b}_i$ and the error term ϵ_i are independent from each other, it is straightforward to obtain the covariance matrix of $\mathbf{y}_i$ as

$$\boldsymbol{\Sigma}_i = \mathbf{R}_i + \mathbf{Z}_i\mathbf{G}\mathbf{Z}_i^T, \tag{6.4}$$

and thus, $\mathbf{y}_i$ follows a multivariate Gaussian distribution as

$$\mathbf{y}_i \sim \mathcal{N}(\mathbf{X}_i\beta, \mathbf{R}_i + \mathbf{Z}_i\mathbf{G}\mathbf{Z}_i^T). \tag{6.5}$$

If we know $\mathbf{G}$ and $\mathbf{R}_i$, we know $\boldsymbol{\Sigma}_i$ from (6.4). Then based on the generalized least squares estimate for linear regression in (5.16), we have

$$\hat{\beta} = \left(\sum_{i=1}^{N}\mathbf{X}_i^T\boldsymbol{\Sigma}_i^{-1}\mathbf{X}_i\right)^{-1}\left(\sum_{i=1}^{N}\mathbf{X}_i^T\boldsymbol{\Sigma}_i^{-1}\mathbf{y}_i\right). \tag{6.6}$$

However, $\mathbf{G}$ and $\mathbf{R}_i$ are unknown in practice. The parameterization and estimation of these covariance matrices become the focus of the LME model estimation. The main estimation methods can be classified into two categories: maximum likelihood estimation and quadratic non-iterative distribution-free estimation. We will focus on the maximum likelihood estimation in Section 6.2.1 and other methods in Section 6.2.2. In the following, we will first introduce several methods for the parameterization of the covariance matrix.

Consider a general multivariate normally distributed random vector $\mathbf{y}$ of dimension m. The unstructured form of the covariance matrix of $\mathbf{y}$ is

$$\boldsymbol{\Sigma} = \begin{pmatrix} \sigma_1^2 & \sigma_{12} & \cdots & \sigma_{1m} \\ \sigma_{21} & \sigma_2^2 & \cdots & \sigma_{2m} \\ \vdots & \vdots & \ddots & \vdots \\ \sigma_{m1} & \sigma_{m2} & \cdots & \sigma_m^2 \end{pmatrix}$$

There are $m(m+1)/2$ parameters in the unstructured form. Clearly, if the dimension m is large, it will be difficult to estimate the unstructured form, particularly considering the constraint that the estimated matrix should be positive semi-definite to be a valid covariance matrix. We can impose some patterns on the covariance matrix to reduce the number of parameters yet capture the essential characteristic of the data. The following patterns are often used in practice.

1. Variance component covariance matrix
This pattern is given as

$$\boldsymbol{\Sigma} = \begin{pmatrix} \sigma_1^2 & 0 & \cdots & 0 \\ 0 & \sigma_2^2 & \cdots & 0 \\ \vdots & \vdots & \ddots & \vdots \\ 0 & 0 & \cdots & \sigma_m^2 \end{pmatrix}$$

This pattern implies that the components in $\mathbf{y}$ vector are independent from each other. There are m parameters in this pattern. The diagonal elements of the covariance matrix are called variance components.

2. Compound symmetry covariance matrix
This pattern is given as

$$\boldsymbol{\Sigma} = \sigma^2 \begin{pmatrix} 1 & \rho & \cdots & \rho \\ \rho & 1 & \cdots & \rho \\ \vdots & \vdots & \ddots & \vdots \\ \rho & \rho & \cdots & 1 \end{pmatrix}. \tag{6.7}$$

There are only two parameters $\{\sigma, \rho\}$ in this pattern. For the LME model in (6.3), if we only have one random effect, i.e., $q=1$ and the covariance matrix of error term is in the form of $\sigma_\epsilon^2 \mathbf{I}_i$, then the covariance matrix of the response $\mathbf{y}_i$ will be in the form of compound symmetry (see page 192 of Fitzmaurice et al., 2004). Compound symmetry is one of the first covariance patterns used in the repeated measurement data.

3. Toeplitz and autoregressive covariance matrix
These covariance matrix patterns are mainly used to characterize the autocorrelation of the temporal repeated observations from the same unit. Thus, these patterns are often imposed on $\mathbf{R}_i$ matrix in the LME model.

As shown in Table 6.1, there are m parameters in the Toeplitz covariance matrix and only two parameters in the autoregressive covariance matrix. It is obvious that the autoregressive pattern is a special case of the Toeplitz pattern. These two covariance matrix patterns are often used to characterize the covariance of evenly sampled time series data. The Toeplitz pattern assumes that the

Table 6.1 Toeplitz and autoregressive covariance matrix

Toeplitz	Autoregressive
$\sigma^2 \begin{pmatrix} 1 & \rho_1 & \rho_2 & \cdots & \rho_{m-1} \\ \rho_1 & 1 & \rho_1 & \cdots & \rho_{m-2} \\ \rho_2 & \rho_1 & 1 & \cdots & \rho_{m-3} \\ \vdots & \vdots & \ddots & \vdots & \\ \rho_{m-1} & \rho_{m-2} & \rho_{m-3} & \cdots & 1 \end{pmatrix}$	$\sigma^2 \begin{pmatrix} 1 & \rho & \rho^2 & \cdots & \rho^{m-1} \\ \rho^1 & 1 & \rho^1 & \cdots & \rho^{m-2} \\ \rho^2 & \rho^1 & 1 & \cdots & \rho^{m-3} \\ \vdots & \vdots & \ddots & \vdots & \\ \rho^{m-1} & \rho^{m-2} & \rho^{m-3} & \cdots & 1 \end{pmatrix}$

variance of all the samples (i.e., the components of $\mathbf{y}$) are the same and the correlations between any two different samples with the same time interval between them are the same. For the autoregressive pattern, in addition to the above constraints, we further assume the correlation between two samples decays exponentially along with the time interval between them.

Both Toeplitz and autoregressive patterns are only applicable to evenly observed data, i.e., the time intervals between two adjacent observations are the same. When the observation intervals are not equal, the exponential pattern can be used. Interested readers can refer to Fitzmaurice et al. [2004].

We introduced several commonly used patterns of covariance matrix for a generic random vector. In the mixed effects model, we need to specify the structure of two covariance matrices, $\mathbf{G}$ and $\mathbf{R}_i$. In principle, we could impose a selected pattern to either $\mathbf{G}$ or $\mathbf{R}_i$. However, because what we observe is an aggregated effect of $\mathbf{b}_i$ and ϵ_i in LME models, we may not be able to separate $\mathbf{G}$ or $\mathbf{R}_i$ if their patterns are not carefully selected. In this book, unless explicitly stated, we assume $\mathbf{G}$ is unstructured and $\mathbf{R}_i$ is in the simplified variance component form $\sigma_\epsilon^2 \mathbf{I}_i$.

6.2 Parameter Estimation for LME Model

6.2.1 Maximum Likelihood Estimation Method

Maximum likelihood estimation is the most widely used method for parameter estimation of LME models. Let η denote the distinct parameters in $\mathbf{G}$ and $\mathbf{R}_i$ and let $\boldsymbol{\theta} = \{\boldsymbol{\beta}, \boldsymbol{\eta}\}$ be the collective model parameters. The likelihood for the observed data $\mathbf{y}_a = (\mathbf{y}_1^T, \cdots, \mathbf{y}_N^T)^T$ is

$$L(\boldsymbol{\theta}|\mathbf{y}_a) = \prod_{i=1}^{N} f(\mathbf{y}_i|\boldsymbol{\theta}), \tag{6.8}$$

where $f(\mathbf{y}_i|\boldsymbol{\theta}) = \frac{1}{[(2\pi)^{n_i} |\boldsymbol{\Sigma}_i|]^{1/2}} \exp\left(-\frac{1}{2}(\mathbf{y}_i - \mathbf{X}_i\boldsymbol{\beta})^T \boldsymbol{\Sigma}_i^{-1}(\boldsymbol{\eta})(\mathbf{y}_i - \mathbf{X}_i\boldsymbol{\beta})\right)$ is the multivariate normal density function with mean and covariance given by (6.5).

Taking the logarithm operation on both sizes of (6.8) and denoting the function $-\ln L(\boldsymbol{\theta}|\mathbf{y}_a)$ as $-l(\boldsymbol{\theta}|\mathbf{y}_a)$, which is the *negative log likelihood function*, we have

$$\begin{aligned} -l(\boldsymbol{\theta}|\mathbf{y}_a) &= \frac{1}{2}\sum_{i=1}^{N}[n_i \ln(2\pi) + \ln(|\boldsymbol{\Sigma}_i(\boldsymbol{\eta})|) \\ &\quad + (\mathbf{y}_i - \mathbf{X}_i\boldsymbol{\beta})^T \boldsymbol{\Sigma}_i^{-1}(\boldsymbol{\eta})(\mathbf{y}_i - \mathbf{X}_i\boldsymbol{\beta})]. \end{aligned} \tag{6.9}$$

Then the maximum likelihood estimation (MLE) of the parameters can be obtained as

$$\hat{\boldsymbol{\theta}} = \arg\min_{\boldsymbol{\theta}} - l(\boldsymbol{\theta}|\mathbf{y}_a). \tag{6.10}$$

The optimization problem in (6.10) is a constrained nonlinear optimization problem because we require $\mathbf{G}$ and $\mathbf{R}_i$ to be valid covariance matrices (i.e., positive semi-definite matrices) when evaluated with the estimated parameters $\hat{\eta}$. It is known that if we let $\mathbf{R}_i = \sigma_\epsilon^2 \mathbf{I}_i$, then the sufficient condition of the existence of an MLE solution is that $\sum_i (n_i - q) > p$ (see Demidenko, 2005). However, this will not guarantee that the estimated $\mathbf{G}$ will be a valid covariance matrix. Such a constraint significantly complicates the optimization problem. Various numerical methods have been used to solve this problem, including the Newton–Raphson algorithm, the Fisher scoring algorithm, and the Expectation–Maximization (EM) algorithm. The detailed discussions of the computational algorithms of solving the optimization problem is out of the scope of the book. Interested readers can refer to Demidenko [2005].

For LME models, the MLE method is biased for finite samples. For example, the maximum likelihood estimate of the variance using i.i.d. samples from a normal distribution $(a_1, a_2, ..., a_n)$ is $\hat{\sigma}^2 = (1/n)\sum_i (a_i - \hat{\mu})^2$, which is biased downwards. The reason for this is that the degree of freedom lost in the estimation of the mean μ is not incorporated in the estimation of variance σ^2. To account for this, the sample variance $\hat{\sigma}^2 = (1/(n-1))\sum_i (a_i - \hat{\mu})^2$ is unbiased and commonly used in practice. Following a similar idea, for LME unbiased estimation is obtained through a method called Restricted or Residual Maximum Likelihood (REML) estimation. The basic idea of REML method is to obtain a likelihood function that only involves the covariance parameters. Such a likelihood function can be obtained through a Bayesian argument as $L_R(\eta|\mathbf{y}_a) = \int L(\boldsymbol{\beta}, \eta|\mathbf{y}_a)\mathrm{d}\boldsymbol{\beta}$, which can be viewed as integrating the mean parameter $\boldsymbol{\beta}$ out of the likelihood function. We can also obtain $L_R(\eta \mid \mathbf{y}_a)$ through the likelihood function of the model residual $\mathbf{e}_i = \mathbf{y}_i - \mathbf{X}_i\widehat{\boldsymbol{\beta}}$, where $\widehat{\boldsymbol{\beta}}$ is given in (6.6) and written as a function of η. Intuitively, we can argue that only residuals contain the information about the covariance parameters. The residual likelihood function can be derived noting the residuals also follow a normal distribution and are independent of $\widehat{\boldsymbol{\beta}}$. The restricted negative log likelihood function is (see page 441 of Pawitan, 2001)

$$-l_R(\eta \mid \mathbf{y}) \propto \frac{1}{2}\sum_{i=1}^{N} [\ln(|\boldsymbol{\Sigma}_i(\eta)|) + \ln|\mathbf{X}_i^T\boldsymbol{\Sigma}_i^{-1}(\eta)\mathbf{X}_i| + (\mathbf{y}_i - \mathbf{X}_i\widehat{\boldsymbol{\beta}})^T\boldsymbol{\Sigma}_i^{-1}(\eta)(\mathbf{y}_i - \mathbf{X}_i\widehat{\boldsymbol{\beta}})], \tag{6.11}$$

where the symbol $\propto$ stands for "is proportional to". Please note that $\widehat{\boldsymbol{\beta}}$ is given in (6.6) and thus is a function of η as well.

By minimizing the negative log likelihood or restricted negative log likelihood function, we can get a point estimate of the parameters. In many cases, we also want to obtain the confidence interval of the estimated parameters. A classical confidence interval for parameter estimated by the likelihood based approach is the Wald confidence interval, which is given as

$$\hat{\theta}_i \pm z_{\alpha/2}\,\hat{\sigma}_{\hat{\theta}_i},$$

where α is the selected significance level, $z_{\alpha/2}$ is the $(1-\alpha/2)\times 100\%$ percentile point of the standard normal distribution, and $\hat{\sigma}_{\hat{\theta}_i}$ is the square root of the (i, i)th component of the Fisher information matrix plugged in the estimated parameters in the MLE procedure [Millar, 2011]. Clearly, the Wald confidence interval is based on the assumption that the MLE estimate follows a normal distribution. This is true asymptotically but for the finite sample case, it is inaccurate, particularly for covariance parameters with small sample size.

Another commonly used confidence interval is the likelihood confidence interval. This confidence interval is based on the approximation that the log likelihood ratio follows a χ^2 distribution. Specifically, for a single parameter model, we have the likelihood confidence interval as

$$\theta \in \left\{\theta \,|\, -2\log\frac{L(\theta)}{L(\hat{\theta})} < \chi^2_{\alpha,1}\right\},$$

where $\chi^2_{\alpha,1}$ is the $(1-\alpha)\times 100\%$ percentile point of χ^2 distribution with degree of freedom 1. In the general case, there are multiple parameters in the model.

For such a case we can use the profile likelihood confidence interval. Instead of using the standard likelihood function, we use the so-called profile likelihood

$$L_p(\theta_i) = \max_{\boldsymbol{\theta}_{-i}} L(\theta_i, \boldsymbol{\theta}_{-i}),$$

where $\boldsymbol{\theta}_{-i}$ represents all the parameters excluding θ_i. Please note that $L_p(\theta_i)$ only depends on a single parameter θ_i and the other parameters are "profiled" out.

Now let us use a simple example to demonstrate the inference of LME models.

Example 6.1 In this example, we use the following toy data set to demonstrate the LME model estimation procedure.

Table 6.2 Current values (mA) from three wafers at two sites of each wafer

y_{11}	y_{12}	y_{21}	y_{22}	y_{31}	y_{32}
0.901	1.032	1.022	1.249	0.933	0.989

This toy data set is extracted from the `Wafer` data set from the `nlme` package in `R`. The value of y_{ij} is the intensity of current collected on a n-channel device located at the jth site on the ith wafer when a voltage of 0.8 V is applied to the device. We want to find out if there is variation across different wafers. We can establish a simple LME model for this purpose as

$$\begin{pmatrix} y_{i1} \\ y_{i2} \end{pmatrix} = \begin{pmatrix} 1 \\ 1 \end{pmatrix} \beta + \begin{pmatrix} 1 \\ 1 \end{pmatrix} b_i + \begin{pmatrix} \epsilon_{i1} \\ \epsilon_{i2} \end{pmatrix}.$$

Corresponding to the LME model in (6.3) we have $\mathbf{y}_i = \begin{pmatrix} y_{i1} \\ y_{i2} \end{pmatrix}$, $\mathbf{X}_i = \begin{pmatrix} 1 \\ 1 \end{pmatrix}$, $\mathbf{Z}_i = \begin{pmatrix} 1 \\ 1 \end{pmatrix}$, and $\boldsymbol{\epsilon}_i = \begin{pmatrix} \epsilon_{i1} \\ \epsilon_{i2} \end{pmatrix}$. This is actually a balanced one way analysis of a variance model with random effects. Because we only have one random effect, the covariance matrix of b_i, $\mathbf{G}$, becomes a scalar, i.e., $b_i \sim \mathcal{N}(0, \sigma_b^2)$. We further assume $\boldsymbol{\epsilon}_i \sim \mathcal{N}(\mathbf{0}, \sigma_\epsilon^2 \begin{pmatrix} 1 & 0 \\ 0 & 1 \end{pmatrix})$ and thus $\mathbf{R}_i = \sigma_\epsilon^2 \mathbf{I}_2$. With these assumptions, we can obtain the covariance matrix of $\mathbf{y}_i$ using (6.4) as

$$\boldsymbol{\Sigma}_i = \sigma_\epsilon^2 \mathbf{I}_2 + \begin{pmatrix} 1 \\ 1 \end{pmatrix} \sigma_b^2 \begin{pmatrix} 1 & 1 \end{pmatrix} = \begin{pmatrix} \sigma_\epsilon^2 + \sigma_b^2 & \sigma_b^2 \\ \sigma_b^2 & \sigma_\epsilon^2 + \sigma_b^2 \end{pmatrix}$$

In this case, the unknown parameters for the covariance matrix is $\boldsymbol{\eta} = \left\{\sigma_\epsilon^2, \sigma_b^2\right\}$. From this result, we can see that because of the random effect, the observations from the same wafer, i.e., y_{i1} and y_{i2} are correlated. Also, if we have more observations from one wafer, the dimension of $\boldsymbol{\Sigma}_i$ will increase but the form of $\boldsymbol{\Sigma}_i$ will be kept the same, i.e., the diagonal elements will be $\sigma_\epsilon^2 + \sigma_b^2$ and all the off-diagonal elements will be σ_b^2. Clearly, this covariance matrix is in the form of compound symmetry as shown in (6.7).

Plugging the expressions of $\mathbf{G}, \mathbf{R}_i, \mathbf{X}_i, \mathbf{Z}_i$ into (6.9) and (6.11), we can get the negative log likelihood function $-l(\beta, \sigma_\epsilon^2, \sigma_b^2 \mid \mathbf{y}_a)$ and the restricted negative log likelihood function $-l_R(\sigma_\epsilon^2, \sigma_b^2 \mid \mathbf{y}_a)$. By minimizing these two functions, we can obtain the corresponding estimates for β, σ_ϵ^2, and σ_b^2. Please note for REML, the β parameter estimation is given by (6.6) once the variance parameters are estimated.

For a general LME model, the MLE and REML estimates have to be obtained numerically. However, for some simple cases, the closed form solution is available. In fact, in this example, for both MLE and REML, we have $\hat{\sigma}_\epsilon^2 = \sum_{i=1}^{3}(\sum_{j=1}^{2}(y_{ij} - \bar{y}_i))/3$, where $\bar{y}_i = (y_{i1} + y_{i2})/2$. The estimation for σ_b^2 is different for MLE and REML: under standard MLE, $\hat{\sigma}_b^2 = \sum_{i=1}^{3}(\bar{y}_i - \bar{\bar{y}})/3 - \hat{\sigma}_\epsilon^2/2$ while under REML, $\hat{\sigma}_b^2 = \sum_{i=1}^{3}(\bar{y}_i - \bar{\bar{y}})/2 - \hat{\sigma}_\epsilon^2/2$, where $\bar{\bar{y}}$ is the grand average of all six observed values. For more general expression of the closed expression of the MLE and REML solution for balanced random coefficient models, please refer to Demidenko [2005].

In R language, the `lmer` function in `lme4` package is a powerful routine to obtain the ML and REML for a general linear mixed effects model. The following sample code illustrates its usage.

```
data<-data.frame(wafer=c(1,1,2,2,3,3),
          current=c(0.90088,1.03200,
                    1.02160,1.24900,
                    0.93266,0.98908))
m1.mle = lmer(current ~ 1+(1|wafer),data,REML=FALSE)
```

The second command calls `lmer` to fit the model with a fixed intercept, and a random intercept clustered by `wafer`. The `REML=FALSE` option indicates standard MLE method is used. Using `summary(m1.mle)`, we have

```
Linear mixed model fit by maximum likelihood ['lmer-
Mod']
Formula: current ~ 1 + (1 | wafer)
   Data: data

    AIC       BIC      logLik      deviance      df.resid
   -3.2      -3.9       4.6          -9.2            3

Scaled residuals:
    Min             1Q          Median          3Q          Max
   -1.05339   -0.63054    -0.16234     0.08709    1.99444

Random effects:
Groups          Name           Variance       Std.Dev.
wafer       (Intercept)      0.0005451      0.02335
Residual                     0.0120144      0.10961
Number of obs: 6, groups: wafer, 3

Fixed effects:
              Estimate Std. Error t value
(Intercept)    1.02087    0.04673   21.84
```

The summary of the fitted model first shows some general summary statistics about the model such as AIC, BIC, the log-likelihood, etc. From the output for the random effects, we have the variance of the random effect associated with `wafer` as $\hat{\sigma}_b^2 = 0.00055$ and the error term variance $\hat{\sigma}_\epsilon^2 = 0.012$. The fitted fixed effect $\hat{\beta} = 1.02$. If we remove the option `REML=FALSE` in `lmer`, we will get REML estimate. The estimates of $\hat{\sigma}_\epsilon^2$ and $\hat{\beta}$ are the same as that in MLE case but the variance of the random effect is bigger as $\hat{\sigma}_b^2 = 0.003821$.

From the summary of the fitted model, we can obtain the point estimates of the parameters. The confidence interval of the parameter estimates can be obtained using the `confint` function. Using the command `confint(m1.mle)`, we can get the profile intervals as

```
                 2.5%          97.5%
.sig01         0.00000000    0.2266927
.sigma         0.05820456    0.2245900
  (Intercept)  0.89039439    1.1513455
```

Please note that due to the small sample size of this toy example, the profile likelihood confidence interval does not perform well.

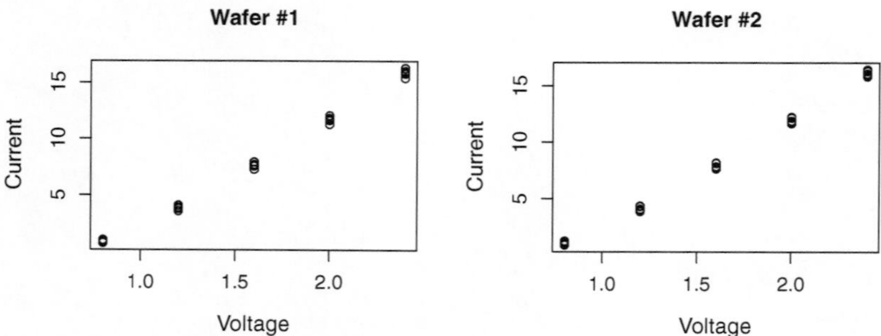

Figure 6.1 Data measured from two wafers

Example 6.2 In this example, we consider the full `Wafer` data set in the `nlme` package. The data set gives the intensity of current at five ascending voltages (i.e., 0.8, 1.2, 1.6, 2.0, 2.4) collected on n-channel devices. The observations were made on eight sites of each of ten wafers. Figure 6.1 shows the observation data from two wafers. Please note that each panel of the figure contains the observations at eight sites.

To study the variation of the current for given voltage, we can use different models. If we use y_{ijk} as the current observation at the ith wafer, the jth site for the kth given voltage, where $i = 1, \ldots, 10$, $j = 1, \ldots, 8$, and $k = 1, \ldots, 5$, and x_k as the kth given voltage (e.g., $x_1 = 0.8$), we can use the following models (the expression in the parenthesis is the corresponding formula expression in `R`):

- Model 1. $y_{ijk} = \beta_0 + x_k\beta_1 + \epsilon_{ijk}$, (current~voltage)
- Model 2. $y_{ijk} = \beta_0 + x_k\beta_1 + b_{0i} + \epsilon_{ijk}$, (current~1+voltage+(1|Wafer))
- Model 3. $y_{ijk} = \beta_0 + x_k\beta_1 + b_{0j} + \epsilon_{ijk}$, (current~1+voltage+(1|Site))
- Model 4. $y_{ijk} = \beta_0 + x_k\beta_1 + b_{0ij} + \epsilon_{ijk}$, (current~1+voltage+(1|Wafer:Site))

The first model is a regular linear regression model that only includes fixed effects and ignores the variation across wafers and sites. Models 2, 3, and 4 are LME models where the random effects model the variation across wafers, sites, and both, respectively. Using the `R` formula expression, we can use `model<-lm(formula,data=Wafer)` function to fit the linear regression model and use `model<-lmer(formula,data=Wafer)` to fit the LME models. We can use the AIC values of the fitted model to compare these models. Using `AIC(model)`, we can obtain the AIC values for models 1 to 4 as 600.14, 483.36, 605.67, 524.30, respectively. Based on these AIC values, we can see that AIC prefers the LME model considering the variation in the intercept across wafers. However, we want to point out that the AIC tends to be in favor of complex models if the sample size is limited. We can see the summary of the fitted Model 2 as follows:

```
Linear mixed model fit by REML ['lmerMod']
Formula: current ~ 1 + voltage + (1 | Wafer)
   Data: Wafer

REML criterion at convergence: 475.4

Scaled residuals:
   Min       1Q   Median      3Q     Max
-1.7814  -0.8019  -0.1197  0.7650  3.2175

Random effects:
 Groups   Name        Variance  Std.Dev.
 Wafer    (Intercept) 0.09346   0.3057
 Residual             0.17535   0.4187
Number of obs: 400, groups: Wafer, 10

Fixed effects:
            Estimate Std. Error t  value
(Intercept) -7.08287      0.11529  -61.44
voltage      9.64866      0.03701  260.69
Correlation of Fixed Effects:
        (Intr)
voltage -0.514
```

6.2.2 Distribution-Free Estimation Methods

The MLE method is a powerful approach for model parameter estimation. However, to apply MLE, we need to specify the parametric distribution of all the random effects in the model, which sometimes is difficult. Here we provide a brief introduction to a few "distribution-free" approaches.

Minimum Norm Quadratic Unbiased Estimator (MINQUE) approach is a popular distribution-free method for parameter estimation of a LME model. MINQUE is proposed by Rao [1971]. The basic idea is to use a quadratic form of the observations to estimate the variance parameters. If we stack the observations $\mathbf{y}_i$, $i = 1,\ldots,N$ together to form a tall vector, then we can get the aggregated form of the linear mixed effects model in (6.3) as

$$\mathbf{y}_a = \mathbf{X}_a\boldsymbol{\beta} + \mathbf{Z}_a\mathbf{b}_a + \epsilon_a, \tag{6.12}$$

where $\mathbf{y}_a = \begin{pmatrix} \mathbf{y}_1 \\ \mathbf{y}_2 \\ \vdots \\ \mathbf{y}_N \end{pmatrix}$, $\mathbf{X}_a = \begin{pmatrix} \mathbf{X}_1 \\ \mathbf{X}_2 \\ \vdots \\ \mathbf{X}_N \end{pmatrix}$, $\mathbf{Z}_a = \begin{pmatrix} \mathbf{Z}_1 & \mathbf{0} & \cdots & \mathbf{0} \\ \mathbf{0} & \mathbf{Z}_2 & \cdots & \mathbf{0} \\ \vdots & \vdots & \ddots & \vdots \\ \mathbf{0} & \mathbf{0} & \cdots & \mathbf{Z}_N \end{pmatrix}$, $\mathbf{b}_a = \begin{pmatrix} \mathbf{b}_1 \\ \mathbf{b}_2 \\ \vdots \\ \mathbf{b}_N \end{pmatrix}$, $\boldsymbol{\epsilon}_a = \begin{pmatrix} \epsilon_1 \\ \epsilon_2 \\ \vdots \\ \epsilon_N \end{pmatrix}$.

Then we use a quadratic form $\mathbf{y}_a^T\mathbf{A}\mathbf{y}_a$ to estimate the variance parameters. For example, to estimate the variance of the error term in (6.3), we let

$$\hat{\sigma}_\epsilon^2 = \mathbf{y}_a^T\mathbf{A}\mathbf{y}_a. \tag{6.13}$$

The rationale of using a quadratic form to estimate σ_ϵ^2 is that if we can directly observe $\boldsymbol{\epsilon}_a$, then a natural estimator for σ_ϵ^2 would be $E(\boldsymbol{\epsilon}_a^T\boldsymbol{\epsilon}_a)$. In reality, $\boldsymbol{\epsilon}_a$ is not observable but is part of $\mathbf{y}_a$. Thus, we may use a general quadratic form of $\mathbf{y}_a$ as shown in (6.13) to approximate $\boldsymbol{\epsilon}_a^T\boldsymbol{\epsilon}_a$. Without loss of generality, we can assume $\mathbf{A}$ is symmetric. Because the estimate should be non-negative to be valid, we require $\mathbf{A}$ to be positive semi-definite.

Now the question is how to select the weighting matrix $\mathbf{A}$ to make it a good approximation. Through some algebraic manipulation, we have

$$\begin{aligned} E(\hat{\sigma}_\epsilon^2) &= E(\mathbf{y}_a^T\mathbf{A}\mathbf{y}_a) \\ &= \boldsymbol{\beta}^T\mathbf{X}_a^T\mathbf{A}\mathbf{X}_a\boldsymbol{\beta} + \sigma_\epsilon^2\text{tr}(\mathbf{A}) + \text{tr}(\mathbf{G}\mathbf{Z}_a^T\mathbf{A}\mathbf{Z}_a) \end{aligned}.$$

Obviously, to make $\hat{\sigma}_\epsilon^2$ unbiased, i.e., $E(\hat{\sigma}_\epsilon^2) = \sigma_\epsilon^2$, we need $\mathbf{X}_a^T\mathbf{A}\mathbf{X}_a = \mathbf{0}$, $\mathbf{Z}_a^T\mathbf{A}\mathbf{Z}_a = \mathbf{0}$, and $\text{tr}(\mathbf{A}) = 1$. Any $\mathbf{A}$ satisfying these conditions will make the estimation of σ_ϵ^2 unbiased. Ideally, we also want to minimize the variance of the estimate, $\text{var}(\hat{\sigma}_\epsilon^2)$. However, the variance of a quadratic term involves the third and fourth moments, which is quite challenging to evaluate. Thus, instead of minimizing the variance, the MINQUE method minimizes the Euclidean norm of the matrix $\mathbf{A}$, which is simply the sum of squares of all elements of $\mathbf{A}$, computed as $\text{tr}(\mathbf{A}\mathbf{A}^T)$. The rationale of minimizing the Euclidean norm of $\mathbf{A}$ is intuitive because generally we observe that large elements of $\mathbf{A}$ lead to a larger variance of $\hat{\sigma}_\epsilon^2$. Thus, it makes sense to minimize the Euclidean norm of $\mathbf{A}$. The final MINQUE formulation for estimating σ_ϵ^2 is $\hat{\sigma}_\epsilon^2 = \mathbf{y}_a^T\mathbf{A}\mathbf{y}_a$, where $\mathbf{A}$ is a symmetric positive semi-definite matrix with minimal $\text{tr}(\mathbf{A}\mathbf{A}^T)$ under restrictions $\mathbf{X}_a^T\mathbf{A}\mathbf{X}_a = \mathbf{0}$, $\mathbf{Z}_a^T\mathbf{A}\mathbf{Z}_a = \mathbf{0}$, and $\text{tr}(\mathbf{A}) = 1$.

For the estimation of $\hat{\sigma}_\epsilon^2$, we actually have a closed form solution of the MINQUE formulation as

$$\hat{\sigma}_\epsilon^2 = \frac{1}{(\sum n_i) - \text{rank}(\mathbf{W})}\mathbf{y}_a^T(\mathbf{I} - \mathbf{W}\mathbf{W}^-)\mathbf{y}_a, \tag{6.14}$$

where $\mathbf{W}$ is the augmented matrix including both $\mathbf{X}_a$ and $\mathbf{Z}_a$ as $\mathbf{W} = [\mathbf{X}_a \quad \mathbf{Z}_a]$, and the superscript '–' represents the Moore–Penrose inverse. If $\mathbf{y}$ follows normal distribution, then the variance of $\hat{\sigma}_\epsilon^2$ is

$$\text{var}(\hat{\sigma}_\epsilon^2) = \frac{2\sigma_\epsilon^4}{(\sum n_i) - \text{rank}(\mathbf{W})}.$$

The detailed derivation of the above results can be found in Demidenko [2005].

The MINQUE estimation of the covariance matrix of the random effects (i.e., the matrix $\mathbf{G}$) follows a similar idea but needs more complicated mathematical manipulation.

Although there is no distributional requirement, the MINQUE method is closely related to REML with normal distribution assumption. In fact, it can be shown that for a balanced random coefficient model, the MINQUE solution is the same as the REML solution [Demidenko, 2005]. It can also be shown that when $\mathbf{G}$ is a diagonal matrix, the MINQUE solution is actually the solution from one step iteration of REML. Nevertheless, if we have strong evidence that the random effects do not follow a normal distribution, the MINQUE method can be used.

Example 6.3 We can use the MINQUE method to estimate the variance parameters in the models listed in Example 6.1. The aggregated model for the data in Example 6.1 can be written as

$$\mathbf{y}_a = \begin{pmatrix} y_{11} \\ y_{12} \\ y_{21} \\ y_{22} \\ y_{31} \\ y_{32} \end{pmatrix} = \begin{pmatrix} 1 \\ 1 \\ 1 \\ 1 \\ 1 \\ 1 \end{pmatrix} \beta + \begin{pmatrix} 1 & 0 & 0 \\ 1 & 0 & 0 \\ 0 & 1 & 0 \\ 0 & 1 & 0 \\ 0 & 0 & 1 \\ 0 & 0 & 1 \end{pmatrix} \begin{pmatrix} b_1 \\ b_2 \\ b_3 \end{pmatrix} + \begin{pmatrix} \epsilon_{11} \\ \epsilon_{12} \\ \epsilon_{21} \\ \epsilon_{22} \\ \epsilon_{31} \\ \epsilon_{32} \end{pmatrix},$$

$\mathbf{W} = [\mathbf{X}_a \quad \mathbf{Z}_a] = \begin{pmatrix} 1 & 1 & 0 & 0 \\ 1 & 1 & 0 & 0 \\ 1 & 0 & 1 & 0 \\ 1 & 0 & 1 & 0 \\ 1 & 0 & 0 & 1 \\ 1 & 0 & 0 & 1 \end{pmatrix}$, and $\text{rank}(\mathbf{W}) = 3$. Substituting the values of $\mathbf{y}_a$ and $\mathbf{W}$ in (6.14), we can get the MINQUE estimate of σ_ϵ^2. After we have assigned the values to $\mathbf{y}_a$ and $\mathbf{W}$, we can compute $\hat{\sigma}_\epsilon^2$ by the following `R` code: `sigma.epsilon.hat <-1/3*t(ya)%*%(diag(6)-W%*%ginv(W))%*%ya` where `ginv` is a function in `MASS` package to calculate the Moore–Penrose inverse of a given matrix. The resulting `sigma.epsilon.hat` is 0.0120144, which is identical to the REML result.

Method of moments is another common parameter estimation method. The basic idea is to first derive the relationship between the population moments and the parameter of interest. Then samples are drawn from the population and sample moments are computed from the samples. Finally the equations are solved for the parameter of interest using the sample moments in place of population moments. In Demidenko [2005], the method of moments is presented for estimating the covariance parameters of the random effects, i.e., $\mathbf{G}$ in mixed effects model (6.3) with the error term covariance matrix assumed to be $\mathbf{R}_i = \sigma_\epsilon^2 \mathbf{I}_{n_i}$ and the unbiased estimate of σ_ϵ^2 given. It is proven that the estimator is unbiased and consistent.

Another related method to the method of moments is the variance least squares estimation method. We first use the ordinary least squares method to estimate the fixed effects, denoted as $\hat{\boldsymbol{\beta}}_{OLS} = (\sum \mathbf{X}_i^T \mathbf{X}_i)^{-1}(\sum \mathbf{X}_i^T \mathbf{y}_i)$, and compute the ordinary least squares residual as $\hat{\mathbf{e}}_i = \mathbf{y}_i - \mathbf{X}_i \hat{\boldsymbol{\beta}}_{OLS}$. Then the basic idea is to pretend $\hat{\boldsymbol{\beta}}_{OLS}$ is the true value of fixed effects $\boldsymbol{\beta}$ and $\hat{\mathbf{e}}_i \hat{\mathbf{e}}_i^T$ is viewed as an estimate for the covariance matrix of the observation from the ith individual unit. And the estimates of $\mathbf{G}$ and σ_ϵ^2 can be obtained through the minimization problem

$$\min_{\sigma_\epsilon^2, \mathbf{G}} \sum_{i=1}^{N} \text{tr}[(\hat{\mathbf{e}}_i \hat{\mathbf{e}}_i^T - \sigma_\epsilon^2 \mathbf{I} - \mathbf{Z}_i \mathbf{G} \mathbf{Z}_i^T)^2]. \tag{6.15}$$

Noting that $\text{tr}(\mathbf{A}^2) = \sum A_{jj}^2$, we can see that the above minimization problem can be viewed as applying the least squares method to the variance parameters. Setting the partial derivative of the cost function in (6.15) to zero we can obtain an analytical solution to the minimization problem. Details can be found in Demidenko [2005]. One advantage of this approach is that it can be easily extended to nonlinear mixed effects models.

6.3 Hypothesis Testing

In previous sections, we presented several approaches for point estimation of the parameters of a LME model. In practice, we are often interested in testing the significance of the parameters in the model in order to generalize the results to the population. In this section, we present some basic results on the hypothesis testing on the fixed effects and the covariance parameters of the random effects in a LME model.

6.3.1 Testing for Fixed Effects

If we know the covariance matrices of the random effects and the error term, under the normal distribution assumption, the maximum likelihood estimate of the fixed effects β is given by (6.6). It is straightforward to obtain

$$\begin{aligned}\text{var}(\hat{\beta}) &= (\sum_{i=1}^{N}\mathbf{X}_i^T\boldsymbol{\Sigma}_i^{-1}\mathbf{X}_i)^{-1}(\sum_{i=1}^{N}\mathbf{X}_i^T\boldsymbol{\Sigma}_i^{-1}\boldsymbol{\Sigma}_i\boldsymbol{\Sigma}_i^{-1}\mathbf{X}_i)(\sum_{i=1}^{N}\mathbf{X}_i^T\boldsymbol{\Sigma}_i^{-1}\mathbf{X}_i)^{-1}\\ &= (\sum_{i=1}^{N}\mathbf{X}_i^T\boldsymbol{\Sigma}_i^{-1}\mathbf{X}_i)^{-1}(\sum_{i=1}^{N}\mathbf{X}_i^T\boldsymbol{\Sigma}_i^{-1}\mathbf{X}_i)(\sum_{i=1}^{N}\mathbf{X}_i^T\boldsymbol{\Sigma}_i^{-1}\mathbf{X}_i)^{-1}\\ &= (\sum_{i=1}^{N}\mathbf{X}_i^T\boldsymbol{\Sigma}_i^{-1}\mathbf{X}_i)^{-1}\end{aligned} \tag{6.16}$$

In practice, the underlying true covariance matrices of the random effects and the error term are unknown. To evaluate $\boldsymbol{\Sigma}_i$, we can plug in the MLE or REML estimates of the covariance parameters, $\hat{\eta}$, as if they are the true parameters. With the variance expression in (6.16), we can establish the approximate Wald tests. In a very general setting, if we want to test

$$H_0 : \mathbf{L}\boldsymbol{\beta} = \mathbf{0}, \quad \text{versus} \quad H_1 : \mathbf{L}\boldsymbol{\beta} \neq \mathbf{0}, \tag{6.17}$$

we can use the following test statistic as

$$\hat{\beta}^T\mathbf{L}^T[\mathbf{L}(\sum_{i=1}^{N}\mathbf{X}_i^T\hat{\boldsymbol{\Sigma}}_i^{-1}\mathbf{X}_i)^{-1}\mathbf{L}^T]^{-1}\mathbf{L}\,\hat{\beta}, \tag{6.18}$$

where $\hat{\boldsymbol{\Sigma}}_i$ represents the estimated covariance matrix of $\mathbf{y}_i$ by using the estimated covariance parameters $\hat{\eta}$. The statistic in (6.18) asymptotically follows a chi-squared distribution with rank($\mathbf{L}$) degrees of freedom under H_0. To reject the null hypothesis, we can check if the test statistic in (6.18) is larger than $\chi^2_{\alpha,\,\text{rank}(\mathbf{L})}$, where α is the specified significance level.

Example 6.4 We can apply the above method to test the fixed effect in the LME in Example 6.1 at a significance level $\alpha = 0.05$. The hypothesis testing problem is formulated as

$$H_0 : \beta = 0, \quad \text{versus} \quad H_1 : \beta \neq 0,$$

Using the MLE result in Example 6.1, we have $\hat{\beta} = 1.02087$, $\hat{\sigma}_b^2 = 0.00055$, $\hat{\sigma}_\epsilon^2 = 0.01201$. Plugging in (6.18), we can get the approximate Wald test statistic.

Here we have $\mathbf{L}=1$, $\boldsymbol{\Sigma}_i = \hat{\sigma}_b^2 + \hat{\sigma}_\epsilon^2 = 0.01256$. We can get the test statistic as 497.8744. The 95% percentile point of chi-squared distribution with degrees of freedom of 1 is 3.84146. The test statistic is well above the critical value so we can reject the null hypothesis.

One weakness of the Wald approximation is that it underestimates the variability of $\hat{\beta}$ because by using (6.16), the variability in $\hat{\eta}$ is ignored. This issue can be compensated by using an approximate F test statistic as

$$F = \frac{\hat{\beta}^T \mathbf{L}^T [\mathbf{L}(\sum_{i=1}^{N} \mathbf{X}_i^T \hat{\boldsymbol{\Sigma}}_i^{-1} \mathbf{X}_i)^{-1} \mathbf{L}^T]^{-1} \mathbf{L}\hat{\beta}}{\text{rank}(\mathbf{L})}, \tag{6.19}$$

which asymptotically follows an F distribution with rank($\mathbf{L}$) as the numerator degrees of freedom of the F distribution under H_0. The denominator degrees of freedom of the F distribution needs to be estimated from the data. A popular method to determine the denominator degrees of freedom is to use Scatterthwaite's method of moment approximation [Hrong-Tai Fai and Cornelius, 1996]. In such a procedure, the first step is to conduct eigendecomposition of the matrix $[\mathbf{L}(\Sigma_{i=1}^N \mathbf{X}_i^T \hat{\boldsymbol{\Sigma}}_i^{-1} \mathbf{X}_i)^{-1} \mathbf{L}^T]^{-1}$ into $\mathbf{PDP}^T$, where $\mathbf{P}$ is an orthogonal matrix and $\mathbf{D}$ is a diagonal matrix. With this decomposition, the numerator of F in (6.19) can be written as a sum of the square of r independent variables with t distributions, where $r = \text{rank}(\mathbf{L})$,

$$rF = \sum_{i=1}^{r} \frac{(\mathbf{PL}\hat{\beta})_i^2}{d_i} = \sum_{i=1}^{r} T^2(v_i),$$

where d_i is the ith diagonal element of $\mathbf{D}$, $(\mathbf{PL}\hat{\beta})_i$ is the ith element of the vector $(\mathbf{PL}\hat{\beta})$, and $T(v_i)$ represents the corresponding t-distributed random variable. The degrees of freedom of each t distribution term are denoted as v_i and can be approximated as

$$v_i = \frac{2d_i}{\mathbf{g}_i^T \mathbf{H} \mathbf{g}_i},$$

where $\mathbf{g}_i$ is the gradient of $(\mathbf{PL})_i(\sum_{i=1}^{N} \mathbf{X}_i^T \hat{\boldsymbol{\Sigma}}_i^{-1} \mathbf{X}_i)^{-1} (\mathbf{PL})_i^T$ with respect to covariance parameters η, $(\mathbf{PL})_i$ denotes the ith row of the matrix $\mathbf{PL}$, and $\mathbf{H}$ is the covariance matrix of $\hat{\eta}$, which can be obtained using the second derivative of the log-likelihood function. With the approximation of v_i, we can have the denominator degrees of freedom of the F test statistic in (6.19) as

$$\nu = \frac{2\sum_{i=1}^{r} \frac{\nu_i}{\nu_i - 2}}{\sum_{i=1}^{r} \frac{\nu_i}{\nu_i - 2} - r}.$$

With the estimated denominator degrees of freedom, to test the hypothesis in (6.17), we can check if the statistic in (6.19) is larger than the $(1-\alpha)\times 100\%$ percentile point of the corresponding F-distribution. The numerical implementations of this testing approach are available in `lmerTest` package in `R` [Kuznetsova et al., 2017].

In addition to the approximate Wald test and F-test methods, the likelihood ratio test (LRT) is another commonly used method. Assume we want to test the following hypothesis:

$$H_0 : \boldsymbol{\beta} \in \boldsymbol{\Omega}^0 \quad \text{versus} \quad H_1 : \boldsymbol{\beta} \in \boldsymbol{\Omega}, \tag{6.20}$$

where $\boldsymbol{\Omega}$ is the space of the parameters of the fixed effects, $\boldsymbol{\Omega}^0$ is a subspace of $\boldsymbol{\Omega}$. A classical result in maximum likelihood estimation theory is that under some regularity conditions, the likelihood ratio statistic in (6.21), under H_0, asymptotically follows a chi-squared distribution,

$$-2\ln\frac{L(\widehat{\boldsymbol{\beta}}^0_{\text{ML}})}{L(\widehat{\boldsymbol{\beta}}_{\text{ML}})}, \tag{6.21}$$

where $L(\widehat{\boldsymbol{\beta}}^0_{\text{ML}})$ and $L(\widehat{\boldsymbol{\beta}}_{\text{ML}})$ are the maximal likelihood values under the submodel parameter space $\boldsymbol{\Omega}^0$ and the full model parameter space $\boldsymbol{\Omega}$, respectively. The degrees of freedom of the chi-squared distribution is the difference between the dimension of the full model parameter space $\boldsymbol{\Omega}$ and the submodel parameter space $\boldsymbol{\Omega}^0$. With this result, to test the hypothesis in (6.20), we can check if the test statistic in (6.21) is greater than the $(1-\alpha)\times 100\%$ percentile point of the corresponding chi-squared distribution. One subtle point that needs to be mentioned is that the likelihood function and the estimated parameters $\widehat{\boldsymbol{\beta}}^0_{\text{ML}}$ and $\widehat{\boldsymbol{\beta}}_{\text{ML}}$ are obtained using regular maximum likelihood estimation, not the restricted maximum likelihood estimation method [Verbeke, 1997].

All above methods are based on asymptotic approximations. Thus, these methods will yield similar results when the sample size is large. When the sample size is small, the results may be inaccurate and quite different.

6.3.2 Testing for Variance–Covariance Parameters

The testing for covariance parameters of random effects is more complicated. Following the result in likelihood theory, we know that under some regularity conditions, the distribution of the MLE and REML estimate of the covariance parameters $\hat{\eta}$ can be approximated by a normal distribution with mean η and covariance matrix given by the inverse of the Fisher information matrix. In addition, the asymptotic distribution result for the likelihood ratio test statistic still holds for covariance parameters. Thus, in principle, the approximated Wald test and the likelihood ratio test will work for the hypothesis testing on the covariance parameters. However, there is an important caveat: the asymptotic results used in the Wald test and the likelihood ratio test require that the parameter values being tested is far from the boundary of the feasible parameter space. In more detail, assume we want to conduct the following hypothesis testing

$$H_0 : \sigma^2_{b_j} = a \quad \text{versus} \quad H_1 : \sigma^2_{b_j} \neq a, \tag{6.22}$$

where $\sigma^2_{b_j}$ is the variance of the jth random effect. Clearly, if $a >> 0$, then the parameter value under H_0 is far from the boundary of the feasible parameter space. Thus, the asymptotic results in Wald test and the likelihood ratio test will hold. Unfortunately, in practice, the most common testing on variance parameters is to check if one or a set of random effects exist. As a result, zero is the most common choice of a in the hypothesis test (6.22). However, zero is exactly the boundary of the feasible space of $\sigma^2_{b_j}$. Consequently, when we try to test the existence of a random effect, the Wald test and the likelihood ratio test cannot be used.

To address this issue, some researchers have shown that when the covariance parameter values are at the boundary of the parameter space, the likelihood ratio statistic is often a mixture of chi-squared distribution, rather than the regular single chi-squared distribution. With this result, various hypothesis testing cases can be addressed, such as no random effects versus one random effect, or q versus $q+k$ random effects. Interested readers can refer to Stram and Lee [1994].

Here we present a commonly used approach to test for the existence of random effects in the model (6.3) [Demidenko, 2005]. The setup of the hypothesis testing is as follows.

$$H_0 : \mathbf{G} = \mathbf{0} \quad \text{versus} \quad H_1 : \mathbf{G} \neq \mathbf{0},$$

The test statistic is given as

$$\frac{S_{OLS} - S_{min} / (r-p)}{S_{min} / [(\sum n_i) - r]},$$

where $S_{OLS} = \sum_{i=1}^{N} || \mathbf{y}_i - \mathbf{X}_i \hat{\beta}_{OLS} ||^2$, $\hat{\beta}_{OLS}$ is the ordinary least squares estimation of β, i.e., $\hat{\beta} = \left(\sum_{i=1}^{N} \mathbf{X}_i^T \mathbf{X}_i \right)^{-1} \left(\sum_{i=1}^{N} \mathbf{X}_i^T \mathbf{y}_i \right)$, $S_{min} = \min_{\mathbf{u}} || \mathbf{y} - \mathbf{W}\mathbf{u} ||^2$, $\mathbf{y}$ is defined in (6.12), $\mathbf{W}$ is defined as $\mathbf{W} = [\mathbf{X} \quad \mathbf{Z}]$, p is the dimension of β, and r is the rank of $\mathbf{W}$. An important result obtained in Demidenko [2005] is that the above test statistic follows the F-distribution with numerator degrees of freedom $(r - p)$ and denominator degrees of freedom of $\left[\left(\sum n_i \right) - r \right]$ under H_0. Thus, to test if $\mathbf{G} = \mathbf{0}$, we can check if the above test statistic is larger than the specified percentile point of the corresponding F-distribution. This is a fairly useful way to test the existence of any random effects.

Example 6.5 In this example, we apply the F-test and the LRT method to test the fixed and random effects in the LME model in Example 6.2, respectively. We are using the `Wafer` data set and consider the model $y_{ijk} = \beta_0 + x_k \beta_1 + b_{0i} + \epsilon_{ijk}$, which is represented by the `R` formula as `(current~1+voltage+(1|Wafer))`. The `lmerTest` package in `R` provides some nice routines to realize the hypothesis testing. We can use the following functions to invoke the test on fixed effects and random effects:

```
library(lmerTest)
m2 = lmer(current ~ 1+voltage+(1|Wafer),data=Wafer,
     REML=FALSE)
anova(m2)
ranova(m2)
```

The output of `anova(m2)` provides the results of the F-test for the fixed effects given below:

```
Type III Analysis of Variance Table with Satterth-
waite's method
        Sum Sq Mean Sq NumDF DenDF F value     Pr(>F)
voltage 11916   11916    1    390    68133 < 2.2e-16 ***
---
Signif.  codes:  0 ***  0.001 ** 0.01  *0.55 .0.1   1
```

It shows that the fixed effect `voltage` is very significant. The numerator degrees of freedom is 1 and the denominator degrees of freedom is 390. The function `ranova` provides the LRT result for the existence of the random effect `(1|Wafer)` as follows:

```
ANOVA-like table for random-effects: Single term
deletions

Model:
current ~ voltage + (1 | Wafer)
               npar logLik AIC LRT Df Pr(>Chisq)
<none>            4 -233.88 475.76
(1 | Wafer)       3 -297.07 600.14 126.38 1 < 2.2e-16
***
---
Signif. codes: 0 *** 0.001 ** 0.01 * 0.05 . 0.1    1
```

The LRT is actually `-2(logLik0-logLik)` where `logLik` is the log likelihood value of the full model (the one corresponding to `<none>` line in the result) and `logLik0` corresponds to the model removing the random effect. The result indicates that the random effect is very significant in the model as well.

Bibliographic Notes

The concept of random effects models was first proposed by R.A. Fisher [1918]. Subsequently, various statistical theories have been developed. The mixed effects model has become an important area for statistical research and finds broad application in medicine, ecology, social science, and industrial areas. There are some good textbooks on mixed effects models. A good introduction level book is Galwey [2014]. The computational aspect of estimation of linear mixed effects model is thoroughly discussed in Pinheiro and Bates [2006]. Good books focusing on the applications of mixed effects models include Zuur et al. [2009], Brown and Prescott [2014], Wu [2009], and Pinheiro and Bates [2006]. The variance component model, which is a special case of mixed effects model, is covered in Searle et al. [2009]. The specific applications of mixed effects models on longitudinal data are presented in Molenberghs and Verbeke [2000], Wu and Zhang [2006]. The statistical testing for mixed effects models are covered in Khuri et al. [2011]. Comprehensive treatments of mixed effects model are provided by Demidenko [2005], Grafarend [2006], and McCulloch and Neuhaus [2001]. Most of the application cases presented in these books are in the biomedical and social science areas. This is not surprising because in those systems, it is very common that significant random effects are associated with each individual under study. Without appropriate consideration of random effects, the effectiveness of the analytics methods will be questionable, even at the population level.

Exercises

1. Consider the following LME model

$$\mathbf{y}_i = \begin{pmatrix} y_{i1} \\ y_{i2} \end{pmatrix} = \begin{pmatrix} 2 & 1 \\ 1 & 3 \end{pmatrix}\begin{pmatrix} \beta_0 \\ \beta_1 \end{pmatrix} + \begin{pmatrix} 1 & 0.5 \\ 0.5 & 2 \end{pmatrix}\begin{pmatrix} b_{i0} \\ b_{i1} \end{pmatrix} + \begin{pmatrix} \varepsilon_{i1} \\ \varepsilon_{i2} \end{pmatrix}$$

We know $\mathbf{b}_i = \begin{pmatrix} b_{i0} \\ b_{i1} \end{pmatrix} \sim \mathcal{N}(0, \mathbf{G})$, $\epsilon_i = \begin{pmatrix} \varepsilon_{i1} \\ \varepsilon_{i2} \end{pmatrix} \sim \mathcal{N}(\mathbf{0}, \mathbf{R})$, and $\mathbf{G} = \begin{pmatrix} 1 & 0 \\ 0 & 0.5 \end{pmatrix}$, $\mathbf{R} = \begin{pmatrix} 7/8 & 0 \\ 0 & 3/4 \end{pmatrix}$.

Further, we have the observations from three units as

$$\begin{aligned} \mathbf{y}_a &= \begin{pmatrix} y_{11} & y_{12} & y_{21} & y_{22} & y_{31} & y_{32} \end{pmatrix}^T \\ &= \begin{pmatrix} 1.2632 & 3.5230 & 1.2718 & 1.6469 & 1.4103 & 4.4497 \end{pmatrix}^T \end{aligned}$$

Please estimate β_0, β_1.

2. Derive the MLE and REML results for the estimation of σ_ϵ^2 and σ_b^2 in Example 6.1.

3. Consider the following LME model with only one fixed effect and one random effect

$$\mathbf{y}_i = \begin{pmatrix} y_{i1} \\ y_{i2} \end{pmatrix} = \begin{pmatrix} 0 \\ 1 \end{pmatrix}\beta + \begin{pmatrix} 1 \\ 0 \end{pmatrix} b + \begin{pmatrix} \epsilon_{i1} \\ \epsilon_{i2} \end{pmatrix}$$

where β is the fixed coefficient, $b \sim \mathcal{N}(0, \sigma_b^2)$, and $\epsilon_{ij} \sim \mathcal{N}(0, \sigma_\epsilon^2)$. We have the observation data from three units as

$$\begin{aligned} \mathbf{y}_a &= \begin{pmatrix} y_{11} & y_{12} & y_{21} & y_{22} & y_{31} & y_{32} \end{pmatrix}^T \\ &= \begin{pmatrix} 0.4772 & 0.2914 & 1.1014 & 0.6172 & -0.2372 & 0.8567 \end{pmatrix}^T \end{aligned}$$

Please answer the following questions.

(a) Derive the negative log-likelihood function $-l(\boldsymbol{\theta}|\mathbf{y}_a)$

(b) Derive restricted negative log-likelihood function $l_R(\sigma_b^2, \sigma_\epsilon^2|\mathbf{y}_a)$

(c) Estimate the model parameters: β, σ_b^2 and σ_ϵ^2.

(d) Conduct the hypothesis testing
$H_0: \sigma_b^2 = 0$ vs $H_1: \sigma_b^2 \neq 0$
(hint: use the `lmerTest` package in `R`)

4. Consider the LME model

$$\mathbf{y}_i = \begin{pmatrix} y_{i1} \\ y_{i2} \end{pmatrix} = \begin{pmatrix} 1 \\ 1 \end{pmatrix}\beta + \begin{pmatrix} 2 \\ 1 \end{pmatrix} b + \begin{pmatrix} \epsilon_{i1} \\ \epsilon_{i2} \end{pmatrix}$$

where β is a fixed coefficient, $b \sim \mathcal{N}\left(0, \sigma_b^2\right)$, and $\epsilon_{ij} \sim \mathcal{N}\left(0, \sigma_\epsilon^2\right)$.
The observation data is given as

$$\mathbf{y} = \begin{pmatrix} y_{11} & y_{12} & y_{21} & y_{22} \end{pmatrix}^T = \begin{pmatrix} -0.0687 & 0.2053 & -2.7898 & -1.0109 \end{pmatrix}^T$$

Please use MLE and REML methods to estimate σ_b^2 and σ_ϵ^2.

5. The `design_matrix.csv` and `obs.csv` files on the book companion website contain the data from a real machining process for a V-6 automotive engine head. The observations are obtained during three stages: (1) milling cover face, (2) milling joint face and drilling datum holes, and (3) tapping holes and milling slots. The coordinates of 15 and 16 points on the joint face and the cover face of engine heads, respectively, are measured to determine the dimensional integrity of these two surfaces. The file `obs.csv` contains the observations from eight engine units. Each row represents the observations from a sensor at a location and each column contains the observations corresponding to an engine unit. `design_matrix.csv` contains the design matrix which is derived from the physical process. In the model, the deviation of the six locating pins are the model predictors, which includes both the fixed and random effects. We want to estimate the mean and the variance of the deviations of the locating pins.
From the given information, we can construct a LME model.

$$\mathbf{y}_i = \mathbf{X}\beta + \mathbf{X}\mathbf{b}_i + \epsilon_i, \quad i = 1, \ldots, 8$$

where $\beta \in \mathcal{R}^6$, $\mathbf{b}_i \in \mathcal{R}^6$ and $\mathbf{b}_i \sim \mathcal{N}(\mathbf{0}, \mathbf{G})$, $\mathbf{G} = \text{diag}(\sigma_1^2, \ldots, \sigma_6^2)$, $\epsilon_i \sim \mathcal{N}(\mathbf{0}, \mathbf{R})$, and $\mathbf{R} = \sigma_\epsilon^2 \mathbf{I}_{31\times31}$.
Please answer the following questions.
   (a) Estimate β and $\mathbf{G}$ by MLE using `R`.
   (b) Estimate β and $\mathbf{G}$ by REML using `R`.
   (c) Is there any difference between the results from MLE and REML? If yes, explain why.
6. The `degradation.csv` file on the book companion website includes the observations of a degradation indicator of 10 batteries. Each column contains the observation from one battery. It is known that the battery degradation indicator follows the function form

$$r_i(t) = b_{i0} + b_{i1} t^{1.2} + b_{i2} t^{1.7} + \epsilon_i$$

Please answer the following questions.
   (a) Plot the observations for the degradation indicator.
   (b) Fit a linear regression model using the above given function form.
   (c) Fit a LME model and estimate the model parameters by REML method.
   (d) Why do we have to fit the LME model? Explain.

Part II

Random Effects Approaches for Diagnosis and Prognosis

7

Diagnosis of Variation Source Using PCA

One important application area of system diagnosis is quality control. In industrial practice, we often need to identify the root causes of the problems in product quality. For example, dimensional integrity is a major quality concern in many discrete-part manufacturing processes, such as assembly processes and machining processes. For a perfect manufacturing process, we want the dimensions of the product to be identical to each other. Obviously, such a perfect process does not exist. There are inherent variations in the dimensional qualities of the final product: large variation indicates bad quality [Montgomery, 2009].

The root causes of the product quality variation are called *variation sources*, which form an important category of process faults. For example, the dimensional variation of a product is affected by many variation sources in the process as shown in Figure 7.1, e.g., positioning variability of fixture locators, alignment variability of machine tools, and random deformation of compliant parts.

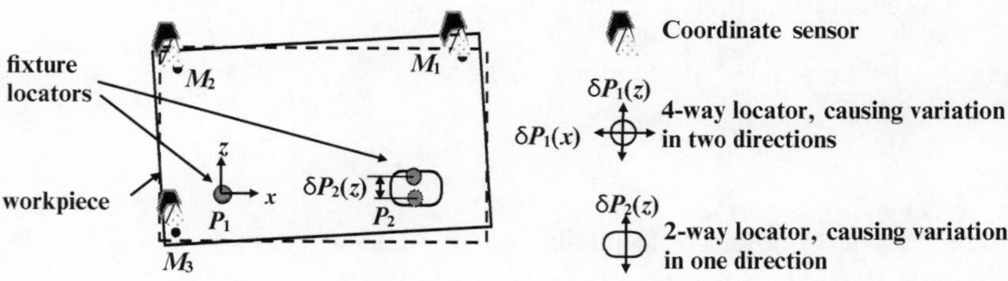

Figure 7.1 An illustration of fixture in an assembly process for a 2-D workpiece

Industrial Data Analytics for Diagnosis and Prognosis: A Random Effects Modelling Approach, First Edition. Shiyu Zhou and Yong Chen.

In this chapter and the following two chapters, we present methods to solve the problem of diagnosing variation sources based on the product quality observations. In this chapter, we focus on an engineering approach for variation source diagnosis using principal component analysis (PCA). The advantage of this method is that it is relatively easy to use and easy to interpret.

7.1 Linking Variation Sources to PCA

In order to diagnose variation sources, the first step is to establish a model that links the product quality observations to the process variation sources. In general, the impact of the process variation sources on the product quality is nonlinear, which can be represented by a general function $\psi(\cdot)$ in (7.1),

$$\mathbf{y}_i = \psi(\mathbf{b}_i) + \epsilon_i, \tag{7.1}$$

where $\mathbf{y}_i$ is the quality observations from the ith product/unit (e.g., the coordinate sensor measurements of the workpiece in Figure 7.1), $\mathbf{b}_i$ represents the process inputs (e.g., the mismatch between the position of a pin and the corresponding part hole as shown in Figure 7.1), ϵ_i is the error term due to measurement noise. Although the relationship between $\mathbf{y}_i$ and $\mathbf{b}_i$ is in general nonlinear, a linear relationship can provide a fairly good approximation because the quality deviations are often small in magnitude. This linear approximation can be obtained through Taylor series expansion around the nominal value $\mathbf{b}_0$ as

$$\mathbf{y}_i \approx \psi(\mathbf{b}_0) + \left.\frac{\partial \psi(\mathbf{b}_i)}{\partial \mathbf{b}_i}\right|_{\mathbf{b}_i=\mathbf{b}_0} \Delta \mathbf{b}_i + \epsilon_i. \tag{7.2}$$

If $\mathbf{b}_i$ is equal to $\mathbf{b}_0$, there is no quality deviation and $\psi(\mathbf{b}_0)$ gives the nominal quality observation. If we denote $\Delta \mathbf{y}_i$ as $\mathbf{y}_i - \psi(\mathbf{b}_0)$ and $\left.\frac{\partial \psi(\mathbf{b}_i)}{\partial \mathbf{b}_i}\right|_{\mathbf{b}_i=\mathbf{b}_0}$ as $\mathbf{Z}$, then we have

$$\Delta \mathbf{y}_i \approx \mathbf{Z} \cdot \Delta \mathbf{b}_i + \epsilon_i.$$

For the sake of notation simplicity, we can replace $\Delta \mathbf{y}_i$ by $\mathbf{y}_i$ and $\Delta \mathbf{b}_i$ by $\mathbf{b}_i$ to obtain the linear model as

$$\mathbf{y}_i = \mathbf{Z} \cdot \mathbf{b}_i + \epsilon_i. \tag{7.3}$$

The model in (7.3) is a quite general quality–fault model that has been used widely in quality control literature. We would like to provide some discussions on this model.

- The output $\mathbf{y}_i$ is the quality observation and is also called the Key Quality Characteristics (KQCs). For the same type of products produced by the same process, the dimensionality of the quality observation vectors for different units is often the same. Thus, we can assume $\mathbf{y}_i \in \mathcal{R}^{n\times1}$ for $\forall i \in \{1, 2, \ldots, N\}$, where N is the number of units.
- In practice the input $\mathbf{b}_i$ is called the process Key Control Characteristics (KCC). It represents the process parameters and conditions that impact on the product quality. For the same type of products and the same process, we can reasonably assume $\mathbf{b}_i$, $\forall i \in 1, 2, \ldots, N$, has the same dimensionality, i.e., $\mathbf{b}_i \in \mathcal{R}^{q\times1}$. In quality control, there are two types of process faults. The first one is that the process input $\mathbf{b}_i$ has a fixed non-zero shift. As a result, the KQCs $\mathbf{y}_i$ will have a fixed non-zero deviation from the specification. Such a quality error is called the "mean-shift" error. This type of error is often relatively easy to be detected and compensated. Another type of process error is called the "variation" error. For this type of error, $\mathbf{b}_i$ takes different values for different products/units and some components of $\mathbf{b}_i$ have excessive variance. As a result, the KQCs $\mathbf{y}_i$ will have an excessive variation, which leads to inconsistencies in the product quality. The detection and diagnosis of "variation" error is relatively more challenging. In this book, we will focus on the "variation" error. Thus, we assume $\mathbf{b}_i$ follows a multivariate normal distribution with zero mean, i.e., $\mathbf{b}_i \sim \mathcal{N}(\mathbf{0}, \mathbf{G})$. Further, we assume the components of $\mathbf{b}_i$ are independent because the process faults are often independent of each other. In other words, we assume $\mathbf{G}$ is a diagonal matrix, i.e., $\mathbf{G} = \text{diag}\{\sigma_1^2\ \sigma_2^2\ \ldots\ \sigma_q^2\}$.
- For the error term ϵ_i, we assume ϵ_i is an n by 1 vector that follows multivariate normal distribution. If ϵ_i only includes measurement noise, it is reasonable to assume ϵ_i follows a simple distribution as $\epsilon_i \sim \mathcal{N}(\mathbf{0}, \sigma_\epsilon^2\mathbf{I})$, where $\mathbf{I}$ is the identity matrix. Please also note that according to (7.2), ϵ_i may also include the linearization error so it may not follow such a simple normal distribution. We will discuss such case in this chapter as well.
- The matrix $\mathbf{Z}$ in (7.3) is the design matrix with dimension $n\times q$. $\mathbf{Z}$ can be obtained from the physical principles of the process. For example, some researchers have established the quality–fault relation for multistage assembly processes and machining processes based on physical analysis of the processes (see Shi and Zhou, 2009 and the references therein). However, for a complex process, it is often quite challenging to obtain $\mathbf{Z}$ from first principles. The cases of identifying variation sources when $\mathbf{Z}$ is unknown will be discussed in Section 7.4 in this chapter.

From the above discussion, it is clear that the model in (7.3) is a special case of the LME model in (6.3) without the fixed effects part. Further, the design matrix $\mathbf{Z}_i$ is identical for different units, i.e., $\mathbf{Z}_i = \mathbf{Z}$, $\forall i \in \{1, 2, \ldots, N\}$. Thus, the model in (7.3) is a balanced random coefficient regression model as discussed in Chapter 6. Such a model is also called a linear replicated model [Rao and Kleffe, 1988] in some literature.

If $\mathbf{Z}$ is known, we can use the estimation methods for LME models introduced in Chapter 6 to estimate the variance parameters of the random effects $\mathbf{b}_i$. The resulting estimates can be directly used to identify the variation sources. Essentially, if the estimated magnitude of the variance of certain component(s) of $\mathbf{b}_i$ is higher than the allowed tolerance, then we can claim the corresponding component(s) is(are) the variance source(s) in the process. Such an approach will be discussed in Chapter 8. In this chapter, we will introduce the pattern matching technique based on principal component analysis. This approach has been used in engineering practice due to the ease of use and its good interpretability.

The pattern matching technique is based on the relationship between the eigenvector of the covariance matrix of $\mathbf{y}_i$ and the column vectors of the design matrix $\mathbf{Z}$. If we denote $\boldsymbol{\Sigma}_\mathbf{y}$ and $\mathbf{R}$ as the covariance matrix of output $\mathbf{y}$ and the error term ϵ, respectively, then from (7.3) we have

$$\boldsymbol{\Sigma}_\mathbf{y} = \mathbf{Z}\mathbf{G}\mathbf{Z}^T + \mathbf{R}. \tag{7.4}$$

Since $\mathbf{G}$ is a diagonal matrix as discussed above and let us further assume that only the jth element of $\mathbf{b}_i$ has non-zero variance, denoted as σ_j^2, then we have

$$\boldsymbol{\Sigma}_\mathbf{y} = \sigma_j^2 \mathbf{z}_j \mathbf{z}_j^T + \mathbf{R}, \tag{7.5}$$

where $\mathbf{z}_j$ is the jth column vector of $\mathbf{Z}$. If we multiply $\mathbf{z}_j$ on both sides of (7.5) and further assume $\mathbf{R}$ is in the simple form as $\sigma_\epsilon^2\mathbf{I}$, we have

$$\begin{aligned}\boldsymbol{\Sigma}_\mathbf{y} \cdot \mathbf{z}_j &= \sigma_j^2 \mathbf{z}_j \cdot (\mathbf{z}_j^T \cdot \mathbf{z}_j) + \mathbf{R} \cdot \mathbf{z}_j \\ &= [\sigma_j^2 (\mathbf{z}_j^T \cdot \mathbf{z}_j) + \sigma_\epsilon^2] \cdot \mathbf{z}_j\end{aligned}.$$

Please note that $[\sigma_j^2(\mathbf{z}_j^T \cdot \mathbf{z}_j) + \sigma_\epsilon^2]$ is a scalar. Thus, $\mathbf{z}_j$ is an eigenvector of $\boldsymbol{\Sigma}_\mathbf{y}$. The corresponding eigenvalue is $\sigma_j^2(\mathbf{z}_j^T \cdot \mathbf{z}_j) + \sigma_\epsilon^2$. We can show that this eigenvalue is the largest eigenvalue of $\boldsymbol{\Sigma}_\mathbf{y}$ under the condition that $\mathbf{G}$ is a diagonal matrix with only one non-zero element and $\mathbf{R}$ is $\sigma_\epsilon^2\mathbf{I}$. The other eigenvalues are identical and equal to σ_ϵ^2. Here we use a simple example to illustrate this result.

Example 7.1 Consider the following quality–fault model

$$\begin{pmatrix} y_1 \\ y_2 \\ y_3 \end{pmatrix} = \begin{pmatrix} 1 & 2 & 1 \\ 3 & 1 & 2 \\ 4 & 3 & 2 \end{pmatrix} \begin{pmatrix} b_1 \\ b_2 \\ b_3 \end{pmatrix} + \begin{pmatrix} \epsilon_1 \\ \epsilon_2 \\ \epsilon_3 \end{pmatrix}.$$

We also know $\sigma_1^2 = 1$, $\sigma_2^2 = \sigma_3^2 = 0$, and $\sigma_\epsilon^2 = 0.2$. With the provided information, we can show the eigenvalue/eigenvector pairs for $\mathbf{\Sigma_y}$ as follows. Using the relationship in (7.4) and noting $\mathbf{Z} = \begin{pmatrix} 1 & 2 & 1 \\ 3 & 1 & 2 \\ 4 & 3 & 2 \end{pmatrix}$, $\mathbf{G} = \begin{pmatrix} 1 & 0 & 0 \\ 0 & 0 & 0 \\ 0 & 0 & 0 \end{pmatrix}$, and $\mathbf{R} = \begin{pmatrix} 0.2 & 0 & 0 \\ 0 & 0.2 & 0 \\ 0 & 0 & 0.2 \end{pmatrix}$, we can get $\mathbf{\Sigma_y} = \begin{pmatrix} 1.2 & 3 & 4 \\ 3 & 9.2 & 12 \\ 4 & 12 & 16.2 \end{pmatrix}$. With the following simple `R` code, we can easily obtain the eigenvalue/eigenvector pairs of $\mathbf{\Sigma_y}$.

```
sigma.y<-Z%*%G%*%t(Z)+R
e<-eigen(sigma.y)
```

The resulting variable `e` contains the eigenvalue/eigenvector pairs as

```
eigen() decomposition
$'values'
[1] 26.2   0.2   0.2

$vectors
              [,1]          [,2]            [,3]
[1,]    -0.1961161     0.8728505      -0.4468451
[2,]    -0.5883484    -0.4692975      -0.6584876
[3,]    -0.7844645     0.1337605       0.6055770
```

The principal eigenvalue is 26.2 and the corresponding eigenvector is the leftmost vector listed in the above `$vectors` section. It is clear that this vector is actually the normalized first column vector of $\mathbf{Z}$ corresponding to the component of $\mathbf{b}_i$ with non-zero variance.

The above analysis indicates that when the jth element of $\mathbf{b}_i$ has excessive variance (we also call such situation "the jth process fault occurs"), the eigenvector associated with the largest eigenvalue of $\mathbf{\Sigma_y}$, known as its *principal eigenvector*, should match the jth column vector of $\mathbf{Z}$. We can call the column vectors of $\mathbf{Z}$ the *fault signatures* or *fault signature vectors* and the principal eigenvector of $\mathbf{\Sigma_y}$ the *fault symptom* or *fault symptom vector*. The diagnosis of variation sources can be achieved through the pattern matching between fault signature and fault symptom.

7.2 Diagnosis of Single Variation Source

The previous section established the relationship between the fault and its symptom. Ideally, fault signature $\mathbf{z}_j$ will exactly match the principal eigenvector of $\mathbf{\Sigma_y}$. However, in reality, the error term covariance matrix $\mathbf{R}$ may not have a simple covariance structure $\sigma_\epsilon^2\mathbf{I}$, particularly when it also includes a linearization error. When $\mathbf{R}$ is not in the form of $\sigma_\epsilon^2\mathbf{I}$, a fault symptom will deviate from the corresponding fault signature. Another difficulty is due to the fact that the population covariance matrix $\mathbf{\Sigma_y}$ is never exactly known and will have to be estimated from the sample covariance matrix $\mathbf{S_y}$ of observation data. The principal eigenvector of $\mathbf{S_y}$ is a random vector, meaning that we will know a random sample of the fault symptom instead of its exact value. In this section, we present a robust pattern matching technique to address these two issues. This technique is first reported in Li et al. [2007].

The assumptions used in the robust matching method are: (*i*) the quality–fault relationship can be adequately described by (7.3) and the matrix $\mathbf{Z}$ is known. The process faults and the system error terms are assumed to be independent of each other. (*ii*) The covariance matrix of the error term, $\mathbf{R}$, is in a general form. $\mathbf{R}$ is unknown but the range of its eigenvalues is assumed to be known. The rationale of this assumption is that if the error term is dominated by measurement noises, then the range of eigenvalues of $\mathbf{R}$ represents the range of sensor accuracy, which can be obtained through the sensor's performance specification. When the error term consists of high order nonlinear residuals, an offline calibration could provide information regarding the range of the eigenvalues of $\mathbf{R}$. (*iii*) One fault occurs at a time, meaning that multiple faults cannot occur in the system simultaneously. This assumption is not very restrictive because the probability of the occurrence of multiple simultaneous faults is small in many cases. The problem of diagnosing multiple simultaneous faults is discussed in Section 7.3.

To establish a robust pattern matching method, we need to consider both the uncertainties caused by the general structured $\mathbf{R}$ and by the sample covariance matrix $\mathbf{S_y}$ when we use the PCA-based pattern matching method. These uncertainties are illustrated by Figure 7.2. Vector $\mathbf{z}_j$ is a column vector of $\mathbf{Z}$, which is the same as the principal eigenvector of $\mathbf{ZGZ}^T$ when the jth fault occurs. With the presence of an unstructured $\mathbf{R}$, the principal eigenvector of $\mathbf{\Sigma_y}$, i.e., $\mathbf{ZGZ}^T+\mathbf{R}$, denoted as $\mathbf{v}_j^1$ (the superscript represents that it is the principal eigenvector), is not the same as $\mathbf{z}_j$. Instead, it will fall in a cone as shown in Figure 7.2. The boundary of the cone can be represented by the angle γ_j. The principal eigenvector of the sample covariance matrix $\mathbf{S_y}$, denoted as $\mathbf{u_s}$, will be different from $\mathbf{v}_j^1$ due to sample uncertainty. The final boundary for $\mathbf{u_s}$ can be illustrated by the dashed-line cone as shown in Figure 7.2. The cone can be viewed

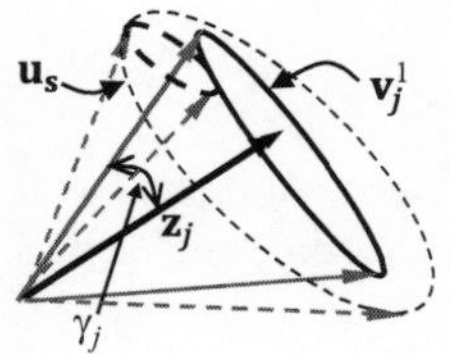

Figure 7.2 Uncertainties in pattern matching for single variation source diagnosis

as the confidence boundary for the fault symptom vector $\mathbf{u_s}$. If the sample principal eigenvector falls in the confidence boundary of a fault signature vector $\mathbf{z}_j$, we can claim that the corresponding fault occurs. Note that Figure 7.2 exaggerates the size of the boundary for the sake of illustration.

To quantify the deviations between $\mathbf{z}_j$ and $\mathbf{u_s}$, we have the following results. We denote $\{\lambda_{\mathbf{b},j}^1, \mathbf{z}_j\}$, $\{\lambda_{\mathbf{y},j}^1, \mathbf{v}_j^1\}$, and $\{l_{\mathbf{y},j}^1, \mathbf{u_s}\}$ as the principal eigenvalue and eigenvector pair of $\mathbf{ZGZ}^T$, $\boldsymbol{\Sigma}_\mathbf{y}$ and $\mathbf{S_y}$, respectively, when the jth fault occurs.

Lemma 7.1 Let θ_j be the angle between $\mathbf{z}_j$ and $\mathbf{v}_j^1$, then the upper bound of θ_j is given as

$$\theta_j \le \sin^{-1}\left(\frac{4}{\lambda_{\mathbf{b},j}^1}\sqrt{\lambda_{\max}^2(\mathbf{R}) - \lambda_{\min}^2(\mathbf{R})}\right), \tag{7.6}$$

where $\lambda_{\max}(\mathbf{R})$ and $\lambda_{\min}(\mathbf{R})$ are the largest and smallest eigenvalues of $\mathbf{R}$.

We denote this boundary $\sin^{-1}\left(\frac{4}{\lambda_{\mathbf{b},j}^1}\sqrt{\lambda_{\max}^2(\mathbf{R}) - \lambda_{\max}^2(\mathbf{R})}\right)$ as γ_j. This result suggests that the perturbation boundary between $\mathbf{z}_j$ and $\mathbf{v}_j^1$ depends on $\lambda_{\mathbf{b},j}^1 / \lambda_{\max}(\mathbf{R})$ and $\lambda_{\max}(\mathbf{R}) / \lambda_{\min}(\mathbf{R})$. The value of $\lambda_{\mathbf{b},j}^1 / \lambda_{\max}(\mathbf{R})$ can be viewed as the signal-to-noise ratio, while $\lambda_{\max}(\mathbf{R}) / \lambda_{\min}(\mathbf{R})$ indicates the deviation of the structure of $\mathbf{R}$ from the simple structure $\sigma_\epsilon^2\mathbf{I}$. With the increase in the signal-to-noise ratio, the perturbation boundary will get smaller and with a larger structural deviation, the perturbation boundary will get larger. The detailed proof of this result can be found in Li et al. [2007].

The uncertainty in the sample covariance matrix $\mathbf{S_y}$ is quantified in the following result.

Lemma 7.2 If λ^1 is a distinct eigenvalue of the n by n covariance matrix $\boldsymbol{\Sigma}_\mathbf{y}$ of normally distributed random vector $\mathbf{y}$, then $N^{1/2}(\mathbf{u_s} - \mathbf{v}^1)$ asymptotically follows a n-variate normal distribution $\mathcal{N}(\mathbf{0}, \boldsymbol{\Gamma})$, where

$$\boldsymbol{\Gamma} = \lambda^1 \sum_{i=2}^{n} \frac{\lambda^i}{(\lambda^1 - \lambda^i)^2} \mathbf{v}^i (\mathbf{v}^i)^T,$$

and $\mathbf{u}_\mathrm{s}$ is the corresponding eigenvector of the sample covariance matrix that is obtained with N samples of $\mathbf{y}$ and $(\lambda^i, \mathbf{v}^i)$, $i = 1, \ldots, n$ are the eigenvalue and eigenvector pairs of $\boldsymbol{\Sigma}_\mathbf{y}$.

The proof of this result can be found in Muirhead [2009]. This result indicates that due to the sampling uncertainty, different realizations of $\mathbf{u}_\mathrm{s}$ will generate an n-dimensional ellipsoid centering around the corresponding vector $\mathbf{v}^1$. There is no upper bound for the angle between $\mathbf{u}_\mathrm{s}$ and $\mathbf{v}^1$. However, we can establish a hypothesis testing procedure to test if $\mathbf{v}^1$ is the underlying vector at the center of the n-dimensional ellipsoid generated by the realizations of $\mathbf{u}_\mathrm{s}$. Specifically, denoting $\mathbf{v}_\mathrm{s}$ as the principal eigenvector of the population covariance matrix corresponding to $\mathbf{S}_\mathrm{y}$ and $\mathbf{v}_k^1$ as the principal eigenvector of $\boldsymbol{\Sigma}_\mathbf{y}$ when the kth fault occurs in the process, $k = 1, 2, \ldots, q$, we can formulate the test as:

$$\begin{aligned} H_0 &: \ \mathbf{v}_\mathrm{s} = \mathbf{v}_k^1 \\ H_1 &: \ \mathbf{v}_\mathrm{s} \neq \mathbf{v}_k^1 \end{aligned}. \tag{7.7}$$

The testing statistic for the above hypothesis can be developed based on Lemma 7.2 as

$$W_k \equiv (N-1)\,(\mathbf{v}_k^1 - \mathbf{u}_\mathrm{s})^T (l_{\mathbf{y},k}^1 \cdot \mathbf{S}_\mathbf{y}^{-1} - 2\mathbf{I} + \frac{1}{l_{\mathbf{y},k}^1} \mathbf{S}_\mathrm{y})\,(\mathbf{v}_k^1 - \mathbf{u}_\mathrm{s}) \tag{7.8}$$

and it is known that W_k asymptotically follows the distribution of χ_{n-1}^2 under H_0. If $W_k > \chi_{\alpha,n-1}^2$, where $\chi_{\alpha,n-1}^2$ is the upper 100α percentile of the χ^2 distribution with degrees of freedom $n-1$, we reject the null hypothesis, meaning that $\mathbf{v}_k^1$ is not the principal eigenvector of current $\boldsymbol{\Sigma}_\mathbf{y}$. Otherwise, we accept the null hypothesis, meaning that $\mathbf{v}_k^1$ is the principal eigenvector of currently observed $\boldsymbol{\Sigma}_\mathbf{y}$, and thus we know the kth fault occurred in the system.

An issue in computing the test statistic W_k is that the values of $\mathbf{v}_k^1$, $k = 1, 2, \ldots, q$, are unknown. As shown in Figure 7.2, we know $\mathbf{z}_k$ and $\mathbf{u}_\mathrm{s}$ but $\mathbf{v}_k^1$ is unknown. We establish a heuristic matching procedure to address this issue. The basic idea is shown in Figure 7.3. Vectors $\mathbf{z}_k$ and $\mathbf{z}_j$ are known column vectors, also called fault signatures, from the matrix $\mathbf{Z}$. The vectors $\mathbf{v}_k^1$ and $\mathbf{v}_j^1$ are the eigenvectors of the population covariance matrix $\boldsymbol{\Sigma}_\mathrm{y}$. Please note that $\mathbf{v}_k^1$ and $\mathbf{v}_j^1$ are unknown because the exact value of $\boldsymbol{\Sigma}_\mathbf{y}$ is unknown. However, from Lemma 7.1, we know that $\mathbf{v}_k^1$ and $\mathbf{v}_j^1$ fall within a boundary around $\mathbf{z}_k$ and $\mathbf{z}_j$, respectively. Such a boundary is illustrated by the cones in Figure 7.3. The vector $\mathbf{u}_\mathrm{s}$ is the principal eigenvector of the sample covariance matrix $\mathbf{S}_\mathbf{y}$. Once observation data are collected, $\mathbf{u}_\mathrm{s}$ can be calculated. In practice, if $\mathbf{u}_\mathrm{s}$ falls in the cone of one signature

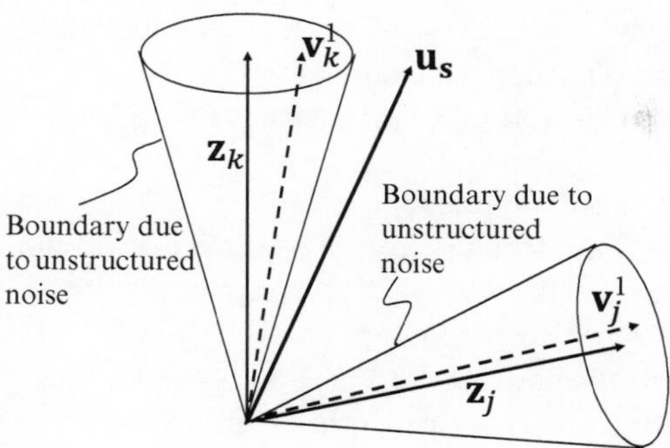

Figure 7.3 Illustration of pattern matching strategy for a single fault

vector $\mathbf{z}_j$, then the decision rule is simple: we can claim that the jth component of process inputs $\mathbf{b}_i$ has excessive variation. There may be cases that multiple signature vectors are too close to each other and their cones overlap. In such cases, if $\mathbf{u}_\mathbf{s}$ falls in the overlap area, then we cannot differentiate which fault occurs. In such cases, we need to select a different observation scheme to change the design matrix $\mathbf{Z}$ to separate these signature vectors.

A challenging case is that $\mathbf{u}_\mathbf{s}$ does not fall in any cones. In such cases, we need to consider the sample uncertainty and use the hypothesis testing in (7.7) to decide which fault occurs. To address the issue of not knowing the eigenvectors of $\boldsymbol{\Sigma}_\mathbf{y}$, we use the following optimization method to find an eigenvector of $\boldsymbol{\Sigma}_\mathbf{y}$ that gives the smallest value of the test statistic (7.8) for the hypothesis testing. Then we can use the obtained value to decide if the corresponding fault occurs,

$$W_k^* = \min_{\mathbf{v}_k^1}(N-1)(\mathbf{v}_k^1 - \mathbf{u}_\mathbf{s})^T(l_{\mathbf{y},k}^1 \cdot \mathbf{S}_\mathbf{y}^{-1} - 2\mathbf{I} + \frac{1}{l_{\mathbf{y},k}^1}\mathbf{S}_\mathbf{y})(\mathbf{v}_k^1 - \mathbf{u}_\mathbf{s})$$
$$\text{subject to} \quad (\mathbf{v}_k^1)^T \mathbf{v}_k^1 = 1 \quad \text{and} \quad \cos^{-1}((\mathbf{v}_k^1)^T \cdot \mathbf{z}_k) \leq \gamma_k, \tag{7.9}$$

where γ_k is the perturbation boundary that can be obtained by Lemma 7.1. The intuition is that since we do not know what the true $\mathbf{v}_k^1$ is, we will find a vector within the feasible range around the fault signature $\mathbf{z}_k$ that gives the smallest value of the test statistic W_k. Then we use the obtained W_k to conduct the hypothesis testing. This is a heuristic approach and the testing result may cause extra false positives. In other words, we may wrongly identify a $\mathbf{z}_k$ as the underlying fault signature.

To summarize, the proposed pattern matching procedure is listed as follows.

1. A quality–fault model as (7.3) is developed. Based on the model, we can obtain the columns of $\mathbf{Z}$ as the fault signature vectors.

2. The multivariate observations $\mathbf{y}_i$, $i = 1, \ldots, N$, of product quality features are obtained during production. The dimensionality of $\mathbf{y}_i$ is n.
3. Based on $\mathbf{y}_i$, $i = 1, \ldots, N$, calculate the sample covariance matrix $\mathbf{S}_\mathbf{y}$ and its principal eigenvector $\mathbf{u}_\mathbf{s}$.
4. Estimate $\lambda_{max}(\mathbf{R})$ and $\lambda_{min}(\mathbf{R})$ from accuracy specification of the observation system. Based on (7.6), we can calculate γ_j. Here we often need to substitute $\lambda^1_{\mathbf{b},j}$ with the tolerance specification of the corresponding process variable if the precise value of $\lambda^1_{\mathbf{b},j}$ is not yet known.
5. Calculate the angles θ_k between $\mathbf{z}_k$ and $\mathbf{u}_\mathbf{s}$, $k = 1 \ldots q$. If all of the angles are larger than the corresponding γ_k, go to next step, otherwise go to Step 7.
6. Solve the optimization problem in (7.9) to find the optimal value of $\mathbf{v}^1_k$ and the corresponding W^*_k, $k = 1 \ldots q$.
7. Find k for $\theta_k \leq \gamma_k$ or $W^*_k \leq \chi^2_{\alpha,n-1}$. We claim the kth fault occurs. If none exists, we claim that none of the known faults occur. If multiple faults are identified, then we claim faults occur but we cannot differentiate them.

One point that needs to be mentioned is that before conducting the fault diagnosis, we assume that excessive variations have been detected in the quality observations. Statistical process control techniques can be used for detecting variance increase.

A comprehensive example of this fault identification procedure for a two stage machining process can be found in Li et al. [2007]. Here we provide a simple example illustrating the uncertainties in the eigenvectors of $\mathbf{\Sigma}_\mathbf{y}$ and $\mathbf{S}_\mathbf{y}$.

Example 7.2 Let us consider a simple two-dimensional model as (7.3) with $\mathbf{Z} = \begin{pmatrix} 1 & 1 \\ 2 & 1 \end{pmatrix}$, $\mathbf{R} = \begin{pmatrix} 0.05 & 0 \\ 0 & 0.01 \end{pmatrix}$, and $\mathbf{G} = \begin{pmatrix} 2 & 0 \\ 0 & 0 \end{pmatrix}$. Using this information, we can obtain the deviation between the fault signature vector, which is the first column of the design matrix $\mathbf{Z}$, and the principal eigenvector of $\mathbf{\Sigma}_\mathbf{y}$.

It is easy to see that $\mathbf{ZGZ}^T = \begin{pmatrix} 2 & 4 \\ 4 & 8 \end{pmatrix}$ with one non-zero eigenvalue of 10. Using (7.6), we can obtain the upper bound of the angle between $\mathbf{z}_1 = \begin{pmatrix} 1 \\ 2 \end{pmatrix}$ and the first eigenvector of $\mathbf{\Sigma}_\mathbf{y}$ as $\theta_1 \leq \sin^{-1}(\frac{4}{10}\sqrt{0.05^2 - 0.01^2}) = \gamma_1$. In this case, we can get $\gamma_1 = 1.12^\circ$. Indeed, we can compute $\mathbf{\Sigma}_\mathbf{y} = \mathbf{ZGZ}^T + \mathbf{R}$ as $\mathbf{\Sigma}_\mathbf{y} = \begin{pmatrix} 2.05 & 4 \\ 4 & 8.01 \end{pmatrix}$ with eigenvalue (10.018 0.042) and corresponding eigenvectors as $(0.4486\ \ 0.8937)^T$ and $(-0.8937\ \ 0.4486)^T$. The principal eigenvector of $\mathbf{\Sigma}_\mathbf{y}$ is $(0.4486\ \ 0.8937)^T$ and its angle slightly differs from the angle of the corresponding column vector of $\mathbf{Z}$, which is $(1\ \ 2)^T$. The angle between $(0.4486\ \ 0.8937)^T$ and $(1\ \ 2)^T$ is 0.0919°. Clearly, this angle is smaller than the upper bound which is 1.12°.

Example 7.3 Using the process model in Example 7.2, we can show the impact of sample uncertainty on the observed principal eigenvector in practice. To show the sample uncertainty, we need to first simulate the observation data $\mathbf{y}_i$ for $i = 1, 2, \ldots, N$ and then use the simulated data to compute the sample covariance matrix $\mathbf{S}_\mathbf{y}$ and its principal eigenvector $\mathbf{u}_\mathbf{s}$. Then, the W_k statistic in (7.8) can be calculated to show the uncertainty in $\mathbf{u}_\mathbf{s}$ due to sampling.

The following R function can be used to simulate the observation data under a given process model.

```
Simulate.vs.data<-function(Z,G,R,num){
#Z,G,R are process parameters. num is the sample size
b<-mvrnorm(n = num, rep(0,ncol(Z)), G);
noise<-mvrnorm(n = num, rep(0,nrow(Z)), R);
y<-Z%*%t(b)+t(noise)
return(y)}
```

With the simulated data, we can repeatedly simulate the observation data and compute the W_k statistic for each simulation run. The distribution of W_k can then be shown by a histogram and compared with its asymptotic distribution. The following R script uses a `for` loop to realize this, where `simu.rep` is the number of simulation runs.

```
w.values <-matrix(0,simu.rep ,1)
sigma.no.noise = Z%*%G%*%t(Z);
sigma.with.noise = sigma.no.noise + R;
e.with.noise <-eigen(sigma.with.noise);
v1 <-e.with.noise$vectors[,1];
for (i in 1:simu.rep){
sample.covariance <-
    cov(t(Simulate.vs.data(Z,G,R,sample.size)))
e.sample.covariance <-eigen(sample.covariance);
u1<-e.sample.covariance$vectors[,1];
l1<-e.sample.covariance$values[1]
#calculate W_k statistic
w.values[i]<-(sample.size-1)*t(v1-u1)%*%
            (l1*solve(sample.covariance)
            -2*diag(nrow(Z))+1/l1*sample.covariance)
            %*%(v1-u1)
}
```

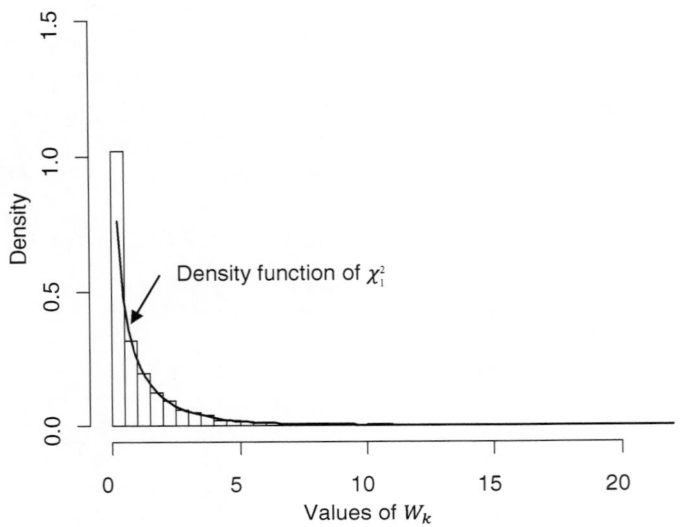

Figure 7.4 Comparison of histogram of simulated W_k statistic and the corresponding χ^2 distribution

If we use 10000 simulation runs with sample size equal to 50, we can obtain Figure 7.4, which compares the histogram of W_k and the theoretical χ^2 distribution with degree of freedom of 1. The histogram fits quite well with the theoretical distribution.

7.3 Diagnosis of Multiple Variation Sources

The method for diagnosing a single variation source as described in Section 7.2 can be extended to the case of diagnosing multiple variation sources. Consider the model in (7.3) and (7.4). Without loss of generality, we assume that there are q' $(1 \leq q' \leq q)$ faults happening in the system, that is, there would exist q' out of q non-zero diagonal entries in $\mathbf{G}$. By changing the sequence of the column vectors of matrices $\mathbf{Z}$ and $\boldsymbol{\Sigma}_{\mathrm{y}}$ according to the magnitude of variation sources, we can get the following:

$$\boldsymbol{\Sigma}_{\mathbf{y}} = \left(\mathbf{z}^{(1)} \ldots\ \mathbf{z}^{(q'+1)}\ \mathbf{z}^{(q'+1)} \ldots\ \mathbf{z}^{(q)}\right) \begin{pmatrix} \sigma^2_{(1)} & & & & \\ & \ddots & & & \\ & & \sigma^2_{(q')} & & \\ & & & 0 & \\ & & & & \ddots & \\ & & & & & 0 \end{pmatrix} \begin{pmatrix} (\mathbf{z}^{(1)})^T \\ \vdots \\ (\mathbf{z}^{(q')})^T \\ (\mathbf{z}^{(q'+1)})^T \\ \vdots \\ (\mathbf{z}^{(q)})^T \end{pmatrix} + \mathbf{R}, \tag{7.10}$$

where $\mathbf{z}_{(1)}$,..., $\mathbf{z}_{(q')}$ are the column vectors of $\mathbf{Z}$ that are associated with the non-zero process variation sources $\sigma^2_{(1)}$,..., $\sigma^2_{(q')}$, where $\sigma^2_{(1)} \geq \sigma^2_{(2)} \geq \ldots \geq \sigma^2_{(q')}$ are the magnitudes of the variation sources. The numbers in the parentheses represent the ordering in the rearranged sequence of these columns. If we multiply an eigenvector corresponding to one of the q' largest eigenvalues of $\boldsymbol{\Sigma}_{\mathbf{y}}$ on both sides of (7.10), it is easy to see that the eigenvector is a linear combination of $\mathbf{z}_{(1)}$,..., $\mathbf{z}_{(q')}$ if $\mathbf{R}$ is in the simple form of $\sigma^2_{\epsilon}\mathbf{I}$. Because the eigenvectors are orthogonal to each other, we can conclude that the vectors $\mathbf{z}_{(1)}$,..., $\mathbf{z}_{(q')}$ span the same linear space of the eigenvectors of $\boldsymbol{\Sigma}_{\mathbf{y}}$ that are associated with the q' largest eigenvalues. The *span* of a set of vectors is the linear space consisting of elements that are linear combinations of the set of vectors. Based on this result, we can establish a pattern matching approach for multiple variation source diagnosis.

We denote the eigenvalue and eigenvector pairs of $\mathbf{ZGZ}^T$, $\boldsymbol{\Sigma}_{\mathbf{y}}$ and $\mathbf{S}_{\mathbf{y}}$ as $\{m^{(i)}, \mathbf{h}^{(i)}\}$, $\{\lambda^{(i)}, \mathbf{v}^{(i)}\}$ and $\{l^{(i)}, \mathbf{u}^{(i)}\}$ $(i = 1, \ldots, q')$, respectively, where the superscript (i) indicates the ith largest eigenvalue and the corresponding eigenvector. Please note that due to the existence of multiple non-zero variation sources, the eigenvectors of $\mathbf{ZGZ}^T$ are not the same as the column vectors of $\mathbf{Z}$. So we use $\mathbf{h}^{(i)}$ to represent them. The relationships between these vectors are illustrated in Figure 7.5. For simplicity, we only show the case of two faults occurring, i.e., two components of $\mathbf{b}_i$ having excessively large variance.

The figure shows the case that two faults, corresponding to $\mathbf{z}^{(1)}$ and $\mathbf{z}^{(2)}$, occur in a process. From the theory stated above, the space spanned by $\mathbf{z}^{(1)}$ and $\mathbf{z}^{(2)}$ is the same as that spanned by the first two eigenvectors of $\mathbf{ZGZ}^T$, $\mathbf{h}^{(1)}$ and $\mathbf{h}^{(2)}$. From the geometric point of view, the two vectors $\mathbf{z}^{(1)}$ and $\mathbf{z}^{(2)}$ will form a plane which can also be formed by $\mathbf{h}^{(1)}$ and $\mathbf{h}^{(2)}$. The plane is denoted as $\mathbf{P}_1$ in Figure 7.5. Similar to the case in single fault diagnosis, if the covariance matrix of the error term is in the simple form $\sigma^2_{\epsilon}\mathbf{I}$, the linear space spanned by the eigenvectors of $\boldsymbol{\Sigma}_{\mathbf{y}}$, $\mathbf{v}^{(1)}$ and $\mathbf{v}^{(2)}$, is the same as the space spanned by the

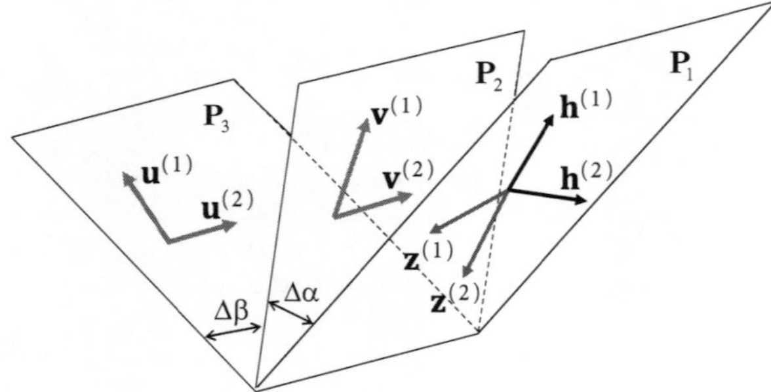

Figure 7.5 Illustration of relationship of principal eigenvectors of $\mathbf{ZGZ}^T$, $\boldsymbol{\Sigma}_\mathbf{y}$ and $\mathbf{S}_\mathbf{y}$

corresponding $\mathbf{z}^{(1)}$ and $\mathbf{z}^{(2)}$. However, with the presence of unstructured error terms, $\mathbf{v}^{(1)}$ and $\mathbf{v}^{(2)}$ will form a plane $\mathbf{P}_2$ that is different from $\mathbf{P}_1$. In practice, we can only get the eigenvectors of $\mathbf{S}_\mathbf{y}$ from the observation data. The two eigenvectors of $\mathbf{S}_\mathbf{y}$, $\mathbf{u}^{(1)}$ and $\mathbf{u}^{(2)}$, will span another plane, denoted as $\mathbf{P}_3$, which will deviate from $\mathbf{P}_1$ and $\mathbf{P}_2$. One point that needs to be mentioned is that the location of $\mathbf{P}_3$ is actually random due to the sampling uncertainty.

In variation source diagnosis, we need to identify the underlying signature vectors (e.g., $\mathbf{z}^{(1)}$ and $\mathbf{z}^{(2)}$ in Figure 7.5) from the observed eigenvectors (e.g., $\mathbf{u}^{(1)}$ and $\mathbf{u}^{(2)}$ in Figure 7.5) through pattern matching. To achieve this, we need to measure the difference between linear spaces $\mathbf{P}_1$ and $\mathbf{P}_2$ and between $\mathbf{P}_2$ and $\mathbf{P}_3$.

The difference between two linear spaces can be measured by the *principal angles* between them [Krzanowski, 1979, 1982].

Definition 7.1 Suppose $\mathbf{L}_1$ and $\mathbf{L}_2$ are two subspaces in $\mathcal{R}^n$, whose dimension satisfy $r = \dim(\mathbf{L}_1) \geq \dim(\mathbf{L}_2) = q \geq 1$, then the principal angles $\theta_1,...,\theta_q \in [0, \pi/2]$ between $\mathbf{L}_1$ and $\mathbf{L}_2$ are defined recursively by $\cos(\theta_k) = \max_{\mathbf{u}\in\mathbf{L}_1} \max_{\mathbf{t}\in\mathbf{L}_2} \mathbf{u}^T\mathbf{t} \equiv \mathbf{u}_k^T\mathbf{t}_k$, subject to $\|\mathbf{u}\| = \|\mathbf{t}\| = 1, \mathbf{u}^T\mathbf{u}_i = \mathbf{t}^T\mathbf{t}_i = 0$, $i = 1, ..., k-1$.

The definition is a bit complicated but we are mainly concerning with the largest principal angle. From the above definition, we can show that the largest principal angle is the largest angle between an arbitrary vector $\mathbf{v}^1$ in $\mathbf{L}_1$ and the vector in $\mathbf{L}_2$ that is closest to $\mathbf{v}^1$. If the dimensionalities of $\mathbf{L}_1$ and $\mathbf{L}_2$ are the same, the largest principal angle between them can be calculated using the following result.

Lemma 7.3 Let $\mathbf{U}$ and $\mathbf{V}$ be the sets of q orthonormal bases of two subspaces $\mathbf{L}_1$ and $\mathbf{L}_2$, respectively, where $q = \dim(\mathbf{L}_1) = \dim(\mathbf{L}_2)$. The largest principal angle between $\mathbf{L}_1$ and $\mathbf{L}_2$ is given by $\cos(\sqrt{\lambda_q})$, where λ_q is the smallest eigenvalue of $\mathbf{U}^T\mathbf{V}\mathbf{V}^T\mathbf{U}$.

The proof of the result can be found in Krzanowski [1979]. We can use the largest principal angle to measure the difference among the subspaces $\mathbf{P}_1$, $\mathbf{P}_2$, and $\mathbf{P}_3$ as shown in Figure 7.5. By checking the magnitude of the difference, we can determine if two subspaces are the same. In the ideal case, the principal angle should be zero if the two subspaces are the same. However, due to the unstructured error term in the system, the principal angle is always larger than zero. Similar to the single fault case, we can obtain an upper bound on the principal angle $\Delta\alpha$ between $\mathbf{P}_1$ and $\mathbf{P}_2$ in Figure 7.5.

Lemma 7.4 Consider $\mathbf{ZGZ}^T$ and $\boldsymbol{\Sigma}_\mathbf{y} = \mathbf{ZGZ}^T + \mathbf{R}$. Suppose there are q' faults occurring in the system, i.e., $m^{(1)} \geq m^{(2)} \geq \ldots \geq m^{(q')} > 0 = m^{(q'+1)} = \ldots = m^{(q)}$, the column vectors in $\mathbf{Z}$ corresponding to non-zero variation sources are $\{\mathbf{z}^{(1)}, \ldots, \mathbf{z}^{(q')}\}$ and the corresponding eigenvectors of $\boldsymbol{\Sigma}_\mathbf{y}$ are $\{\mathbf{v}^{(1)}, \ldots, \mathbf{v}^{(q')}\}$. Then under some non-restrictive conditions, the upper bound on the principal angles between the space spanned by $\{\mathbf{z}^{(1)}, \ldots, \mathbf{z}^{(q')}\}$ and $\{\mathbf{v}^{(1)}, \ldots, \mathbf{v}^{(q')}\}$ is

$$\Delta\alpha \leq \sin^{-1}\left(\frac{4}{m^{(q')}}\sqrt{\lambda^2_{\max}(\mathbf{R}) - \lambda^2_{\min}(\mathbf{R})}\right). \tag{7.11}$$

The detailed proof and some discussions of this result can be found in Li and Zhou [2006]. The value of $m^{(q')}$ is unknown in practice. To overcome this issue, an approximation of the upper bound of the largest principal angle, denoted as γ_1, is developed in Li and Zhou [2006] as

$$\gamma_1 \approx \frac{4}{[l^{(q')} - \lambda_{\max}(\mathbf{R})] - [l^{(q'+1)} - \lambda_{\min}(\mathbf{R})]}\sqrt{\lambda^2_{\max}(\mathbf{R}) - \lambda^2_{\min}(\mathbf{R})}, \tag{7.12}$$

where l represents the eigenvalues of the sample covariance matrix $\mathbf{S}_\mathbf{y}$. Equation (7.12) gives the boundary of the difference between $\mathbf{P}_1$ and $\mathbf{P}_2$ as shown in Figure 7.5.

Because of the sample uncertainty, we also need to quantify the difference between $\mathbf{P}_2$ and $\mathbf{P}_3$. Due to the complexity in computing the principal angle between two subspaces, there is no closed form expression to quantify the uncertainty between $\mathbf{P}_2$ and $\mathbf{P}_3$. Instead, Monte Carlo simulation can be used for this purpose. In such a simulation, for a given system with a known design matrix $\mathbf{Z}$, we can sample N realizations of the random vector $\mathbf{b}_i$ and the error term ϵ_i and then generate N samples of the observation $\mathbf{y}_i$, $i = 1, \ldots, N$, according to (7.3). Then we can compute $\mathbf{S}_\mathbf{y}$ and its eigenvalues and eigenvectors. On the other hand, in the simulation, $\boldsymbol{\Sigma}_{\mathbf{y}}$ is known and so are its eigenvalues and eigenvectors. We can pick up the set of eigenvectors of $\mathbf{S}_{\mathbf{y}}$ and $\boldsymbol{\Sigma}_{\mathbf{y}}$ and compute the largest principal angle ($\Delta\beta$ in Figure 7.5) between them. We can repeat the process M times to get the empirical distribution of $\Delta\beta$. From the empirical distribution, we can get its 95% (or other specified value such as 99%) quantile value as the boundary of $\Delta\beta$. We denote the boundary as γ_2.

One issue that needs to be addressed in the simulation procedure is how to select the parameters for the variances of the components of $\mathbf{b}_i$ (i.e., $\sigma_1^2, \ldots, \sigma_{q'}^2$, noting the rest $\sigma_{q'+1}^2, \ldots, \sigma_q^2$ are zeros) and $\boldsymbol{\epsilon}_i$ (i.e., $\sigma_{\max}^2(\boldsymbol{\epsilon})$ and $\sigma_{\min}^2(\boldsymbol{\epsilon})$). Based on some studies [Anderson et al., 1963; Flury, 1987; Krzanowski, 1982], it is known that the eigenvectors of the sample covariance matrix $\mathbf{S}_{\mathbf{y}}$ under the model in (7.3) are sensitive to $\sigma_{q'}^2 / \sigma_{\max}^2(\boldsymbol{\epsilon})$ and $\sigma_1^2 / (\text{tr}\,(\mathbf{R}) + \sum_{i=1}^{q'} \sigma_i^2)$. In summary, we select the parameters in simulation including sample size N, the number of faults q', the above two ratios, and the number of replicates M. Then we can use the simulation procedure to obtain a bound on $\Delta\beta$, which is caused by sample uncertainty.

We can summarize the steps for diagnosis of multiple variation sources as follows.

1. Similar to the single fault diagnosis case, a quality–fault model as (7.3) is developed. Based on the model, fault signature vectors $\mathbf{z}_j$, $j = 1, 2, \ldots q$, are obtained from the columns of $\mathbf{Z}$.
2. The multivariate observations $\mathbf{y}_i, i = 1, \ldots, N$, of product quality features are obtained during production. The dimension of observations is n. Based on the quality observation, we can calculate the sample covariance matrix $\mathbf{S}_{\mathbf{y}}$.
3. From $\mathbf{S}_{\mathbf{y}}$, determine the number of significant variation sources, q'. This problem is equivalent in selecting the number of principal components that should be held in a PCA model to represent the data set. We can use the scree plot introduced in Chapter 4. A few more sophisticated methods can be found in Valle et al. [1999].
4. Based on the sample covariance matrix $\mathbf{S}_{\mathbf{y}}$, calculate the q' pairs of principal eigenvalues and eigenvectors $\{l^{(i)}, \mathbf{u}^{(i)}\}$ $(i = 1, \ldots, q')$ using PCA. There are $C_q^{q'}$ combinations in total if we select q' column vectors from all q column vectors in matrix $\mathbf{Z}$. We denote a selected group of column vectors as $\mathbf{Z}_k$, $k = 1, 2, .., C_q^{q'}$.
5. Estimate $\lambda_{\max}(\mathbf{R})$ and $\lambda_{\min}(\mathbf{R})$ from the specifications of the measurement system. Then we can calculate γ_1 based on (7.12). We can check each group of column vectors $\mathbf{Z}_k$ and calculate the largest principal angles $\Delta\theta_k$

between span $\{\mathbf{u}^{(1)}, \ldots, \mathbf{u}^{(q')}\}$ and range($\mathbf{Z}_k$), $k = 1 \ldots C_q^{q'}$. The range of a matrix is the linear space spanned by its columns. If all of the angles are larger than γ_1, go to next step. Otherwise we can stop and claim the q' components of $\mathbf{b}_i$ corresponding to the $\mathbf{Z}_k$ that gives the smallest $\Delta\theta$ are the variation sources.

6. Using the Monte Carlo simulation discussed above, determine the bound for γ_2. If all θ_k are larger than $(\gamma_1 + \gamma_2)$, we can claim that *unknown* variation sources occur.
7. If we can find a θ_k that is smaller that $(\gamma_1 + \gamma_2)$, we can claim the corresponding q' variation sources occur.

The above procedure is a heuristic procedure. In the following we will show an example to illustrate the key concepts used in this section. A comprehensive case study for diagnosing multiple variation sources can be found in Li and Zhou [2006].

Example 7.4 Consider the following process model

$$\begin{pmatrix} y_1 \\ y_2 \\ y_3 \\ y_4 \end{pmatrix} = \begin{pmatrix} 9 & 0 & 10 \\ 1 & 3 & 2 \\ 9 & 5 & 10 \\ 6 & 10 & 10 \end{pmatrix} \begin{pmatrix} b_1 \\ b_2 \\ b_3 \end{pmatrix} + \begin{pmatrix} \epsilon_1 \\ \epsilon_2 \\ \epsilon_3 \\ \epsilon_4 \end{pmatrix}.$$

Assume we know $\mathbf{G} = \begin{pmatrix} 2 & 0 & 0 \\ 0 & 1 & 0 \\ 0 & 0 & 0 \end{pmatrix}$ and $\mathbf{R} = \begin{pmatrix} 0.5 & 0 & 0 & 0 \\ 0 & 0.2 & 0 & 0 \\ 0 & 0 & 0.3 & 0 \\ 0 & 0 & 0 & 0.1 \end{pmatrix}$. We can compute the principal angle between the subspace spanned by the two fault signature vectors (i.e., the left two column vectors of $\mathbf{Z}$) and the two eigenvectors corresponding to the two largest eigenvalues of $\mathbf{\Sigma}_\mathbf{y}$. Using an `R` code that is similar to that in Example 7.2, we can obtain the eigenvalues and eigenvectors of $\mathbf{ZGZ}^T$ and $\mathbf{\Sigma}_\mathbf{y}$. Then we can pick two signature vectors from the $\mathbf{Z}$ matrix and compute the principal angles between the linear space spanned by the signature vectors and the eigenvectors. The function `Principal.angle` below gives the principal angle between two subspaces.

```
Principal.angle <-function(v.signature ,u){
  a<-qr(v.signature)
  Qa<-qr.Q(a)
  temp <-t(Qa)%*%u%*%t(u)%*%Qa
  e.temp <-eigen(temp)
  return (acos(sqrt(min(e.temp$values))))
  }
```

In the function `Principal.angle`, we utilize the QR matrix decomposition to compute the principal angles between two linear spaces. The first input `v.signature` contains the signature vectors that do not need to be normalized, while the second input `u` needs to contain a set of orthonormal vectors such as the eigenvectors of a matrix. Using this routine, we can obtain the principal angle between the space spanned by the signature vectors $\begin{pmatrix} 9 & 0 \\ 1 & 3 \\ 9 & 5 \\ 6 & 10 \end{pmatrix}$ and the space spanned by the two eigenvectors corresponding the two largest eigenvalues of $\mathbf{ZGZ}^T$ and $\Sigma_\mathbf{y}$ are $\theta_1 = 0$ and $\theta_2 = 0.11^\circ$, respectively. Also, using the result in (7.11), we can get the perturbation boundary as approximately 1.75°.

7.4 Data Driven Method for Diagnosing Variation Sources

One common requirement for the single or multiple variation sources diagnosis methods introduced in the previous sections is that the design matrix **Z** in (7.3) is known. Although in general, **Z** can be obtained through an analysis of process physics, it may be quite challenging to derive **Z** for a complex process. In this section, we introduce a data driven approach for variation source identification without knowing **Z**. This material is based on Jin and Zhou [2006a].

The basic idea is to learn the column vectors of **Z** through the observed quality data **y** and then use the learned fault signatures for variation source diagnosis. Specifically, we can transform the original quality–fault model in (7.3) to a new form

$$\mathbf{y}_i = \mathbf{Z}^* \mathbf{b}_i^* + \epsilon_i, \tag{7.13}$$

where $\mathbf{b}_i^* = \mathbf{G}^{-1/2}\mathbf{b}_i$ and $\mathbf{Z}^* = \mathbf{Z}\mathbf{G}^{1/2}$, $\mathbf{G}$ is the covariance matrix of $\mathbf{b}_i$. With this form, we can easily see that the covariance matrix of $\mathbf{b}_i^*$ is an identity matrix. One caveat in the transformation is that $\mathbf{G}^{-1/2}$ should exist. For a diagonal matrix $\mathbf{G}$ with one or more zero diagonal elements, $\mathbf{G}^{-1/2}$ does not exist. However, this issue can be easily addressed by reducing the dimension of the model by removing the elements in $\mathbf{b}_i$ with zero variance and the corresponding columns in $\mathbf{Z}$. Thus, the transformation for obtaining (7.13) is generally applicable.

The maximum likelihood estimation of $\mathbf{Z}^*$ can be obtained using eigenvalues and eigenvectors of $\mathbf{S}_\mathbf{y}$ [Tipping and Bishop, 1999]. Assume there are q' faults occurring in the system and the eigenvalues of $\mathbf{S}_\mathbf{y}$ are $l_1 \geq l_2 \cdots \geq l_{q'} \geq l_{q'+1} \cdots \geq l_p$. Denote $\mathbf{U}_{q'}$ as a $n \times q'$ matrix consisting of the q' eigenvectors of $\mathbf{S}_\mathbf{y}$ that are associated with $l_1, \ldots, l_{q'}$, $\mathbf{\Gamma}_{q'}$ as a $q' \times q'$ diagonal matrix with diagonal elements of $l_1, \ldots, l_{q'}$, and $\mathbf{Q}$ as an arbitrary $q' \times q'$ orthogonal matrix. Then the MLE of $\mathbf{Z}^*$

$$\hat{\mathbf{Z}}^* = \mathbf{U}_{q'}(\mathbf{\Gamma}_{q'} - \hat{\sigma}_\epsilon^2 \mathbf{I})^{1/2}\mathbf{Q}, \tag{7.14}$$

where $\hat{\sigma}_\epsilon^2 = \sum_{j=q'+1}^{n} l_j / (n - q')$ is the MLE of the variance of the error term. Please note we assume $\mathbf{R}$ has the simple structure of $\sigma_\epsilon^2 \mathbf{I}$.

Clearly, the estimation of $\mathbf{Z}^*$ is not unique. It possesses rotational indeterminacy because of the existence of $\mathbf{Q}$. From (7.14), we have the MLE of $\mathbf{Z}$ as

$$\hat{\mathbf{Z}} = \mathbf{U}_{q'}(\mathbf{\Gamma}_{q'} - \hat{\sigma}_\epsilon^2 \mathbf{I})^{1/2}\mathbf{Q}\mathbf{G}^{-1/2}.$$

Under the multiple faults case, i.e., $q' > 1$, the above equation provides limited information on $\mathbf{Z}$ because $\mathbf{Q}$ is an arbitrary orthogonal matrix and $\mathbf{G}$ is unknown. However, if $q' = 1$, then it is clear that $\hat{\mathbf{Z}}$ becomes a vector and is proportional to $\mathbf{U}_{q'}$. However, the indeterminacy of the scale of $\hat{\mathbf{Z}}$ still exists: we cannot uniquely identify the length of $\hat{\mathbf{Z}}$ and the magnitude of the fault. This indeterminacy can be removed by putting a constraint on the length of $\hat{\mathbf{Z}}$ as 1. With this constraint and when $q' = 1$ and thus $\hat{\mathbf{Z}}$ becomes a vector, we have

$$\tilde{\mathbf{z}}_f = \hat{\mathbf{Z}} / \|\hat{\mathbf{Z}}\| \quad \text{and} \quad \tilde{\sigma}_f^2 = l_1 - \hat{\sigma}_\epsilon^2, \tag{7.15}$$

where we denote that $\tilde{\mathbf{z}}_f$ and $\tilde{\sigma}_f^2$ are the estimates of the direction of the column vector of $\mathbf{Z}$ corresponding to the current single fault and the fault's magnitude (variance), respectively. The tilde sign indicates the estimated value actually differs from the true value by a constant multiplicative scalar.

Equation (7.15) sheds light on the fault diagnosis without knowing the design matrix **Z**. First, based on (7.15), we can estimate the fault signature vector when a single fault happens. Then fault signature vectors can be collected to form a library. The fault diagnosis can be achieved through a matching between current variation patterns and the fault signature vectors in the library. The steps are summarized as follows.

1. The multivariate product quality observations are obtained in the data collection step. The sample size is assumed to be N and the dimension of the observation is n.
2. Based on the quality observation, we can calculate the sample covariance matrix $\mathbf{S}_\mathrm{y}$. From this covariance matrix, the number of significant variation sources can be estimated in Step 3 introduced in Section 7.3. The corresponding eigenvectors and eigenvalues can be obtained.
3. From Step 2, if only one variation source exists in the system, then the eigenvector associated with the largest eigenvalue of $\mathbf{S}_\mathrm{y}$ is an estimate of the signature vector associated with the current variation source. If there are already some signature vectors in the library, then we need to test if the current signature vector is the same as one of them. If yes, we can determine that the current fault is the same as a previous fault. If not, we need to inspect the process and locate the physical root cause of the new fault. If the new fault is located, we will put the eigenvector associated with the largest eigenvalue of $\mathbf{S}_\mathrm{y}$ into the library as the signature vector of the new fault.
4. From Step 2, if q' ($q' > 1$) faults exist in the system, we can use the learned signature vectors in the library to identify which faults have happened. First, we can randomly select q' signature vectors in the library. If the total number of signature vectors in the current library is smaller than q', then we can only say that some unknown faults have happened. If the total number of signature vectors is greater than or equal to q', we can select q' vectors from the library and check if the q' selected signature vectors and the current variation patterns match. If there is a match, the same set of faults associated with the selected signature vectors have happened. If not, we can select another set of signature vectors from the library and check again. If all the potential combinations of signature vectors have been tested and none spans the same space as current significant eigenvectors of $\mathbf{S}_\mathrm{y}$, we cannot identify the faults in this case and can only claim that some unknown faults occurred.

In this procedure, statistical matching between the observed principal eigenvector(s) and the signature vector(s) in the library is needed. Different from that in Sections 7.2 and 7.3, the signature vectors in the library here are not deterministic vectors; rather they are also eigenvectors of the sample covariance matrix obtained from previous quality observations. To address this issue, we could adopt the testing approach for common eigenspace. The testing of common eigenspace is to test if the eigenvectors of the covariance matrix of two set of observations $\{\mathbf{y}_{11},\ldots,\mathbf{y}_{1n}\}$ and $\{\mathbf{y}_{21},\ldots,\mathbf{y}_{2n}\}$ associated with the first q' largest eigenvalues span the same subspace. Several techniques have been developed [Flury, 1987; Schott, 1988, 1991]. However, one critical limitation of those approaches is that it cannot combine multiple signature vectors in the library to form a new "multiple fault signature". They require that all the eigenvectors come from the same sample set. As a result, the flexibility of those methods is limited.

To overcome this limitation, we could adopt the methods presented in Section 7.3 with some modifications in the simulation procedure. Instead of using the deterministic column vectors of $\mathbf{Z}$, the signature vectors should also be simulated and then the bound of the principal angle γ_2 can be obtained. The other steps can be kept the same.

The major characteristic of the method presented in this section is that it does not require a predefined model that links the system observations and the variation sources. Instead, this method identifies the unique signature vector for each fault only based on quality observation data. With the accumulation of observation data and scenarios of fault occurrences, the fault signature library will gradually increase. This method can be used in quick root cause diagnosis of a manufacturing process, which will lead to product quality improvement, production downtime reduction, and hence a remarkable cost reduction in manufacturing systems.

In the above data driven diagnosis method, the library of the fault signatures is learned when a single fault occurs. However, when multiple faults occur, some information can still be extracted through the analysis of the linear spaces spanned by the principal eigenvectors of the sample covariance matrix of the quality observations. The detailed procedure can be found in Jin and Zhou [2006b].

Bibliographic Notes

PCA, as a very powerful tool to identify the variation patterns in multivariate data, has been used widely in practice for industrial process monitoring, fault detection, and diagnosis. The applications of PCA in process industry can be found in Ge and Song [2012], Qin [2012]. For discrete manufacturing processes, using PCA pattern matching method for fixture fault diagnosis in automotive body assembly processes was first proposed in Ceglarek and Shi [1996]. Various aspects of this diagnosis method are studied in subsequent works, including the impact of error terms on the pattern matching performance, simultaneous matching for multiple faults, and learning the fault signature from data directly instead of from physical model [Rong et al., 2000; Li et al., 2007; Li and Zhou, 2006; Ding et al., 2002a; Jin and Zhou, 2006a,b].

Exercises

1. Consider the following simple quality–fault model

$$\begin{pmatrix} y_1 \\ y_2 \end{pmatrix} = \begin{pmatrix} 1 & 1 \\ 2 & 3 \end{pmatrix} \begin{pmatrix} b_1 \\ b_2 \end{pmatrix} + \begin{pmatrix} \epsilon_1 \\ \epsilon_2 \end{pmatrix}.$$

Please answer the following questions.

(a) If the covariance matrix of $\mathbf{b}$ is $\mathbf{G} = \begin{pmatrix} 1 & 0 \\ 0 & 0 \end{pmatrix}$ and the error term covariance matrix is $\mathbf{R} = \begin{pmatrix} 0.1 & 0 \\ 0 & 0.1 \end{pmatrix}$, then what is $\mathbf{\Sigma}_\mathbf{y}$, what are the eigenvalue/eigenvector pairs of $\mathbf{\Sigma}_\mathbf{y}$? What are the angles between the eigenvectors of $\mathbf{\Sigma}_\mathbf{y}$ and the column vectors of $\mathbf{Z}$?

(b) If $\mathbf{R} = \begin{pmatrix} 0.1 & 0 \\ 0 & 0.3 \end{pmatrix}$ and other conditions are the same, answer the questions in (a) again.

(c) If $\mathbf{R} = \begin{pmatrix} 0.1 & 0 \\ 0 & 0.5 \end{pmatrix}$ and other conditions are the same, answer the questions in (a) again.

(d) Discuss the results in (a), (b), (c).

2. Consider the following quality–fault model

$$\begin{pmatrix} y_1 \\ y_2 \end{pmatrix} = \begin{pmatrix} 1 & -1 \\ 3 & 5 \end{pmatrix} \begin{pmatrix} b_1 \\ b_2 \end{pmatrix} + \begin{pmatrix} \epsilon_1 \\ \epsilon_2 \end{pmatrix}.$$

Through an analysis of the measurement systems, we know the covariance matrix of the error term is in the form as $\mathbf{R} = \begin{pmatrix} 0.1 & 0 \\ 0 & 0.4 \end{pmatrix}$.

We collect the observation data from five units as follows:

Table 7.1 Observation data

Unit	1	2	3	4	5
y_1	−0.19	−2.38	−0.28	1.21	−0.67
y_2	0.27	−4.97	−1.40	2.68	−0.58

Using the observation data, please identify which one among b_1 and b_2 has excessive variation.

3. Consider the following quality–fault model

$$\begin{pmatrix} y_1 \\ y_2 \end{pmatrix} = \begin{pmatrix} 1 & 1 \\ 3 & 5 \end{pmatrix} \begin{pmatrix} b_1 \\ b_2 \end{pmatrix} + \begin{pmatrix} \epsilon_1 \\ \epsilon_2 \end{pmatrix}.$$

Through an analysis of the measurement systems, we know the covariance matrix of the error terms is in the form as $\mathbf{R} = \begin{pmatrix} 0.1 & 0 \\ 0 & 0.4 \end{pmatrix}$. Please answer the following questions.
We collect the observation data from five units as follows:

Table 7.2: Observation data

Unit	1	2	3	4	5
y_1	1.75	1.40	−0.51	−0.73	0.85
y_2	6.89	2.03	−0.16	−2.33	3.00

Using the observation data, can you identify if any one of b_1 and b_2 has excessive variation? If yes, which one? If not, why?

4. Consider the following quality–fault model

$$\begin{pmatrix} y_1 \\ y_2 \\ y_3 \end{pmatrix} = \begin{pmatrix} 8 & 9 & 2 \\ 9 & -6 & 5 \\ 1 & 0 & 9 \end{pmatrix} \begin{pmatrix} b_1 \\ b_2 \\ b_3 \end{pmatrix} + \begin{pmatrix} \epsilon_1 \\ \epsilon_2 \\ \epsilon_3 \end{pmatrix}.$$

(a) If the covariance matrix of $\mathbf{b}$ is $\mathbf{G} = \begin{pmatrix} 1 & 0 & 0 \\ 0 & 1.3 & 0 \\ 0 & 0 & 0 \end{pmatrix}$ and the error term covariance matrix is $\mathbf{R} = \begin{pmatrix} 0.1 & 0 & 0 \\ 0 & 0.4 & 0 \\ 0 & 0 & 0.5 \end{pmatrix}$, then what is $\mathbf{\Sigma_y}$, what are the eigenvalue/eigenvector pairs of $\mathbf{\Sigma_y}$? What are the angles between the eigenvectors of $\mathbf{\Sigma_y}$ and the column vectors of $\mathbf{Z}$?

(b) What are the principal angles between the two eigenvectors of $\boldsymbol{\Sigma}_{\mathbf{y}}$ associated with the two largest eigenvalues and two column vectors of $\mathbf{Z}$? Please note that we have three choices of two column vectors of $\mathbf{Z}$. Based on these principal angles, what conclusion can we draw?

5. From a manufacturing process, we obtain the following four groups of observation data. Each group consists of the observations from four units.

Table 7.3 Observation data

	Unit	1	2	3	4
	y_1	−1.55	2.19	−0.21	0.09
Group 1	y_2	−1.03	3.05	−1.75	0.20
	y_3	0.45	0.44	0.35	0.68
	y_1	3.00	−0.63	0.41	2.41
Group 2	y_2	1.60	−1.08	−0.22	2.15
	y_3	−0.26	0.25	1.08	−0.45
	y_1	0.17	0.23	−3.03	−0.84
Group 3	y_2	0.09	0.18	−3.09	−1.67
	y_3	−0.52	−0.04	−0.40	−0.14
	y_1	0.55	2.40	−0.52	1.48
Group 4	y_2	0.60	2.08	−1.04	0.28
	y_3	0.78	0.60	0.37	−0.61

(a) Please use a scree plot to identify the number of variation sources with excessive variation for each group of observations.

(b) For these four group of observations, do any of them come from similar faulty condition?

8

Diagnosis of Variation Sources Through Random Effects Estimation

In Chapter 7, we presented several variation source diagnosis methods using PCA. Those methods are easy to be implemented and interpreted. However, those methods are generally heuristic in nature and not very effective for the diagnosis of multiple faults. In this chapter, we introduce a few methods for variation source diagnosis through direct parameter estimation of the random effects model.

From Chapter 7, we can cast the quality–fault model as a special form of the linear mixed effects model as (7.3). For the sake of convenience, we repeat the model below

$$\mathbf{y}_i = \mathbf{Z}\mathbf{b}_i + \epsilon_i, i = 1, 2, \dots, N, \tag{8.1}$$

where $\mathbf{y}_i \in \mathcal{R}^{n\times 1}$ represents the quality observations, $\mathbf{b}_i \in \mathcal{R}^{q\times 1}$ represents the random effect of q process variation sources, and ϵ_i represents the additive error term due to measurement noises. Under the context of variation source diagnosis, we can often assume the variation sources and the error term follow normal distribution $\mathbf{b}_i \sim \mathcal{N}(\mathbf{0}, \mathbf{G})$, $\epsilon_i \sim \mathcal{N}(\mathbf{0}, \mathbf{R})$, where $\mathbf{G} = \text{diag}\{\sigma_1^2 \ \ \sigma_2^2 \ \ \dots \ \ \sigma_q^2\}$, $\mathbf{R} = \sigma_\epsilon^2\mathbf{I}$. With these assumptions, the variance–covariance relationship between quality observations (which have zero mean) and process variation sources can be obtained from model (8.1) as

$$\boldsymbol{\Sigma}_\mathrm{y} = \mathbf{Z}\mathbf{G}\mathbf{Z}^T + \sigma_\epsilon^2\mathbf{I}_n = \sum_{j=1}^{q+1}\sigma_j^2\,\mathbf{V}_j, \tag{8.2}$$

where $\mathbf{V}_j = \mathbf{z}_j\mathbf{z}_j^T$ for $j = 1, \dots, q$, $\mathbf{z}_j$ is the jth column vector of $\mathbf{Z}$, $\mathbf{V}_{q+1} = \mathbf{I}_n$, and $\sigma_{q+1}^2 = \sigma_\epsilon^2$. The $\{\sigma_j^2\}_{j=1}^{q+1}$ are the variances of process variation sources

Industrial Data Analytics for Diagnosis and Prognosis: A Random Effects Modelling Approach, First Edition. Shiyu Zhou and Yong Chen.

(including the variance of the error terms). In this chapter, we also call them *variance components*. The problem of variation source diagnosis is to estimate the variance components based on observations $\{\mathbf{y}_i\}_{i=1}^N$.

Example 8.1 Consider a system with $n=6$, $q=3$, and the matrix $\mathbf{Z}$ given by

$$\mathbf{Z}^T = \begin{pmatrix} 1 & 1 & 0 & 0 & 0 & 0 \\ -1 & -1 & 1 & 1 & -1 & -1 \\ 0 & 0 & 0 & 0 & 1 & 1 \end{pmatrix}. \tag{8.3}$$

We can use the following R codes to find $\mathbf{V}_j$, $j=1,\ldots,q$ for this example:

```
n <- 6; q <- 3
Z.vec<-c(1, 1, 0, 0, 0, 0,
         -1, -1, 1, 1, -1, -1,
         0, 0, 0, 0, 1, 1)
Z <- matrix(Z.vec, n, q)
V <- array(0, c(n, n, q+1))
for (j in 1:q){V[,,j] = Z[,j]%*% t(Z[,j])}
V[,,q+1] = diag(n)
```

In the above codes, the matrices $\mathbf{V}_j$ are obtained in `V[, , j]` ($j=1, \ldots, q+1$). The obtained $\mathbf{V}_j$ are:

$$\mathbf{V}_1 = \begin{pmatrix} 1 & 1 & 0 & 0 & 0 & 0 \\ 1 & 1 & 0 & 0 & 0 & 0 \\ 0 & 0 & 0 & 0 & 0 & 0 \\ 0 & 0 & 0 & 0 & 0 & 0 \\ 0 & 0 & 0 & 0 & 0 & 0 \\ 0 & 0 & 0 & 0 & 0 & 0 \end{pmatrix} \mathbf{V}_2 = \begin{pmatrix} 1 & 1 & -1 & -1 & 1 & 1 \\ 1 & 1 & -1 & -1 & 1 & 1 \\ -1 & -1 & 1 & 1 & -1 & -1 \\ -1 & -1 & 1 & 1 & -1 & -1 \\ 1 & 1 & -1 & -1 & 1 & 1 \\ 1 & 1 & -1 & -1 & 1 & 1 \end{pmatrix}$$

$$\mathbf{V}_3 = \begin{pmatrix} 0 & 0 & 0 & 0 & 0 & 0 \\ 0 & 0 & 0 & 0 & 0 & 0 \\ 0 & 0 & 0 & 0 & 0 & 0 \\ 0 & 0 & 0 & 0 & 0 & 0 \\ 0 & 0 & 0 & 0 & 1 & 1 \\ 0 & 0 & 0 & 0 & 1 & 1 \end{pmatrix},$$

and $\mathbf{V}_4 = \mathbf{I}_6$. Based on (8.1), for the system of Example 8.1, we have

$$\boldsymbol{\Sigma}_{\mathbf{y}} = \sigma_1^2\mathbf{V}_1 + \sigma_2^2\mathbf{V}_2 + \sigma_3^2\mathbf{V}_3 + \sigma_\epsilon^2\mathbf{I}_6. \tag{8.4}$$

From (8.3), it can be seen that the first and third variation sources in this system only affect the first two and the last two observations, respectively, which explains why $\mathbf{V}_1$ has non-zero elements only in the top 2×2 submatrix and $\mathbf{V}_3$ has non-zero elements only in the bottom 2×2 submatrix.

Example 8.2 We consider an example of the automotive body assembly process. A linear model is derived and presented in Section 4 of Apley and Shi, [1998] for a process similar to the one illustrated in Figure 7.1. In this particular model, there are nine observations ($n = 9$) and three independent variation sources ($q = 3$). The matrix Z is given in Apley and Shi [1998] as

$$\mathbf{Z}^T = \begin{pmatrix} 0.093 & 0 & -0.093 & 0.093 & 0 & 0.647 & -0.370 & 0 & 0.647 \\ 0.577 & 0 & 0 & 0.577 & 0 & 0 & 0.577 & 0 & 0 \\ -0.120 & 0 & 0.843 & -0.120 & 0 & -0.120 & 0.482 & 0 & -0.120 \end{pmatrix}.$$

By replacing the first three lines of the R codes in Example 8.1 with the following R codes, we can obtain $\mathbf{V}_j$, $j = 1, 2, 3$ for Example 8.2.

```
n<-9; q<-3
Z.vec<-c(.093,0,-.093,.093,0,.647,-.370,0,.647,
.577, 0, 0, .577, 0, 0, .577, 0, 0,
-.120, 0, .843, -.120, 0, -.120, .482, 0, -.120)
```

Equation (8.4) is still applicable to Example 8.2. The exact $\mathbf{V}_j$, $j = 1, 2, 3$ for Example 8.2 are omitted here.

8.1 Estimation of Variance Components

One estimator of the variance components in (8.2) is based on the MLE method as introduced in Chapter 6. Denote $\boldsymbol{\sigma}^2 := (\sigma_1^2 \dots \sigma_{q+1}^2)^T$ and $\hat{\boldsymbol{\sigma}}^2$ as its estimated value. It can be shown that for the special linear mixed model in (8.1), MLE is the solution to the following nonlinear equation [Anderson, 1969]:

$$\{\mathrm{tr}(\hat{\boldsymbol{\Sigma}}_{\mathbf{y}}^{-1}\mathbf{V}_i\hat{\boldsymbol{\Sigma}}_{\mathbf{y}}^{-1}\mathbf{V}_j)\}_{i,j=1}^{q+1}\hat{\boldsymbol{\sigma}}^2 = \{\mathrm{tr}(\hat{\boldsymbol{\Sigma}}_{\mathbf{y}}^{-1}\mathbf{V}_i\hat{\boldsymbol{\Sigma}}_{\mathbf{y}}^{-1}\mathbf{S}_{\mathbf{y}})\}_{i=1}^{q+1}. \tag{8.5}$$

In addition to the MLE method, there are two other types of online variance estimators introduced in the literature. These online estimators are suitable for online diagnosis applications because they are easy to compute. These two estimators will be introduced as follows.

1. Least-Squares (LS) Fit Estimator

The basic idea of LS fit estimator is to minimize the sum of the squared difference between the estimate $\hat{\boldsymbol{\Sigma}}_{\mathbf{y}} := \sum_{j=1}^{q+1} \hat{\sigma}_j^2 \mathbf{V}_j$ and the observation covariance matrix $\mathbf{S}_{\mathbf{y}}$, where $\mathbf{S}_{\mathbf{y}}$ is defined as

$$\mathbf{S}_{\mathbf{y}} := \frac{1}{N} \sum_{i=1}^{N} (\mathbf{y}_i \mathbf{y}_i^T).$$

Therefore, we need to solve

$$\min \; \| \hat{\boldsymbol{\Sigma}}_{\mathbf{y}} - \mathbf{S}_{\mathbf{y}} \|^2,$$

where $\| \cdot \|$ is the Euclidean norm of a matrix (i.e., the Frobenius norm) and the minimization is performed with respect to the estimates of variance components, $\{\hat{\sigma}_j^2\}_{j=1}^{q+1}$. d'Assumpcao [1980] and Böhme [1986] derived the LS fit estimator as

$$\hat{\boldsymbol{\sigma}}^2 = [\{\mathrm{tr}(\mathbf{V}_j \mathbf{V}_i)\}_{i,j=1}^{q+1}]^{-1} \{\mathrm{tr}(\mathbf{V}_i \mathbf{S}_{\mathbf{y}})\}_{i=1}^{q+1}, \tag{8.6}$$

where $\{\cdot\}_{i,j=1}^{q+1}$ is a $(q+1)\times(q+1)$ matrix, and $\{\cdot\}_{i=1}^{q+1}$ is a $(q+1)\times 1$ column vector.

2. Two-Step (2S) Estimator

In Apley and Shi [1998], a two-step procedure is used to estimate the variance components for a system in which the matrix $\mathbf{Z}$ has full column rank. The two-step procedure is described as follows:

Step 1: Estimate $\hat{\mathbf{b}}_i = (\mathbf{Z}^T\mathbf{Z})^{-1}\mathbf{Z}^T\mathbf{y}_i$ for $i = 1, \ldots, N$.

Step 2: Estimate σ_ϵ^2 by $\hat{\sigma}_\epsilon^2 = \frac{1}{N(n-q)} \sum_{i=1}^{N} \mathbf{e}_i^T \mathbf{e}_i$, where $\mathbf{e}_i = \mathbf{y}_i - \mathbf{Z}\hat{\mathbf{b}}_i$; and estimate $\{\sigma_j^2\}_{j=1}^{q}$ by

$$\hat{\sigma}_j^2 = \frac{1}{N} \hat{\mathbf{b}}_j^T \hat{\mathbf{b}}_j - \hat{\sigma}_\epsilon^2 (\mathbf{Z}^T\mathbf{Z})_{j,j}^{-1}, \text{ for } j = 1, \ldots, q, \tag{8.7}$$

where $(\mathbf{Z}^T\mathbf{Z})^{-1}_{j,j}$ is the (j, j)th element in $(\mathbf{Z}^T\mathbf{Z})^{-1}$.

In the above two-step procedure, the first step is used to find the least squares estimation of the random effect $\mathbf{b}_i$, for each unit i, i=1, ..., N. In step 2, the estimated $\mathbf{b}_i$ is then used to estimate the variance components.

It is worth pointing out the relation between the two-step estimator and an estimator from Stoica and Nehorai [1995], which is given by

$$\widehat{\mathbf{G}} = \mathbf{Z}^+\mathbf{S}_\mathbf{y}(\mathbf{Z}^+)^T - \hat{\sigma}_\epsilon^2(\mathbf{Z}^T\mathbf{Z})^T. \tag{8.8}$$

$$\hat{\sigma}_\epsilon^2 = \mathrm{tr}((\mathbf{I}_n - \mathbf{Z}\mathbf{Z}^+)\mathbf{S}_\mathbf{y})/(n-q), \tag{8.9}$$

where $\mathbf{Z}^+ =: (\mathbf{Z}^T\mathbf{Z})^{-1}\mathbf{Z}^T$. We will see that the two-step estimator is exactly the diagonal elements of the estimator in (8.8) and (8.9). Substituting $\hat{\mathbf{b}}_i = \mathbf{Z}^+\mathbf{y}_i$ into (8.7), we have

$$\begin{aligned}
\hat{\sigma}_\epsilon^2 &= \frac{1}{N(n-q)}\sum_{i=1}^{N}\mathbf{y}_i^T(\mathbf{I}_n - \mathbf{Z}\mathbf{Z}^+)^T(\mathbf{I}_n - \mathbf{Z}\mathbf{Z}^+)\mathbf{y}_i \\
&= \frac{1}{N(n-q)}\mathrm{tr}((\mathbf{I}_n - \mathbf{Z}\mathbf{Z}^+)\sum_{i=1}^{N}\mathbf{y}_i\mathbf{y}_i^T) \qquad , \\
&= \frac{1}{n-q}\mathrm{tr}((\mathbf{I}_n - \mathbf{Z}\mathbf{Z}^+)\mathbf{S}_\mathbf{y}),
\end{aligned}$$

which is the same as $\hat{\sigma}_\epsilon^2$ in (8.9). Further,

$$\begin{aligned}
\hat{\sigma}_j^2 &= \frac{1}{N}\sum_{i=1}^{N}(\mathbf{Z}_j^+\mathbf{y}_i)(\mathbf{Z}_j^+\mathbf{y}_i)^T - \hat{\sigma}_\epsilon^2(\mathbf{Z}^T\mathbf{Z})^{-1}_{j,j} \\
&= \mathbf{Z}_j^+(\frac{1}{N}\sum_{i=1}^{N}\mathbf{y}_i\mathbf{y}_i^T)(\mathbf{Z}_j^+)^T - \hat{\sigma}_\epsilon^2(\mathbf{Z}^T\mathbf{Z})^{-1}_{j,j}, \\
&= \mathbf{Z}_j^+\mathbf{S}_\mathbf{y}(\mathbf{Z}_j^+)^T - \hat{\sigma}_\epsilon^2(\mathbf{Z}^T\mathbf{Z})^{-1}_{j,j},
\end{aligned}$$

where $\mathbf{Z}_j^+$ is the jth row vector of $\mathbf{Z}^+$. It is clear that the above equation is equal to the (j, j)th diagonal element of $\widehat{\mathbf{G}}$ in (8.8).

Example 8.3 In this example, we calculate the MLE and the two online estimators for the system in Example 8.1 using simulated data. For Example 8.1, we assume that $\mathbf{G} = \mathbf{I}_3$, $\sigma_\epsilon = 1$, and the sample size $N = 50$. First, we use the following `R` code to generate quality observations $\mathbf{y}_i$, i=1, 2, ..., N, for Example 8.1 based on (8.1).

```
library(MASS)
sigma.b <- diag(c(1, 1, 1))
sigma <- 1
N <- 50
set.seed(10)
b <- t(mvrnorm(N, mu=rep(0, q), Sigma=sigma.b))
e <- matrix(rnorm(n*N, sd=sigma), n, N)
y <- Z%*%b+e
```

In the R codes above, the ith column of the matrix variable y corresponds to $\mathbf{y}_i$, $i = 1, 2, \ldots, N$. Now we find the LS fit estimator using the following codes:

```
s.y <- (1/N)*y%*%t(y)
lhs <- matrix(0, q+1, q+1)
for (i in 1:(q+1))
for (j in 1:(q+1))
lhs[i,j]=sum(diag(V[,,j]%*%V[,,i]))
rhs <- rep(0, q+1)
for (i in 1: (q+1))
rhs[i] <- sum(diag(V[,,i]%*%s.y))
var.hat <- solve(lhs)%*%rhs
```

The vector var.hat in the R codes above corresponds to $\hat{\boldsymbol{\sigma}}^2$ in (8.6). Based on the R outputs, the estimation of $\boldsymbol{\sigma}^2$ using the LS fit estimator is $\hat{\sigma}_1^2 = 0.742$, $\hat{\sigma}_2^2 = 1.046$, $\hat{\sigma}_3^2 = 0.740$, and $\hat{\sigma}_\epsilon^2 = 0.897$. Next, we calculate the two-step estimator based on (8.8) and (8.9) using the following codes.

```
Z.n <- solve(t(Z)%*%Z)%*%t(Z)
var.e = sum(diag(( diag(n)-Z%*%Z.n)%*%s.y))/(n-q)
G.hat = Z.n%*%s.y%*%t(Z.n)-var.e*solve(t(Z)%*%Z)
```

In the R codes above, var.e corresponds to $\hat{\sigma}_\epsilon^2$ and G.hat gives $\widehat{\mathbf{G}}$ in (8.8). Since the two-step estimator is exactly the diagonal elements of $\widehat{\mathbf{G}}$, we can use the R code diag(G.hat) to extract $\hat{\sigma}_i^2$, $i = 1, 2, 3$. Based on the R outputs, the estimation of $\boldsymbol{\sigma}^2$ using the two-step estimator is $\hat{\sigma}_1^2 = 0.788$, $\hat{\sigma}_2^2 = 1.080$,

$\hat{\sigma}_3^2 = 1.176$, and $\hat{\sigma}_\epsilon^2 = 0.880$. To find the MLE for this example, we use the `lmer` function in the `lme4` package as in Chapter 6. In order to use the `lmer()` function, we need to stack up the vectors $\mathbf{y}_i$, $i = 1, 2, \ldots, N$ using the `melt()` function from the `reshape2` package. The `R` codes for the MLE is given as follows.

```
library(reshape2)
library(lme4)
data.mle <- data.frame(cbind(melt(y)[,2:3],
                          rep(1, N)%x%Z))
names(data.mle)[1] <- "group"
data.mle$group <- as.factor(data.mle$group)
m1 <- lmer(value ~ 0 + (0+X1|group) +
             (0+X2|group) + (0+X3|group),
             data <-data.mle , REML=F)
as.data.frame(VarCorr(m1))
```

Since the model in (8.1) contains random effects only, no fixed effect is included in the `lmer()` function. The `VarCorr()` function is used to extract the estimates of all variance components from the fitted random effects model. Based on the `R` outputs, the estimation of $\boldsymbol{\sigma}^2$ using the MLE is $\hat{\sigma}_1^2 = 0.622$, $\hat{\sigma}_2^2 = 1.026$, $\hat{\sigma}_3^2 = 0.869$, and $\hat{\sigma}_\epsilon^2 = 0.914$.

Example 8.4 Now we revisit the automotive body assembly process in Example 8.2. To illustrate the application of the variance estimators, we perform a simulation study with variance components given as $\boldsymbol{\sigma}^2 = \left(0.0011\ 0.0025\ 0.0044\ 0.0006\right)^T$ (including that of the error term) and a sample size of $N = 25$. The following `R` codes are needed to set the parameters correspondingly.

```
Sigma.b <- diag(c(0.0011 , 0.0025 , 0.0044))
sigma <- sqrt (0.0006)
N <- 25
```

The remaining codes are the same as those used in Example 8.3 to obtain estimation of variance components based on LS fit, two-step, and MLE. The estimated variance components for the three estimators are listed in Table 8.1. Comparing the estimated values and the true values of variance components for Example 8.3 and Example 8.4, it can be seen that there are various degrees

Table 8.1 Estimated variance components of three estimators for Example 8.4

	$\hat{\sigma}^2$
LS fit	$(0.00011\ 0.00200\ 0.00331\ 0.00060)^T$
Two-Step	$(0.00020\ 0.00199\ 0.00328\ 0.00060)^T$
MLE	$(0.00018\ 0.00197\ 0.00327\ 0.00060)^T$

of estimation error for all three estimators. The properties of these estimators, including their estimation accuracy, will be studied in Section 8.2.

8.2 Properties of Variation Source Estimators

In this section we study some key properties of the variation source estimators introduced in Section 8.1, including existence condition, bias, and variance. Existence condition affects the applicability of the estimators. Bias and variance are two aspects of the accuracy of the estimators.

1. Existence of Variance Components Estimators
The condition for the two-step estimator to exist is that $\mathbf{Z}^T\mathbf{Z}$ should be of full rank and $n \geq q + 1$, where $q + 1$ is the number of independent variance components, including the variance of the error term. The existence condition of an LS fit estimator is that $\{\mathrm{tr}(\mathbf{V}_j\mathbf{V}_i)\}_{i,j=1}^{q+1}$ should be of full rank. The condition that $\mathbf{Z}^T\mathbf{Z}$ should be of full rank and $n \geq q + 1$ is a stronger condition than that $\{\mathrm{tr}(\mathbf{V}_j\mathbf{V}_i)\}_{i,j=1}^{q+1}$ is of full rank. It can be shown that the existence of the two-step estimator can guarantee the existence of an LS fit estimator [Ding et al., 2005]. However, the reverse is not true, i.e., $\{\mathrm{tr}(\mathbf{V}_j\mathbf{V}_i)\}_{i,j=1}^{q+1}$ may be of full rank even if $\mathbf{Z}^T\mathbf{Z}$ is not. An MLE exists if $\mathbf{S}_\mathrm{y}$ is positive definite and $\mathbf{V}_1$, $\mathbf{V}_2,\ldots,\mathbf{V}_{q+1}$ are linearly independent [Rao and Kleffe, 1988]. It can be seen that $\mathbf{V}_1$, $\mathbf{V}_2,\ldots,\mathbf{V}_{q+1}$ are linearly independent if and only if $\{\mathrm{tr}(\mathbf{V}_j\mathbf{V}_i)\}_{i,j=1}^{q+1}$ is of full rank. Therefore, the existence condition of an MLE is the same as that of the LS fit estimator when $\mathbf{S}_\mathrm{y}$ is positive definite.

Example 8.5 We can use the `is.singular.matrix()` function in `R` package `matrixcalc` to check if a square matrix is singular. For example the following codes check if the existence condition of the two-step estimator is satisfied.

```
library(matrixcalc)
is.singular.matrix(t(Z)%*%Z)
```

For the **Z** matrices of both Example 8.1 and Example 8.2, the output of `is.singular.matrix()` in the above codes is `FALSE`, indicating that $\mathbf{Z}^T\mathbf{Z}$ is of full rank and the two-step estimator exists for both examples. Similarly we can check the existence condition of the LS fit estimator by checking if $\{\text{tr}(\mathbf{V}_j\mathbf{V}_i)\}_{i,j=1}^{q+1}$ is of full rank. It can be seen that the LS fit estimator exists for both Example 8.1 and Example 8.2, too. In fact, based on the discussions earlier, the existence of the two-step estimator always guarantees the existence of the LS fit estimator. An example where the LS fit estimator exists while a two-step estimator does not can be found in question 3 of the exercises of this chapter.

2. Bias of Variance Component Estimators

Both the LS fit estimator and two-step estimator are unbiased. The LS fit estimator is unbiased since $\mathbf{S}_\mathbf{y}$ is the unbiased estimate of $\boldsymbol{\Sigma}_\mathbf{y}$, i.e., $E(\mathbf{S}_\mathbf{y}) = \boldsymbol{\Sigma}_\mathbf{y}$. The unbiasedness of the two-step estimator is proven in Ding et al. [2005]. An MLE is generally biased but it is asymptotically unbiased. The online estimators would have an advantage over the MLE in terms of unbiasedness when the sample size is small. The bias and variance of an estimator are two components of estimation errors. Since the two online estimators are both unbiased, the comparison of their performance should be based on the estimation variance, which is discussed next.

3. Estimation Variance of Variation Source Estimators

The closed-form solution of the variance of the two online estimators, as well as the approximate result on the variance of MLE based on the inverse of the Fisher information matrix, are derived in Ding et al. [2005]. Since the closed-form solution is in a complicated mathematical form, here we study the variance of the estimators using simulations.

Example 8.6 We apply the LS fit estimator, two-step estimator, and MLE to 10,000 independently generated data sets following the parameter settings in Example 8.1 and Example 8.2 and use the sample mean and sample variance of the estimation to approximate its mean and variance. The results for the system in Example 8.1 are shown in Table 8.2. For this system, the overall accuracy of LS fit and two-step estimators are comparable, while MLE is more accurate than both online estimators.

The results for the system in Example 8.2 are shown in Table 8.3. For this system, the two-step estimator has similar accuracy to the MLE estimator, while both of them are more accurate than the LS fit estimator. More general discussions on performance comparison of the estimators are provided in Section 8.3.

Table 8.2 Mean and variance of $\hat{\boldsymbol{\sigma}}^2 = (\hat{\sigma}_1^2\ \hat{\sigma}_2^2\ \hat{\sigma}_2^2\ \hat{\sigma}_\epsilon^2)^T$ for the system in Example 8.1

	LS fit	Two-Step	MLE
sample mean	$(1.00\ 1.00\ 1.00\ 1.00)^T$	$(1.00\ 1.00\ 1.00\ 1.00)^T$	$(1.00\ 1.00\ 1.00\ 1.00)^T$
sample variance	$(0.185\ 0.076\ 0.191\ 0.020)^T$	$(0.171\ 0.095\ 0.178\ 0.013)^T$	$(0.129\ 0.071\ 0.135\ 0.012)^T$

Table 8.3 Mean and variance of $\hat{\boldsymbol{\sigma}}^2 = (\hat{\sigma}_1^2\ \hat{\sigma}_2^2\ \hat{\sigma}_2^2\ \hat{\sigma}_\epsilon^2)^T$ for the system in Example 8.2

	LS fit	Two-Step	MLE
sample mean	$(0.0011\ 0.0025\ 0.0044\ 0.0006)^T$	$(0.0011\ 0.0025\ 0.0044\ 0.0006)^T$	$(0.0011\ 0.0025\ 0.0044\ 0.0006)^T$
sample variance $(\times 10^{-7})$	$(4.7\ 8.1\ 22\ 0.09)^T$	$(2.8\ 7.5\ 22\ 0.05)^T$	$(2.8\ 7.5\ 21\ 0.05)^T$

8.3 Performance Comparison of Variance Component Estimators

In this section, we compare the performance of the estimators, especially the differences between the two online variation source estimators. Since two-step and LS fit estimators are both unbiased, we can focus on the variance of estimators in performance comparison. One purpose of this comparison is to find the condition under which the LS fit and two-step estimators may be good alternatives for the MLE during online variance estimation. LS fit and two-step estimators can be computed through their closed-form expressions. Consequently they are faster to be computed than the MLE. The major drawback of either the LS fit or the two-step estimator is that they may have unacceptably higher variances than an MLE. In order to characterize the overall variance of an estimator for all variance components, we first define the dispersion of a variance estimator by the trace of its covariance matrix, i.e., $D_{ML,2S\ or\ LS} := \text{tr}(\text{cov}(\hat{\boldsymbol{\sigma}}^2_{ML,2S\,or\,LS}))$. In Ding et al. [2005], it is shown that the dispersion of all three estimators is inversely proportional to the sample size, N. Therefore, we expect that the dispersion comparison of the estimators is independent of N. Using D_{ML} as the benchmark, relative difference between a two-step (or LS fit) estimator and an ML estimator can be characterized by the percentage difference (*Dif*), defined as

$$Dif_{2S\,or\,LS} := \frac{D_{2S\,or\,LS} - D_{ML}}{D_{ML}} \times 100\%.$$

The LS fit and two-step estimators become an MLE under some special conditions. The two-step estimator is an MLE in a noise-free environment, i.e., when $\sigma_\epsilon^2 = 0$. In this case, when a two-step estimator exists, observing $\mathbf{y}_i$ is equivalent to directly observing $\mathbf{b}_i$. The sample variances computed from direct observations of $\mathbf{b}_i$, $i = 1, \ldots, N$ are the maximum likelihood estimators of $\{\sigma_j^2\}_{j=1}^q$. Under this noise-free circumstance, the diagonal elements of $\widehat{\mathbf{G}}$ as defined in (8.8) are the same as the sample variances of $\mathbf{b}_i$. Therefore, a two-step estimator is the MLE of $\{\sigma_j^2\}_{j=1}^q$. In contrast, the LS fit estimator becomes an MLE when the random effects $\mathbf{b}_i$ do not have any randomness. The randomness of $\mathbf{y}$ is purely from the error term, i.e., $\mathbf{G} = \mathbf{0}$ and $\sigma_\epsilon^2 \neq 0$. This equivalence can be easily seen by substituting $\widehat{\boldsymbol{\Sigma}}_{\mathbf{y}} = \hat{\sigma}_\epsilon^2 \mathbf{I}_n$ into (8.5), which becomes equivalent to (8.6).

We have seen that the LS fit and two-step estimators are two extreme cases of the MLE. Based on this observation, we expect that a two-step estimator will perform similarly well to an MLE when the error term is relatively small, while an LS fit estimator will perform similarly well to an MLE when the error term dominates in the process. In order to characterize the dominance of the error term, we define an average signal-to-noise (SNR) ratio as follows

$$SNR := \frac{\mathrm{tr}(\mathbf{G})}{q\sigma_\epsilon^2},$$

where $\mathrm{tr}(\mathbf{G})/q = \sum_{i=1}^q \sigma_i^2/q$ represents the average strength of signals and σ_ϵ^2 the strength of the error term. In Table 8.4, we summarize the general properties of the two online variance component estimators.

Table 8.4 Summary of properties of online estimators

	Two-Step Estimator	**LS fit Estimator**
Pros	Closed-form solution	Closed-form solution
	Unbiased	Unbiased
	Close to MLE when SNR is large	Close to MLE when SNR is small
Cons	Performance poor when SNR is small	Performance poor when SNR is large
Existence condition	$\mathbf{ZZ}^T$ full rank and $n \geq q+1$ stronger than LS fit estimator	$\{\mathrm{tr}(\mathbf{V}_j\mathbf{V}_i)\}_{i,j=1}^{q+1}$ full rank may exist even if $\mathbf{ZZ}^T$ is singular

Example 8.7 Consider the system in Example 8.1, the approximated variance of the estimators based on simulation is given in Table 8.2. The corresponding dispersion and *Dif* can then be found, as shown in Table 8.5. In this case, it is clear that the variance of the online estimators is at least 32% larger than that of MLE.

In Ding et al. [2005], the *Dif* is obtained for *SNR* in the range of [0.01,100]. This range of *SNR* is realized by fixing σ_ϵ^2 at unit variance, while changing the value of tr(**G**). Certainly, the variance components σ_i^2 s in **G** can take different values for a given tr(**G**). In Ding et al. [2005], σ_i^2 are set to have equal values, i.e., $\sigma_i^2 = \mathrm{tr}(\mathbf{G})/q$. For instance, if $SNR = 0.1$, given $\sigma_\epsilon^2 = 1$, we have $\mathrm{tr}(\mathbf{G}) = 0.1 \times 3$. Then $\sigma_i^2 = 0.1$ for $i = 1, 2, 3$. Based on Ding et al. [2005], when $SNR < 0.2$, the LS fit estimator performs similarly to MLE, with $Dif < 10\%$. When $SNR > 2$, the two-step estimator performs similarly to MLE, with $Dif < 10\%$. When $SNR = 1$, which is the case for Table 8.5, the two online estimators have the largest *Dif*.

Table 8.5 Comparison of the estimators for the system in Example 8.2

	LS fit	Two-Step	MLE
dispersion	0.472	0.457	0.347
Dif	36%	32%	—

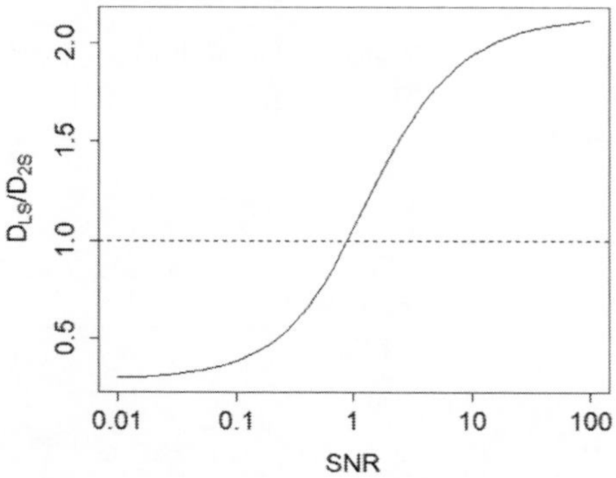

Figure 8.1 D_{LS}/D_{2S} for different *SNR*.

Based on simulation, we plot D_{LS}/D_{2S} versus *SNR* in the range of [0.01,100] in Figure 8.1. From the plot, we can see that the two online estimators perform similarly when *SNR* is close to 1. When $SNR < 1$, the LS fit estimator outperforms the two-step estimator. When $SNR > 1$, the two-step estimator outperforms the LS fit estimator. This is consistent with our general understandings discussed earlier in this section.

Example 8.8 For the system in Example 8.2, the approximated variance of the estimators based on simulation is given in Table 8.3. The corresponding dispersion and *Dif* can then be found, as shown in Table 8.6. Using the parameters given in Example 8.4, the *SNR* for this system can be calculated as equal to $\frac{0.0011+0.0025+0.0044}{3\times0.0006} = 4.44$, which is relatively large. Based on the discussion in this section, we expect that the two-step estimator would perform similarly to the MLE and outperform the LS fit estimator. This is supported by the results in Table 8.6. In practice, the likely range of the *SNR* values can often be obtained from engineering design specifications. For example, in an automotive body assembly process as described and modeled in Example 8.2, the sensor used is one type of non-contact coordinate sensor. The specification of sensor accuracy is $6\sigma_\epsilon = 0.1mm$. On the other hand, the tolerance of the pinhole contact is about $0.2mm$. If the tolerance is approximated by the six-sigma value, then $6\sigma_{locator} = 0.2mm$. This implies that the repeatability of a malfunctioning locator will have a six-sigma value larger than 0.2 mm. Based on this understanding, it is reasonable to assume that $SNR \geq (\frac{6\sigma_{locator}}{6\sigma_\epsilon})^2 = 4$. Given that the *SNR* is greater than four, the two-step estimator is expected to perform similarly to the MLE in terms of estimation variance. On the other hand an LS fit estimator may have significantly larger variance. Therefore the two-step estimator is a better choice for such systems.

Table 8.6 Comparison of the estimators for the system in Example 8.2

	LS fit	Two-Step	MLE
dispersion	3.54×10^{-6}	3.21×10^{-6}	3.16×10^{-6}
Dif	12%	1.7%	—

Bibliographic Notes

In order to diagnose the variation source in industrial processes, it is natural to cast the quality–fault model as a mixed effects model. Then the variation sources can be viewed as the variance components to be estimated in this mixed model. Based on the mixed model, several methods have been proposed to estimate the variance components for the purpose of diagnosing variation sources. The least squares method is used in Chang and Gossard [1998]. The two-step procedure is proposed in Apley and Shi [1998]. A formal maximum-likelihood estimator and the confidence intervals of the estimated variance components are provided in Zhou et al. [2004]. Several methods are compared in Ding et al. [2005]. Recently, some new approaches for variation source diagnosis for an under-determined system are proposed. These methods adopt Bayesian [Bastani et al., 2018; Lee et al., 2020] and compressed sensing approaches [Bastani et al., 2016].

Exercises

1. Consider a system with $n = 6$, $q = 3$, and the matrix $\mathbf{Z}$ given by

$$\mathbf{Z}^T = \begin{pmatrix} 1 & 1 & 0 & 0 & 0 & 0 \\ -2 & -2 & 2 & 2 & -2 & -2 \\ 0 & 0 & 0 & 0 & 1 & 1 \end{pmatrix}.$$

   (a) Find $\mathbf{V}_j$, $j = 1, \ldots, q$ for this system.
   (b) Calculate the LS fit estimator, two-step estimator, and MLE for this system using simulated data. Assume that $\mathbf{G} = \mathbf{I}_3$, $\sigma_\epsilon = 1$, and the sample size $N = 100$.
   (c) Apply the LS fit estimator, two-step estimator, and MLE to 1000 independently generated data sets following the parameter settings in part (b) to estimate the mean and variance of each estimator for this system.
   (d) Find D_{LS}, D_{2S}, and D_{ML}, as well as *Dif* for the LS fit estimator and two-step estimator based on the simulation results from part (c).
2. Generate the plot as in Figure 8.1 for the system in Example 8.2. For what ranges of *SNR* values, is the LS fit estimator (two-step estimator) better?

3. A two-station automotive body assembly process with $n = 18$ and $q = 9$ has design matrix given below.

$$\mathbf{Z}_{18\times 9} = [\mathbf{Z}_1\ \mathbf{Z}_2] = \begin{pmatrix} 0 & 0 & 0 & 0.1215 & -0.3846 & 0 & 0 & 0 & 0.2632 \\ 0 & 0 & 0 & 0.0221 & -0.0699 & 0 & 0 & 0 & 0.0478 \\ 0 & 0 & 0 & 0.1215 & -0.3846 & 0 & 0 & 0 & 0.2632 \\ 0 & 0 & 0 & -0.1817 & 0.5944 & 0 & 0 & 0 & -0.4067 \\ 0 & 0 & 0 & -0.0773 & 0.2448 & 0 & 0 & 0 & -0.1675 \\ 0 & 0 & 0 & -0.3379 & 1.0699 & 0 & 0 & 0 & -0.7321 \\ 0 & 0 & 0 & 0.1656 & -0.5245 & 0 & 0 & 0 & 0.3589 \\ 0 & 0 & 0 & -0.3379 & 1.0699 & 0 & 0 & 0 & -0.7321 \\ 0 & 0 & 0 & 0 & 0 & 0 & 0 & 0 & 0 \\ 0 & 0 & 0 & 0.2054 & 0.6503 & 0 & 0 & 0 & -0.445 \\ -1 & 1 & 0 & -0.3110 & 0 & 0.4 & -0.4 & 0 & 0.311 \\ 0 & 0 & 0 & 0.0574 & 0 & -0.24 & 1.24 & 0 & -1.0574 \\ -1 & 1 & 0 & -0.2153 & 0 & 0 & 0 & 0 & 0.2153 \\ 0 & 0 & 0 & -0.2392 & 0 & 1 & 0 & 0 & -0.7608 \\ -1 & 0 & 1 & -0.0957 & 0 & 0 & 0 & 0.4 & -0.3043 \\ 0 & 0 & 0 & 0.0574 & 0 & 0 & 0 & -0.24 & 0.1826 \\ -1 & 0 & 1 & 0 & 0 & 0 & 0 & 0 & 0 \\ 0 & 0 & 0 & -0.2392 & 0 & 0 & 0 & 1 & -0.7608 \end{pmatrix}.$$

The submatrix $\mathbf{Z}_1$ is a 18×3 matrix corresponding to the first three variation sources. The submatrix $\mathbf{Z}_2$ is a 18×6 matrix corresponding to the last six variation sources.

(a) If only the first three variation sources are considered, check if the respective existence conditions for the LS fit estimator and two-step estimator are satisfied. Based on the results, comment on which estimator has a stronger existence condition.

(b) Repeat part(a) for the last six variation sources.

9

Analysis of System Diagnosability

In Chapters 7 and 8, we presented techniques to diagnose the variation sources in a manufacturing process through the product quality observations. To achieve satisfactory diagnosis performance, the question on system diagnosability needs to be answered. In other words, do the observed quality observations contain sufficient information so that we can identify the variation sources in the process? We present some useful results on system diagnosability based on the analysis of the quality–fault model in this chapter. In Section 9.1, the basic formulation of diagnosability for linear mixed effects model along with the diagnosability criteria will be introduced. The concept of a minimal diagnosable group for a partially diagnosable system will be introduced in Section 9.2. In Section 9.3, we provide methods to evaluate the effectiveness of a measurement system in terms of system diagnosability.

9.1 Diagnosability of Linear Mixed Effects Model

Consider a typical quality–fault model described by a linear mixed effects model as in (9.1) (the same as (6.3)):

$$\mathbf{y}_i = \mathbf{X}_i\beta + \mathbf{Z}_i\mathbf{b}_i + \epsilon_i, \tag{9.1}$$

where $\mathbf{y}_i$ is the quality observation from the ith product unit, β represents the fixed bias in the system and $\mathbf{b}_i$ represents the random variation sources in the system that takes different values for different product units. Similar to Chapters 7 and 8, we assume the variation sources are independent from each other, which means the elements of $\mathbf{b}_i$ are independent random variables.

Industrial Data Analytics for Diagnosis and Prognosis: A Random Effects Modelling Approach, First Edition. Shiyu Zhou and Yong Chen.

Without loss of generality, we can assume the measurement scheme for different product units is identical. In other words, we assume $\mathbf{X}_i$ and $\mathbf{Z}_i$ are fixed for different product unit i. With these assumptions, we can drop the index i in (9.1) for the sake of simplicity:

$$\mathbf{y} = \mathbf{X}\beta + \mathbf{Z}\mathbf{b} + \epsilon, \tag{9.2}$$

where $\mathbf{y} \in \mathcal{R}^{n\times 1}$ is the quality observations, $\mathbf{X} \in \mathcal{R}^{n\times p}$ and $\mathbf{Z} \in \mathcal{R}^{n\times q}$ are design matrices that are determined by the product/process design and the measurement scheme, $\mathbf{b}$ is a random vector and $\mathbf{b} \sim \mathcal{N}(\mathbf{0}, \mathbf{G})$, and $\epsilon \sim \mathcal{N}(\mathbf{0}, \mathbf{R})$ is the error term due to measurement noise. Considering the assumption of independent variation sources and noise, we further have $\mathbf{G} = \begin{pmatrix} \sigma_1^2 & 0 & \cdots & 0 \\ 0 & \sigma_2^2 & \cdots & 0 \\ \vdots & \vdots & \ddots & \vdots \\ 0 & 0 & \cdots & \sigma_q^2 \end{pmatrix}$ and $\mathbf{R} = \sigma_\epsilon^2 \mathbf{I}$, where $\mathbf{I}$ is the identity matrix with dimension of $n \times n$.

For the model in (9.2), the quality observations $\mathbf{y}$ and the design matrices $\mathbf{X}$ and $\mathbf{Z}$ are known. The elements of β and the variance components $\sigma_1^2, \ldots, \sigma_q^2$ are the unknown parameters representing process faults that we want to diagnose. The diagnosability problem is to answer the question: based on the quality observations from multiple product units, i.e., we have multiple random samples of $\mathbf{y}$, can we identify the process faults?

First, we define the concept of parameter identifiability of a general statistical model as follows [Demidenko, 2005].

Definition 9.1 Let a statistical model be defined by a family of distributions for observed data $\mathbf{Y}$ parameterized by the vector $\boldsymbol{\theta}$. The feasible space of $\boldsymbol{\theta}$ is denoted as Θ and the probability distribution of $\mathbf{Y}$ is denoted as $\mathbf{Y} \sim P_{\boldsymbol{\theta}}(\mathbf{y})$, $\boldsymbol{\theta} \in \boldsymbol{\Theta}$. The model is identifiable on Θ if $P_{\boldsymbol{\theta}_1}(\mathbf{y}) = P_{\boldsymbol{\theta}_2}(\mathbf{y})$ implies that $\boldsymbol{\theta}_1 = \boldsymbol{\theta}_2$.

This definition is quite intuitive. Indeed, if $\boldsymbol{\theta}_1 \neq \boldsymbol{\theta}_2$ but the distribution of the observations are the same, then we will not be able to differentiate $\boldsymbol{\theta}_1$ and $\boldsymbol{\theta}_2$. Based on the above general definition of identifiability, we define the specific diagnosability concept under model (9.2) as follows.

Definition 9.2 Under model (9.2), a linear parametric function $\mathbf{u}^T\beta$, $\mathbf{u} \in \mathcal{R}^{p\times 1}$ is said to be diagnosable if for any β_1 and β_2, $\mathbf{u}^T\beta_1 \neq \mathbf{u}^T\beta_2 \Rightarrow E(\mathbf{y})|_{\beta=\beta_1} \neq E(\mathbf{y})|_{\beta=\beta_2}$. Similarly, a linear parametric function $\mathbf{w}^T\boldsymbol{\alpha}$, $\mathbf{w} \in \mathcal{R}^{(q+1)\times 1}$ is said to be diagnosable if for any $\boldsymbol{\alpha}_1$ and $\boldsymbol{\alpha}_2$, $\mathbf{w}^T\boldsymbol{\alpha}_1 \neq \mathbf{w}^T\boldsymbol{\alpha}_2 \Rightarrow \mathrm{cov}(\mathbf{y})|_{\boldsymbol{\alpha}=\boldsymbol{\alpha}_1} \neq \mathrm{cov}(\mathbf{y})|_{\boldsymbol{\alpha}=\boldsymbol{\alpha}_2}$ where $\boldsymbol{\alpha}$ is the vector of variance components and the variance of the error term, i.e., $\boldsymbol{\alpha} = (\sigma_1^2 \ldots \sigma_q^2\ \sigma_\epsilon^2)^T$.

In model (9.2), we are only concerned with the mean and variance of process faults because all random variables follow normal distribution and the first and the second order moments completely define the distribution of $\mathbf{y}$. Intuitively, the above definition means that a fault or a combination of faults is called diagnosable if the change in the mean $\boldsymbol{\beta}$ or variance components $\boldsymbol{\alpha}$ causes a change in the mean or variance of observation $\mathbf{y}$. This is not surprising because obviously we will not be able to infer the unknown parameters if the change in these parameters does not show up in the observations. Thus, this definition does not depend on any specific diagnosis algorithm. Please also note that the definition is quite flexible. By selecting different $\mathbf{u}$ and $\mathbf{w}$, the diagnosability of different faults or combinations of faults can be evaluated. For example, by selecting $\mathbf{u}$ or $\mathbf{w} = (1\,0\ldots0)^T$, we can check if the mean or variance of the first fault is diagnosable. If yes, we say the mean or variance of this fault can be uniquely identified or diagnosed.

Definition 9.2 does not provide a way to check if a fault or a combination of faults is diagnosable. We can use the following necessary and sufficient condition of fault diagnosability to check the system diagnosability. The proof can be found in the Appendix at the end of this chapter.

Lemma 9.1 For model (9.2),
(i) $\mathbf{u}^T\boldsymbol{\beta}$ is diagnosable if and only if $\mathbf{u} \in \text{Range}(\mathbf{X}^T)$
(ii) $\mathbf{w}^T\boldsymbol{\alpha}$ is diagnosable if and only if $\mathbf{w} \in \text{Range}(\mathbf{H})$, where $\mathbf{H}$ is a symmetric matrix given by $\mathbf{H} = \begin{pmatrix} (\mathbf{z}_1^T\mathbf{z}_1)^2 & \cdots & (\mathbf{z}_1^T\mathbf{z}_q)^2 & \mathbf{z}_1^T\mathbf{z}_1 \\ \vdots & \vdots & \vdots & \vdots \\ (\mathbf{z}_q^T\mathbf{z}_1)^2 & \cdots & (\mathbf{z}_q^T\mathbf{z}_q)^2 & \mathbf{z}_q^T\mathbf{z}_q \\ \mathbf{z}_1^T\mathbf{z}_1 & \cdots & \mathbf{z}_q^T\mathbf{z}_q & n \end{pmatrix}$, where Range$(\cdot)$ denotes the the range space (i.e., linear space spanned by the column vectors) of a matrix and $\mathbf{z}_i$ is the ith column vector of $\mathbf{Z}$ in (9.2).

Lemma 9.1 gives us a powerful tool to test if some combinations of faults are diagnosable. From this lemma, it is clear that the means of all the faults are uniquely diagnosable if and only if $\mathbf{X}^T\mathbf{X}$ is of full column rank. The variances of all the faults are uniquely diagnosable if and only if $\mathbf{H}$ is of full rank. In the above criterion, the diagnosability of the variance of process faults includes the variance of the error term σ_ϵ^2. This means that even if a fault can be distinguished from other faults, it could still be non-uniquely diagnosable if it is tangled with the error term. In many practical cases, we can often assume the error term is small and thus we can ignore its effects when exploring the diagnosability of process faults. Through reducing $\boldsymbol{\alpha}$ to include only $(\sigma_1^2 \cdots \sigma_q^2)$, we can reduce the $\mathbf{H}$ matrix in Lemma 9.1 to $\mathbf{H}_r$, where $\mathbf{H}_r$ is a sub-matrix of $\mathbf{H}$ as $\mathbf{H}_r = \begin{pmatrix} (\mathbf{z}_1^T\mathbf{z}_1)^2 & \cdots & (\mathbf{z}_1^T\mathbf{z}_q)^2 \\ \vdots & \vdots & \vdots \\ (\mathbf{z}_q^T\mathbf{z}_1)^2 & \cdots & (\mathbf{z}_q^T\mathbf{z}_q)^2 \end{pmatrix}$.

Now let us look at several examples of checking the system diagnosability.

Example 9.1 Consider a very simple system in which the quality–fault relationship is given as $\mathbf{y} = \begin{pmatrix} 1 & 0 \\ 0 & 0 \end{pmatrix}\boldsymbol{\beta} + \begin{pmatrix} 1 & 0 \\ 0 & 0 \end{pmatrix}\mathbf{b} + \epsilon$. Due to the special structure of the design matrix $\mathbf{X}$ and $\mathbf{Z}$ in this example, it is obvious that we cannot identify the mean change of the second element of $\boldsymbol{\beta}$ and the variance change of the second element of $\mathbf{b}$ as their values do not contribute to the observation $\mathbf{y}$ at all. For this simple case, we can also use Definition 9.2 to directly check the diagnosability. We have $E(\mathbf{y}) = \begin{pmatrix} \beta_1 \\ 0 \end{pmatrix}$ and $\text{var}(\mathbf{y}) = \begin{pmatrix} \sigma_1^2 + \sigma_\epsilon^2 & \\ & \sigma_\epsilon^2 \end{pmatrix}$. Thus, from the definition, we can diagnose

the mean β_1, and variance components σ_1^2 and σ_ϵ^2 and any of their combinations. To use the result in Lemma 9.1 to check the diagnosability, we can first get $\mathbf{X}^T = \begin{pmatrix} 1 & 0 \\ 0 & 0 \end{pmatrix}$ and $\mathbf{H} = \begin{pmatrix} 1 & 0 & 1 \\ 0 & 0 & 0 \\ 1 & 0 & 2 \end{pmatrix}$. Then we can check if a selected $\mathbf{u}$ or $\mathbf{w}$ is in the range space of $\mathbf{X}^T$ or $\mathbf{H}$, respectively. From linear algebra, we know this can be checked by checking if Rank $(\mathbf{X}^T) = \text{Rank}(\mathbf{X}^T|\mathbf{u})$ or if $\text{Rank}(\mathbf{H}) = \text{Rank}(\mathbf{H}\,|\mathbf{w})$. If yes, then the corresponding vector is in the range space of the corresponding matrix and thus the corresponding $\mathbf{u}^T\boldsymbol{\beta}$ or $\mathbf{w}^T\boldsymbol{\alpha}$ is diagnosable. This rank testing can be easily achieved using the following R function:

```
check.diagnosability <-function(u,X){
  return(R(t(X))==R(cbind(t(X),u)))}
```

Please note that because $\mathbf{H}$ is a symmetric matrix. The above `check.diagnosability` function can also be used to check the variance diagnosability. Using this function, we can obtain `TRUE` for $\mathbf{u} = (1\,0)^T$, which means β_1 is diagnosable, while for $\mathbf{u} = (0\,1)^T$, we get `FALSE`, which means β_2 is not diagnosable. In fact, for any $\mathbf{u}$ with a non-zero second element, we will get FALSE as we will not get any information on β_2 from the observations. Similarly, we can check $\mathbf{w} = (a\,0\,0)^T$, $\mathbf{w} = (0\,0\,b)^T$, $\mathbf{w} = (a\,0\,b)^T$ for any $a \neq 0$ and $b \neq 0$ and we will get TRUE. For any $\mathbf{w}$ with a non-zero second element, we will get FALSE.

To get more insights, we can try a more complicated system with quality–fault model $\mathbf{y} = \begin{pmatrix} 1 & 2 \\ 0 & 0 \end{pmatrix} \boldsymbol{\beta} + \begin{pmatrix} 1 & 2 \\ 0 & 0 \end{pmatrix} \mathbf{b} + \epsilon$. Using Lemma 9.1, we will find that β_1 is no longer diagnosable. Rather, we will only be able to diagnose $a\left(\beta_1 + 2\beta_2\right)$ for any $a \neq 0$. Similarly, σ_1^2 is no longer diagnosable. We can only diagnose $a(\sigma_1^2 + 4\sigma_2^2)$ for any $a \neq 0$ and σ_ϵ^2. This result is not surprising if we show the expression of $E(\mathbf{y})$ and $\mathrm{var}(\mathbf{y})$. One point we want to mention is that for very simple systems as shown in this example, the diagnosability can be directly checked using the diagnosability definition. However, for a complex system with high dimensional observations and a large number of process faults, the results in Lemma 9.1 will be very useful.

Example 9.2 In this example, we show that there are cases that even when a design matrix $\mathbf{Z}$ does not have full column rank, the variance faults in the system may still be uniquely diagnosable. This is because the rank of the $\mathbf{H}$ matrix in Lemma 9.1 could be higher than that of $\mathbf{Z}$. Consider the following system with quality–fault model as $\mathbf{y} = \begin{pmatrix} 1 & 1 & 2 \\ 2 & 3 & 5 \\ 1 & 1 & 2 \end{pmatrix} \boldsymbol{\beta} + \begin{pmatrix} 1 & 1 & 2 \\ 2 & 3 & 5 \\ 1 & 1 & 2 \end{pmatrix} \mathbf{b} + \epsilon$. It is obvious that the column rank of $\mathbf{X}$ and $\mathbf{Z}$ are both two because the third column is the summation of the first two columns. Thus, we will not be able to uniquely diagnose β_1, β_2, or β_3. However, we can get $\mathbf{H}$ matrix as $\mathbf{H} = \begin{pmatrix} 36 & 64 & 196 & 6 \\ 64 & 121 & 361 & 11 \\ 196 & 361 & 1089 & 33 \\ 6 & 11 & 33 & 3 \end{pmatrix}$, which has a rank of four. This means we can uniquely diagnose each variance component σ_1^2, σ_2^2, σ_3^2, and σ_ϵ^2. This seems to be surprising. However, we can write out the expression of the covariance matrix of $\mathbf{y}$,

$$\mathrm{cov}(\mathbf{y}) = \begin{pmatrix} \sigma_1^2 + \sigma_2^2 + 4\sigma_3^2 + \sigma_\epsilon^2 & 2\sigma_1^2 + 3\sigma_2^2 + 10\sigma_3^2 & \sigma_1^2 + \sigma_2^2 + 4\sigma_3^2 \\ 2\sigma_1^2 + 3\sigma_2^2 + 10\sigma_3^2 & 4\sigma_1^2 + 9\sigma_2^2 + 25\sigma_3^2 + \sigma_\epsilon^2 & 2\sigma_1^2 + 3\sigma_2^2 + 10\sigma_3^2 \\ \sigma_1^2 + \sigma_2^2 + 4\sigma_3^2 & 2\sigma_1^2 + 3\sigma_2^2 + 10\sigma_3^2 & \sigma_1^2 + \sigma_2^2 + 4\sigma_3^2 + \sigma_\epsilon^2 \end{pmatrix}.$$

We can see that in the expression $\text{cov}(\mathbf{y})$, we do have four independent expressions of σ_1^2, σ_2^2, σ_3^2, and σ_ϵ^2 so that we can solve for each of them if we know the values of $\text{cov}(\mathbf{y})$. Particularly, we will not only use the variances of each element of $\mathbf{y}$, but also use the covariance between different elements. This is the reason why even $\mathbf{Z}$ is not of full rank, we may still be able to diagnose all the variance components.

9.2 Minimal Diagnosable Class

Using Lemma 9.1, we can easily check if a process fault is uniquely diagnosable. However, Lemma 9.1 alone is not effective in analyzing a partially diagnosable system when not all faults are uniquely diagnosable. It is not obvious from Lemma 9.1 what the other faults that we need to know before we can identify a non-uniquely diagnosable fault. To analyze the partially diagnosable system, we propose the concept of minimal diagnosable class [Zhou et al., 2003]. We first introduce the concept of diagnosable class, and then present the definition of minimal diagnosable class.

Definition 9.3 A nonempty set of faults form a mean or variance diagnosable class if a nontrivial linear combination of their means or variances are diagnosable. "Nontrivial" means at least one coefficient of the linear combination is non-zero.

Definition 9.4 A nonempty set of faults form a minimal mean or variance diagnosable class if no strict subset of the set is mean or variance diagnosable.

Intuitively, a minimal diagnosable class represents a set of faults that cannot be further differentiated based on the observations. We can only identify a linear combination of them, but cannot identify any of its strict subsets. With this information, we can show the coupling relationship among faults and learn what additional information is needed to identify certain faults.

The minimal diagnosable class can be obtained from the Reduced Row Echelon Form (RREF) of the transpose of testing matrix $\mathbf{X}$ for mean and $\mathbf{H}$ for variance diagnosability. This result is stated in the following lemma.

Lemma 9.2 Let $\boldsymbol{\delta}$ denote a row of the RREF of $\mathbf{X}$ with non-zero elements and $\boldsymbol{\beta}[\boldsymbol{\delta}]$ as the set of elements of $\boldsymbol{\beta}$ that correspond to the non-zero elements of $\boldsymbol{\delta}$, then $\boldsymbol{\beta}[\boldsymbol{\delta}]$ is a minimal mean diagnosable class. Similarly, $\boldsymbol{\alpha}[\boldsymbol{\delta}]$ is a minimal variance diagnosable class if $\boldsymbol{\delta}$ is a row of the RREF of $\mathbf{H}$ with non-zero elements.

The proof of this lemma utilizes the properties of the RREF of a matrix, including if the RREF of a matrix is unique, its row space is the same as the row space of the original matrix, and in a column of RREF with a pivot position, it can only have one non-zero element. The interested reader can refer to Zhou et al. [2003] for details of the proof. The RREF of a matrix can be directly obtained through the `rref` function in `pracma` package in `R`. Considering the system in Example 9.2, we can get

$$\text{RREF}(\mathbf{X}) = \begin{pmatrix} 1 & 0 & 1 \\ 0 & 1 & 1 \\ 0 & 0 & 0 \end{pmatrix} \text{ and } \text{RREF}(\mathbf{H}) = \begin{pmatrix} 1 & 0 & 0 & 0 \\ 0 & 1 & 0 & 0 \\ 0 & 0 & 1 & 0 \\ 0 & 0 & 0 & 1 \end{pmatrix}.$$

According to Lemma 9.2, we can get that for mean diagnosability, $\{\beta_1, \beta_3\}$ and $\{\beta_2, \beta_3\}$ are minimal diagnosable classes, which means we cannot uniquely identify any of β_1, β_2, and β_3. For the variance diagnosability, we have $\{\sigma_1^2\}$, $\{\sigma_2^2\}, \{\sigma_3^2\}$, and $\{\sigma_\epsilon^2\}$ are minimal diagnosable classes, which means all the variance components including σ_ϵ^2 can be uniquely identified.

Lemma 9.2 provides an easy method to obtain some minimal diagnosable classes. However, we cannot guarantee that the obtained minimal diagnosable classes are exhaustive. In other words, the minimal diagnosable classes obtained from $\mathbf{X}$ or $\mathbf{H}$ may not be the complete list of all the minimal diagnosable classes in the system. The following lemma can be used to obtain the complete list of minimal diagnosable classes. We first introduce some notations.

Let $\mathbf{x}_j$ be the jth column of $\mathbf{X}$. We define the jth mean fault in model (9.1) as the bias in β_j. Then we have a total of p mean faults indexed as $\{1, 2,\ldots, p\}$ and the jth fault corresponds to $\mathbf{x}_j$. Similarly, we can index the $q+1$ variance faults as $\{1, 2,\ldots, q+1\}$ and the jth fault corresponds to $\mathbf{h}_j$, the jth column vector of $\mathbf{H}$ that is defined in Lemma 9.1. With these notations, we have the following lemma.

Lemma 9.3 Denote a minimal mean diagnosable class including a number of faults s as $i_1, i_2,\ldots, i_s$, where $i_1, i_2,\ldots, i_s$ are the indices of the faults. Then we can re-arrange the columns of $\mathbf{X}$ and denote it as $\mathbf{X}' = (\mathbf{x}_{i_{s+1}} \mathbf{x}_{i_{s+2}} \ldots, \mathbf{x}_{i_p}, \mathbf{x}_{i_1}, \ldots, \mathbf{x}_{i_s})$. Then the minimal diagnosable class is identical to $\beta[\delta]$, where δ is the last non-zero row of the RREF of $\mathbf{X}'$. Similarly, a minimal variance diagnosable class $i_1, i_2,\ldots, i_s$ is identical to $\alpha[\delta]$, where δ is the last non-zero row of the RREF of $\mathbf{H}' = (\mathbf{h}_{i_{s+1}} \mathbf{h}_{i_{s+2}} \ldots, \mathbf{h}_{i_{q+1}}, \mathbf{h}_{i_1}, \ldots, \mathbf{h}_{i_s})$.

The detailed proof of this lemma can be found in Zhou et al. [2003]. We can demonstrate the lemma using the quality–fault model in Example 9.2. Considering the mean diagnosability and indexing the mean faults as {1, 2, 3}, we know {1, 3} and {2, 3} are two minimal diagnosable classes. For the minimal diagnosable class {2, 3}, we do not need to re-arrange the columns of **X** and the last non-zero row of RREF(**X**) is (0 1 1) (see the result given below Lemma 9.2), which is consistent with the above lemma. For the minimal diagnosable class {1, 3}, we need to re-arrange the columns of **X** by switching the first and the second column of **X** as $\mathbf{X}' = \begin{pmatrix} 1 & 1 & 2 \\ 3 & 2 & 5 \\ 1 & 1 & 2 \end{pmatrix}$. The RREF of $\mathbf{X}'$ is given as $\text{RREF}\,(\mathbf{X}') = \begin{pmatrix} 1 & 0 & 1 \\ 0 & 1 & 1 \\ 0 & 0 & 0 \end{pmatrix}$, which indicates fault {1, 3} is a minimal diagnosable class. Please note that the non-zero elements of RREF($\mathbf{X}'$) now correspond to the re-arranged index, i.e., the 1 in the second column of RREF($\mathbf{X}'$) actually indicates the first fault now. The variance diagnosability for this example is also consistent with the lemma. Because we can uniquely identify each variance fault in this example, regardless of how we re-arrange the columns of **H**, the last non-zero row of $\text{RREF}\left(\mathbf{H}'\right)$ will always be (0 0 … 0 1), which indicates that the variance fault corresponding to the last column of the $\mathbf{H}'$ is diagnosable.

Lemma 9.3 tells us that a complete list of minimal diagnosable classes can be obtained by thoroughly permuting the columns of the testing matrix **X** or **H** because for any minimal diagnosable class, we can always find a column permutation that makes the columns corresponding to the faults in the class the last columns in the permuted matrix.

Example 9.3 Let us consider a quality–fault model with design matrices $\mathbf{X} = \mathbf{Z} = \begin{pmatrix} 1 & -1 & 0 & 2 \\ 0 & 1 & 1 & -1 \\ 1 & 0 & 1 & 1 \end{pmatrix}$. Using Lemma 9.2 and 9.3, we can get all the minimal mean diagnosable classes {1, 3, 4}, {2, 3, 4}, {1, 2, 4}, {1, 2, 3} and all the minimal variance diagnosable classes {5} (i.e., σ_ϵ^2), {1, 4}, {1, 3}, {1, 2}, {2, 4}, {2, 3}, and {3, 4}.

9.3 Measurement System Evaluation Based on System Diagnosability

To evaluate the effectiveness of a measurement scheme, we need several easy-to-interpret indices to characterize the system diagnosability under the given measurement system. Different measurement schemes will lead to different design matrices and thus different system diagnosability. In Zhou et al. [2003], three criteria are used for the evaluation of a measurement system: information quantity, information quality, and system flexibility.

The information quantity refers to how much information we can get from the observation data for diagnosing system faults. When two measurement systems are used for the same manufacturing system, the number of potential faults is the same. However, for two different measurement systems, the number of faults that we need to know to ensure full diagnosability (i.e., all the faults are uniquely identifiable) could be different. In general, for the model given in (9.2), we will need to know at least $p - \text{Rank}(\mathbf{X})$ number of elements of β to fully identify all the components of β and at least $q + 1 - \text{Rank}(\mathbf{H})$ number of variance components to fully identify all the variance components [Zhou et al. 2003]. Thus, we can use $\text{Rank}(\mathbf{X})$ and $\text{Rank}(\mathbf{H})$ as a measure of information quantity.

The second criterion is about information quality. Even if two measurement systems provide the same amount of information per the rank of $\mathbf{X}$ and $\mathbf{H}$, the detailed information content could be different. In practice, it is always desirable to be able to uniquely diagnose a fault so that a corrective action can be undertaken right away to eliminate the fault and to restore the system to its normal condition. The decision regarding corrective action cannot be made if a fault is coupled with others (i.e., there are multiple elements in the minimal diagnosable class). In such cases, additional investigation or observations are needed. We can use the number of uniquely identifiable faults to benchmark the quality of information. The uniquely identifiable faults can be found by counting the number of the minimal diagnosable classes that contain only a single fault. For the same system, if a measurement scheme provides a larger number of minimal diagnosable classes with a single element, then the measurement scheme is superior in terms of information quality.

The third criterion is the flexibility provided by the current measurement system when we try to achieve the full diagnosability. Some measurement systems could be rigid in the sense that certain faults or fault combinations, which may be difficult to measure in practice, have to be known to achieve a fully diagnosable system. Some other measurement systems may provide information in a flexible way, i.e., many combinations of faults can be selected to make the system fully diagnosable. In Zhou et al. [2003], a concept of *minimal complementary class* is proposed to evaluate flexibility. A minimal complementary class is such a minimal set of faults that if they are known, all the faults of the system can be uniquely identified. Consider a system with four faults and three minimal diagnosable classes as $\{1, 2\}$, $\{1, 3, 4\}$, and $\{2, 3, 4\}$. One can verify that a minimal complementary class for this system is $\{1, 3\}$. Clearly, if fault 1 is known, we will be able to identify fault 2; and if both fault 1 and fault 3 are known, we will be able to identify fault 4 and thus all the faults are uniquely identifiable. We can further show that $\{1, 4\}$, $\{2, 3\}$, $\{2, 4\}$, and $\{3, 4\}$ are also minimal complementary classes. The number of minimal complementary classes for this system is five. A detailed computational procedure to find all the minimal complementary classes for a given quality–fault model can be found in Zhou et al. [2003]. A measurement system with more minimal complementary classes is considered to be more flexible.

Bibliographic Notes

The diagnosability problem discussed in this chapter is a special case of the identifiability problem of a general statistical model. The systematic treatment of this problem can be found in Rao [1992], Cole [2020]. The identifiability of the variance components of a mixed effects model is discussed in Rao and Kleffe [1988]. The diagnosability of variation sources in a discrete manufacturing process is first investigated in Ding et al. [2002b]. The materials introduced in this chapter are based on Zhou et al. [2003]. The concept of Matroid has also been used for diagnosability of variation sources in manufacturing systems [Chen, 2006].

Exercises

1. Consider the following quality–fault model $\mathbf{y} = \begin{pmatrix} 1 & 2 \\ 2 & 4 \end{pmatrix}\begin{pmatrix} \beta_1 \\ \beta_2 \end{pmatrix} + \begin{pmatrix} 1 & 2 \\ 2 & 4 \end{pmatrix}\begin{pmatrix} b_1 \\ b_2 \end{pmatrix} + \epsilon$
   (a) Can we identify the values of β_1 and β_2 based on multiple observations of $\mathbf{y}$? If not, can we identify any combinations of them?
   (b) Can we identify the values of the variance of b_1 and b_2? If not, can we identify any combinations of them?
2. Consider the following quality–fault model

$$\mathbf{y} = \begin{pmatrix} 3 & 2 & 2.5 \\ 2 & 4 & 3 \\ 1 & 6 & 3.5 \end{pmatrix}\begin{pmatrix} \beta_1 \\ \beta_2 \\ \beta_3 \end{pmatrix} + \begin{pmatrix} 3 & 2 & 2.5 \\ 2 & 4 & 3 \\ 1 & 6 & 3.5 \end{pmatrix}\begin{pmatrix} b_1 \\ b_2 \\ b_3 \end{pmatrix} + \epsilon$$

(a) Can we identify the values of β_1, β_2, β_3 based on multiple observations of $\mathbf{y}$? If not, can we identify any combinations of them?

(b) Can we identify the values of the variance of b_1, b_2, b_3? If not, can we identify any combinations of them?

3. Consider the system given in Problem 2. What are the possible minimal diagnosable classes?
4. Consider a multistage assembly manufacturing processes. The quality observations are the dimensional quality measurements at different locations of assembled products. The process faults are the locations of the pins of the fixture system. The quality–fault model is described by (9.2) and we know the $\mathbf{X}$ and $\mathbf{Z}$ are the same 10 by 18 matrix as

$$
\left(\begin{array}{cccccccccc}
1 & 0.786 & -0.786 & 0 & 0 & 0 & 0 & 0 & 0 & 0 \\
0 & 1.143 & -0.143 & 0 & 0 & 0 & 0 & 0 & 0 & 0 \\
0 & 0 & 0 & 1 & 1.1 & -1.1 & 0 & 0 & 0 & 0 \\
0 & 0 & 0 & 0 & 2.26 & -1.26 & 0 & 0 & 0 & 0 \\
0 & 0.401 & -0.786 & 0 & 0 & 0.385 & 1 & 0.385 & -0.385 & 0 \\
0 & 0.073 & -0.143 & 0 & 0 & 0.07 & 0 & 1.07 & -0.070 & 0 \\
0 & 0 & 0 & 0 & 0 & 0 & 0 & 0 & 0 & 1 \\
0 & 0 & 0 & 0 & 0 & 0 & 0 & 0 & 0 & 0 \\
0 & 0.401 & -0.786 & 0 & 0 & 0.385 & 0 & 0.122 & -0.385 & 0 \\
0 & 0.073 & -0.143 & 0 & 0 & 0.07 & 0 & 0.022 & -0.7 & 0
\end{array}\right.
$$

$$
\left.\begin{array}{cccccccc}
0 & 0 & 0 & 0 & 0 & 0 & 0 & 0 \\
0 & 0 & 0 & 0 & 0 & 0 & 0 & 0 \\
0 & 0 & 0 & 0 & 0 & 0 & 0 & 0 \\
0 & 0 & 0 & 0 & 0 & 0 & 0 & 0 \\
0 & 0 & 0 & 0 & 0 & 0 & 0 & 0 \\
0 & 0 & 0 & 0 & 0 & 0 & 0 & 0 \\
0.6 & -0.6 & 0 & 0 & 0 & 0 & 0 & 0 \\
2.48 & -1.48 & 0 & 0 & 0 & 0 & 0 & 0 \\
0 & 0 & 0 & 0 & 0.263 & 1 & 0.263 & -0.263 \\
0 & 0 & 0 & 0 & 0.048 & 0 & 1.048 & -0.048
\end{array}\right)
$$

Ignoring the error term, please find the minimal variance diagnosable classes.

Appendix

Proof of Lemma 9.1

This lemma is a special case of a result in Rao and Kleffe [1988], which is stated as: consider a general linear mixed effects model $\mathbf{y}=\mathbf{X}\beta+\epsilon$, where β represents the fixed effects and ϵ is the random effect with zero mean and $\text{var}(\epsilon)=\sigma_1^2\mathbf{V}_1+\sigma_2^2\mathbf{V}_2+\ldots+\sigma_r^2\mathbf{V}_r$. In Rao and Kleffe [1988], it is shown that $\mathbf{u}^T\beta$ is identifiable if and only if $\mathbf{u}\in\text{Range}(\mathbf{X}^T)$; $\mathbf{w}^T\alpha$ is identifiable if and only if $\mathbf{w}\in\text{Range}(\mathbf{H}')$, where $\mathbf{H}'=(\text{tr}(\mathbf{V}_i\mathbf{V}_j))$, where $1\leq i\leq r$ and $1\leq j\leq r$. Under the model in (9.2), we have $r=q+1$ and $\mathbf{V}_i=\mathbf{z}_i\mathbf{z}_i^T$ for $1\leq i\leq q$ and $\mathbf{V}_{q+1}=\mathbf{I}$. It is easy to show that $\mathbf{H}'$ is identical to the $\mathbf{H}$ defined in Lemma 9.1. Thus, Lemma 9.1 is a special case of the result in Rao and Kleffe [1988].

10

Prognosis Through Mixed Effects Models for Longitudinal Data

Longitudinal data, also known as panel data, is a collection of repeated observations of the same subject, taken from a population, over a period of time. The defining characteristic of longitudinal data is that the data is collected from multiple units at multiple time instances. The classical time series data can be viewed as a special case of longitudinal data, where it is a long series from a single unit. Another subtle difference between time series data and longitudinal data is that we often require the interval between two consecutive observations in a time series be identical while that often is not required for longitudinal data.

Longitudinal data are widely used in biological and health sciences, economic and financial studies, and social sciences. For example, the repeated measures of individual household income of families within a town over a period of time constitute a longitudinal dataset that can be used to study economic growth and income inequality. Repeated measures of certain health status of patients within a control group and a comparison group over a period of time form a longitudinal dataset and are commonly used in clinical trial studies. With the fast development of the Internet of Things and information technology, longitudinal data become commonly available in industrial practices. For example, as shown in Figure 1.1, repeated measures over time from multiple cars on the road can be collected. These data form a longitudinal dataset and can be used for system condition monitoring and prognosis.

This chapter describes statistical models and methods for analyzing longitudinal data and their industrial applications. Specifically, we will focus on the forecasting and failure event prediction based on longitudinal data collected from multiple units. In industrial applications, the reliability of a critical

Industrial Data Analytics for Diagnosis and Prognosis: A Random Effects Modelling Approach, First Edition. Shiyu Zhou and Yong Chen.

unit is crucial to guarantee the overall functional capabilities of the entire system. Failure of such a unit can be catastrophic. Turbine engines of airplanes, power supplies of computers, and batteries of automobiles are typical examples where failure of the unit would lead to breakdown of the entire system. For these reasons, assessing the degradation status and predict the Remaining Useful Life (RUL) of an asset or unit is an important area in industrial data analytics.

RUL prediction is used to predict the remaining useful life of a unit at a particular time of operation. An important method of RUL prediction is to measure one or more physical characteristic(s) of the system at multiple time instances. Such a measure is often called a condition monitoring (CM) signal [Si et al., 2011] and it should be strongly associated with the failure of the unit and contains important information about the health status of the unit. Such a condition monitoring signal is also called degradation signal . With the observation of the temporal progression of condition monitoring signals, we can then forecast its future value and predict the RUL. However, the temporal progression data from a specific unit are often very limited, particularly at the beginning stage of the life of the unit. With the fast development of sensing and information technology, we can often observe the condition monitoring signals from many units. The rich data provides great opportunities to address the above challenge and the mixed effects models can help us to harvest the information in the data. By using the mixed effects modeling approach for longitudinal data, we can, in addition to using the progression data of the unit currently under study, borrow information from the progression data of other similar units. The basic premise is that the degradation paths will behave similarly for similar units and conditions.

This chapter is organized as follows. In Section 10.1, we introduce the mixed effects models for longitudinal data, including the model setup, estimation, and properties. In Section 10.2, we focus on the forecasting of the future value of longitudinal data for an individual unit. The method of using the forecasts of condition monitoring data to get the RUL prediction is introduced in Section 10.3. Section 10.4 and 10.5 present some extensions to the base method.

10.1 Mixed Effects Model for Longitudinal Data

Let y_{ij} represent an observation of a degradation signal at t_{ij}, where $i = 1 \dots m$ is the unit index and $j = 1 \dots n_i$ is the observation index. y_{ij} can be viewed as the response variable and we want to model its relationship with other variables and its time progression characteristics. Correspondingly, we let $\mathbf{z}_{ij}$ be a vector of length q of explanatory variables observed at t_{ij}. To describe the time progression of the response variable, $\mathbf{z}_{ij}$ contains time t and/or the functions of time t and may also contain the observations of related process and environmental variables. For example, if we let $\mathbf{z}_{ij} = (1\ t_{ij}\ t_{ij}^2)^T$, then the temporal progression of y_{ij} will follow a quadratic polynomial form.

To identify the relationship between the response variable and the exploratory variable, a linear regression is often used

$$\begin{aligned} y_{ij} &= b_{i1}z_{ij1} + b_{i2}z_{ij2} + \ldots + b_{iq}z_{ijq} + \epsilon_{ij} \\ &= \mathbf{z}_{ij}^T \mathbf{b}_i + \epsilon_{ij}, \end{aligned}$$

where $\mathbf{b}_i = (b_{i1} \ldots b_{iq})^T$ is the coefficient vector and ϵ_{ij} is often an assumed zero-mean normally distributed random variable representing the error term.

In matrix notation, the linear regression for the ith unit takes the form

$$\mathbf{y}_i = \mathbf{Z}_i \mathbf{b}_i + \epsilon_i, \tag{10.1}$$

where $\mathbf{y}_i = (y_{i1} \cdots y_{in_i})^T$ and $\mathbf{Z}_i$ is a $n_i \times q$ matrix with $\mathbf{z}_{ij}^T$ being the jth row and $\epsilon_i = (\epsilon_{i1} \cdots \epsilon_{in_i})^T$. The model in (10.1) is for an individual unit. If we have data from multiple similar units, then it is reasonable to assume the data exhibit some common characteristic. Thus, we can further specify $\mathbf{b}_i$ as a random vector following a multivariate Gaussian distribution $\mathbf{b}_i \sim \mathcal{N}(\boldsymbol{\mu}_\mathbf{b}, \boldsymbol{\Sigma}_\mathbf{b})$, where $\boldsymbol{\mu}_\mathbf{b}$ and $\boldsymbol{\Sigma}_\mathbf{b}$ are the mean and covariance matrix of $\mathbf{b}_i$, respectively. The error term is often assumed to follow independent normal distribution as $\epsilon_i \sim \mathcal{N}(\mathbf{0}, \sigma_\epsilon^2 \mathbf{I}_{n_i})$, where $\mathbf{I}_{n_i}$ is a n_i by n_i identity matrix. For the sake of convenience in derivations, we also often define $\boldsymbol{\Sigma}_\mathbf{b} = \sigma_\epsilon^2 \mathbf{D}$. Please note that for different units, the number of observations and the interval between consecutive observations might be different. So the number of rows in $\mathbf{Z}_i$ might be different for different units. However, the number of columns of $\mathbf{Z}_i$ are the same for different i.

With the above model specification, it is clear that the model in (10.1) is the special case of the general linear mixed effects model in (6.3) called regression model with random coefficients. Indeed, if we let $\mathbf{b}_i = \boldsymbol{\mu}_\mathbf{b} + \tilde{\mathbf{b}}_i$, where $\tilde{\mathbf{b}}_i \sim \mathcal{N}(\mathbf{0}, \sigma_\epsilon^2 \mathbf{D})$, we can put (10.1) in the form of a general linear mixed effects model as in (6.3) as

$$\mathbf{y}_i = \mathbf{Z}_i \boldsymbol{\mu}_\mathbf{b} + \mathbf{Z}_i \tilde{\mathbf{b}}_i + \epsilon_i.$$

Thus, the general estimation methods and model selection methods for mixed effects models can be used for this model without modification. We want to mention that for the sake of simplicity, we slightly abused the notation here: in previous chapters, we use $\mathbf{b}_i$ to represent the random effects with zero mean, while in this chapter $\mathbf{b}_i$ has non-zero mean $\boldsymbol{\mu}_\mathbf{b}$.

Below we give two examples of mixed effects modeling for longitudinal data.

Example 10.1 The dataset in Figure 1.3 can be viewed as longitudinal data and we can use mixed effects model to describe it. The data contains the observations of internal resistance at different time instances from 14 batteries. The data are re-drawn in Figure 10.1 for the sake of convenience.

One characteristic of this data is that the time instances of the available observations for different batteries are different. Some batteries fail early and thus no observations are available after failure events. If in a longitudinal dataset, the time instances of the observations are the same for all the individuals, the dataset will be called balanced; otherwise, it is called unbalanced. The battery resistance dataset is obviously unbalanced.

Using the mixed effects model estimation technique introduced in Chapter 6, we can fit a mixed effects model for the data. Assume we arrange the data as shown in Table 10.1. If we want to build a mixed effects model in which the battery resistance is linear to time, and use the notation of (10.1), then from the data for the first unit, we will have $\mathbf{y}_1 = (3.5\ \ 3.6\ \ 3.7)^T$ and $\mathbf{Z}_1 = \begin{pmatrix} 1 & 1 \\ 1 & 2 \\ 1 & 3 \end{pmatrix}$, where the first column of $\mathbf{Z}_1$ corresponds to the intercept of the model and the second column corresponds to the regressor t. The dimension of $\mathbf{b}_1$ will be 2×1 including the intercept and slope of the model. Similarly, if we want to build a model in which the response is quadratic with time t, then we need to add a

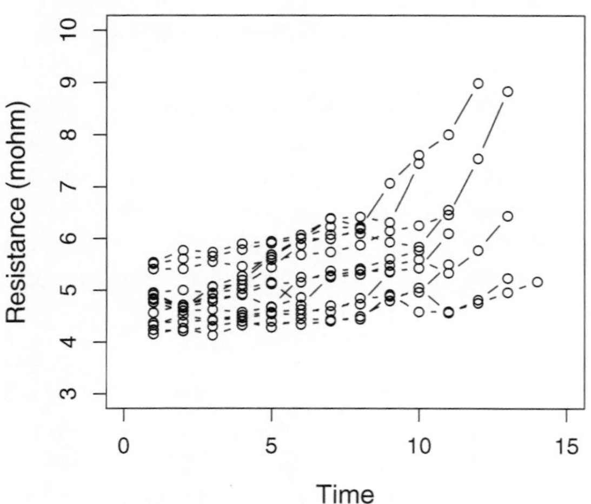

Figure 10.1 Internal resistance measures from multiple batteries

Table 10.1 Illustration of the battery resistance data set

ID	Time	Resistance
1	1	3.5
1	2	3.6
1	3	3.7
2	1	3.45
2	2	3.28
⋮	⋮	⋮

column $(1^2 \quad 2^2 \quad 3^2)^T$ in $\mathbf{Z}_1$, where 1, 2, 3 are the observation times. In this case, the dimension of $\mathbf{b}_1$ will be 3×1 and the last element of $\mathbf{b}_1$ is the coefficient of the quadratic term. For another unit i, $\mathbf{y}_i$ and $\mathbf{Z}_i$ are arranged similarly.

If the data in Table 10.1 is saved in a `csv` file, named `bdata.csv`, then we can use the following `R` code to estimate a mixed effects model

```
bdata <- read.table("bdata.csv",sep=",",header=TRUE)
blme2 <- lme(r ~ obstime+I(obstime^2),
         random= ~ obstime+I(obstime^2)|battid,
            data=bdata)
summary(blme2)
```

In the above code, we use a quadratic function in time to fit the data. We can fit a linear and cubic function to the data using the following command, respectively:

```
blme1 <- lme(r~obstime, random= obstime|battid,
               data=bdata)
blme3 <- lme(r~obstime+I(obstime^2)+I(obstime^3),
          random=~obstime+I(obstime^2)+I(obstime^3)|
           battid, data=bdata)
```

Table 10.2 Estimated parameters under different polynominal degree

	blme1	**blme2**	**blme3**
$\hat{\mathbf{b}}$	$\begin{pmatrix}4.46 & 0.14\end{pmatrix}^T$	$\begin{pmatrix}4.81 & -0.01 & 0.01\end{pmatrix}^T$	$\begin{pmatrix}4.63 & 0.12 & -0.01 & 0.00\end{pmatrix}^T$
$\hat{\sigma}_\epsilon^2 \widehat{\mathbf{D}}$	$\begin{pmatrix}0.067 & 0.016 \\ 0.016 & 0.004\end{pmatrix}$	$\begin{pmatrix}0.321 & -0.051 & 0.006 \\ -0.053 & 0.018 & -0.001 \\ 0.006 & -0.001 & 0.000\end{pmatrix}$	$\begin{pmatrix}0.234 & -0.000 & -0.003 & 0.000 \\ -0.000 & 0.000 & 0.000 & -0.000 \\ -0.003 & 0.000 & 0.000 & -0.000 \\ 0.0000 & -0.000 & -0.000 & 0.000\end{pmatrix}$
$\hat{\sigma}_\epsilon^2$	0.082	0.039	0.029
BIC	145	99	110

The function `lme` is a `R` routine to estimate the parameters of a mixed effects model. The usage of this function can be found in Chapter 6. Through the fitted model `blme1,blme2` and `blme3`, we can obtain the estimated model parameters, including the mean and covariance of the random effects using the following code (using `blme1` as an example),

```
sigma.epsilon.square <-  (blme1$sigma)^2
Mu.b <- blme1$coefficients$fixed
Sigma.b <- var(blme1$coefficients$random$battid)
```

The values are listed in Table 10.2. The symbols $\hat{\mathbf{b}}$, $\widehat{\mathbf{D}}$, and $\hat{\sigma}_\epsilon^2$ represent the estimated values of parameters $\mathbf{b}$, $\mathbf{D}$, and $\hat{\sigma}_\epsilon^2$ as defined in model (10.1). The values of `sigma.epsilon.square`, `Mu.b`, and `Sigma.b` correspond to $\hat{\sigma}_\epsilon^2$, $\hat{\boldsymbol{\mu}}_{\mathbf{b}}$, and $\hat{\sigma}_\epsilon^2 \widehat{\mathbf{D}}$, respectively.

Three polynomial mixed effects models are fitted to the same set of data. Information criteria can be used to select among these models. In the above table, we listed the values of the Bayesian Information Criteria (BIC) and we can see that the quadratic model is the best among the three. With the estimated parameters, we can simulate the progression of the battery resistance. Some simulation results are shown in Figure 10.2.

The `R` function `simulate.lme` can be used to simulate a fitted mixed effects model. However, the standard `simulate.lme` function will not return the simulated response data. A slight modification of the `simulate.lme`

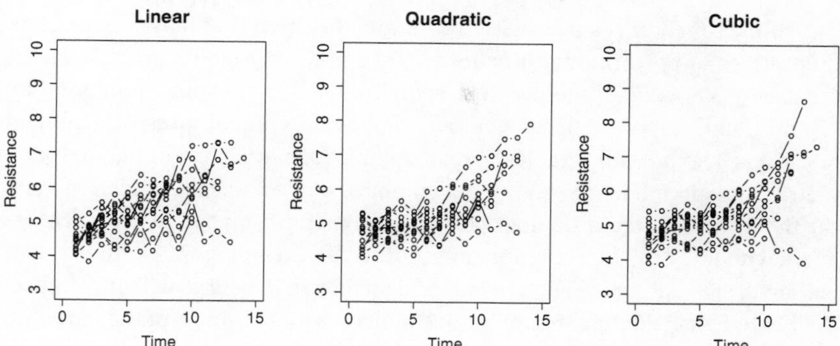

Figure 10.2 Simulated battery internal resistance based on the fitted mixed effects models

function can be used to obtain the raw simulated response data: we can use `trace("simulate.lme",edit=TRUE)` and then add `return(base2)` at the end of the function. Please note that this change will only be effective for the current session and will not permanently change the `simulate.lme`. With this change, we can use

```
sdata<-simulate.lme(blme1)
```

to obtain the simulated data, where `blme1` is the fitted mixed effects model object. The returned data `sdata` has the same size of the original data used to fit `blme1` and we can plot it accordingly to obtain Figure 10.2.

We can see that although visually the scatteredness of the simulated curves is similar to the observed data shown in Figure 10.1, the shapes of curve of some of the simulated data are somewhat different from the observed data. A more flexible mixed effects model can be used. Instead of using simple power functions, spline functions can be used as the regressors. In the following example, we demonstrate this approach.

Example 10.2 In this example, we illustrate the usage of flexible B-spline functions as the regression functions in the mixed effects model. The data we want to model is the same battery resistance data shown in Figure 10.1. In the previous example, simple power functions are used as the regressors. More specifically, for the quadratic model, the design matrix $\mathbf{Z}_i$ in (10.1) contains three columns corresponding to 1 (the intercept), t, and t^2, respectively. It is obvious that more complex regressors can be used to make the model more flexible and expressive. Indeed, B-spline bases are popular choices of regressors in practice thanks to their smoothness and flexibility.

A spline function is a piecewise polynomial function that may contain multiple pieces of polynominal functions. The connecting points of these pieces are called "knots". B-spline, or basis spline, is one type of spline function that has minimal support with respect to a given degree, smoothness, and domain partition. B-spline can find very broad applications in curve fitting, nonparametric statistic modeling, computer graphics, etc. An excellent reference on the theory of B-spline is De Boor [1978]. B-spline function can fit very complex curve with low degree of polynomial. Another key advantage of B-spline is that we can design a B-spline such that a local change in the curve will only impact the related local domain, but will not change the global shape of the curve. To specify B-spline basis functions, we need to specify the degree of the polynomial, the range of the independent variable t, and the knots. In practice, third degree polynomial is often selected because it provides a good trade-off between the complexity and the smoothness of the fitted curve. Instead of specifying the knot values, degree of freedom can be specified. The knot values can be determined based on the degree of freedom with some pre-determined rules. Figure 10.3 shows cubic B-spline bases in the range (0,1). It can be seen that different degrees of freedom will lead to different number of bases. A high degree of freedom can make the model more flexible but at the expense of a larger number of parameters that need to be estimated. For example, if we select the five basis functions in the right panel of Figure 10.3, then the design matrix $\mathbf{Z}_i$ will have five columns (each column corresponds to each basis) and $\mathbf{b}_i$ will be a 5×1 vector.

Limited by the scope of this book, we will not discuss the details of the properties of B-spline but just provide some guidelines on how to use B-spline as the regressor in a mixed effects model. Fitting a mixed effects model using B-spline functions as the regressors is actually quite straightforward in `R`. The `R` function `bs(t,df,degree)` can directly generate the B-spline basis

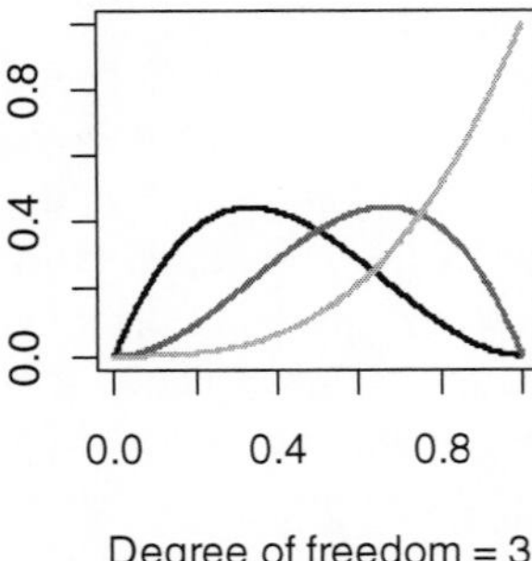

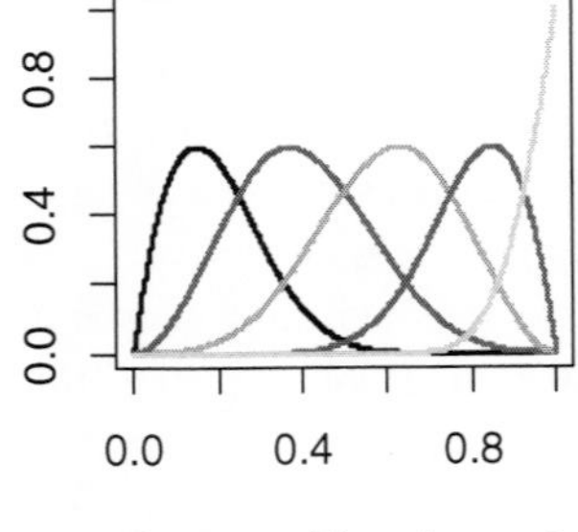

Figure 10.3 Cubic B-spline basis functions with different degrees of freedom

functions evaluated at given values of independent variable t with specified number of basis `df` and degree of the basis `degree`. Using `bs()`, we can fit the mixed effects model using `lme()` function in `R` as

```
lmeControl(returnObject=TRUE,opt="optim",
          optimMethod = "SANN")
blme_bs3 <- lme(r~bs(obstime,df=5,degree=3),
       random= ~bs(obstime,df=5,degree=3)|battid,
           data=bdata)
```

Please note that we do not use the default optimization algorithm for fitting the model, rather, we specified the `"SANN"` method as the optimization algorithm. The default method has convergence issues when dealing with the B-spline based mixed effects model. Using the above code, we are fitting the data in Figure 10.1 using B-spline basis functions as regressors. The B-spline basis functions used are cubic polynomial with five degrees of freedom. Including the intercept in the mixed effects model, $\mathbf{b}_i$ will be a 6×1 vector. The `lme()` function will provide the estimation of the mean and covariance matrix of $\mathbf{b}_i$, and the variance of the error term. In this particular example, the estimated mean of $\mathbf{b}_i$ is $(4.81\ -0.03\ 0.18\ 0.89\ 1.02\ 3.60)^T$. With the estimated parameters, we can again simulate the observations for multiple units as shown in Figure 10.4. In Figure 10.4, we included the B-spline mixed effects model with degree 1, 2, and 3, respectively. The simulated data looks more similar to the original data in Figure 10.1.

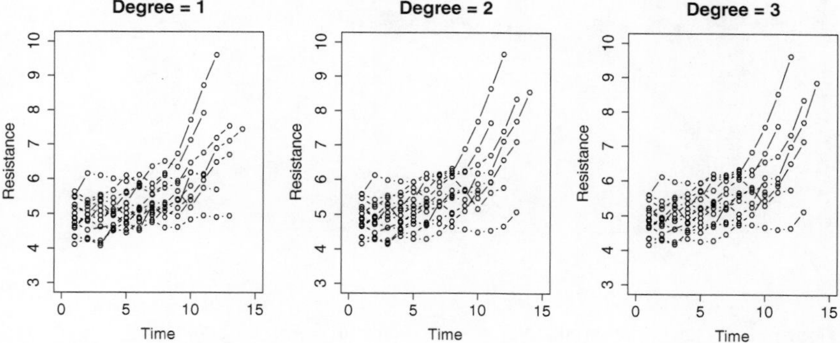

Figure 10.4 Simulated battery resistance data from the fitted linear mixed effects model using B-spline with five degrees of freedom

Specifically, the simulated data show an important feature that is exhibited in the original data, that is, the increase of the signal is slow at the early stage but becomes quicker at the later stages. This feature is not observed in the simple polynomial based mixed effects model as shown in Figure 10.2.

Example 10.3 In this example, we illustrate the use of a B-spline based mixed effects model on a different dataset. The data we want to model is shown in Figure 10.5, which are the temperatures measured at the evaporator outlet of an ice-making machine. Each curve corresponds to one ice-making cycle. We can see that although the temperature measurements from different cycles are similar, there are some variations among cycles. Mixed effects models can be used to describe such variation.

Using similar code of `lme()` and `bs()`, we can fit the data in Figure 10.5 using B-spline basis functions as regressors. The B-spline basis functions used are cubic polynomial but we tried both three and five degrees of freedom. With the estimated parameters, we can again simulate the temperature measurements for each cycle as shown in Figure 10.6. Both simulated data with three and five degrees of freedom look very similar to the original data in Figure 10.5. However, the case with three bases shows some trend between time six and ten that is not in the original data, while the case with five bases does not have such features.

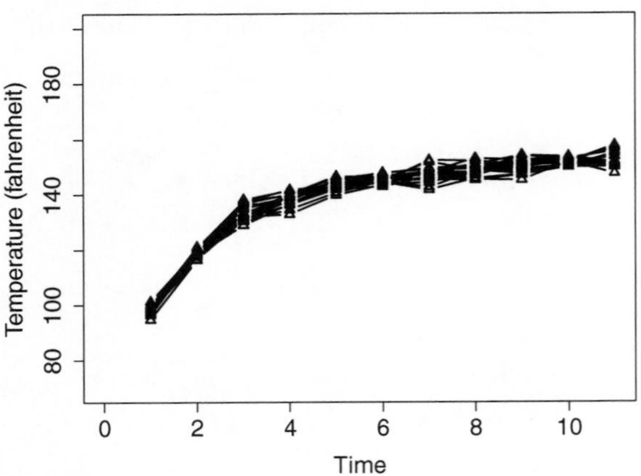

Figure 10.5 Local temperature observations for an ice-making machine

10.2 Random Effects Estimation and Prediction for an Individual Unit

One interesting property of the mixed effects model is that we can make predictions for a specific unit based on the available observations from that unit. Assume we have a unit p. The basic idea of predicting the value of y_p at a future time is to estimate the value of the random effects $\mathbf{b}_p$ and then plug the estimated value into the regression (10.1) to predict the value of the response at a future time. To realize this idea, we use a Bayesian update scheme, which is conceptually intuitive. This method has been adopted in various recent degradation prognosis works and shows satisfactory performance in practical applications [Gebraeel et al., 2005; Zhou et al., 2014; Son et al., 2015a, 2016; Kontar et al., 2017a].

Assume we want to make predictions for a new unit p at time t^*. This new unit is similar to the units from which the historical dataset were collected. These units can be considered as individuals sampled from the same population. At the time instant t^* when the prediction is to be made, assume there are n_p observations from unit p: $\mathbf{y}_i^* = (y_p(t_{p1}) \dots y_p(t_{pn_p}))^T$, where $t_{pn_p} \leq t^*$. In matrix notation, the regression for this unit takes the form of (10.1) as

$$\mathbf{y}_p^* = \mathbf{Z}_p^* \mathbf{b}_p + \epsilon_p^*. \tag{10.2}$$

The asterisk indicates the variables are dependent upon the time instant t^*. Under the mixed effects model, we know that $\mathbf{b}_p$ follows a normal distribution. Using the Bayesian framework, we can naturally choose the distribution of the

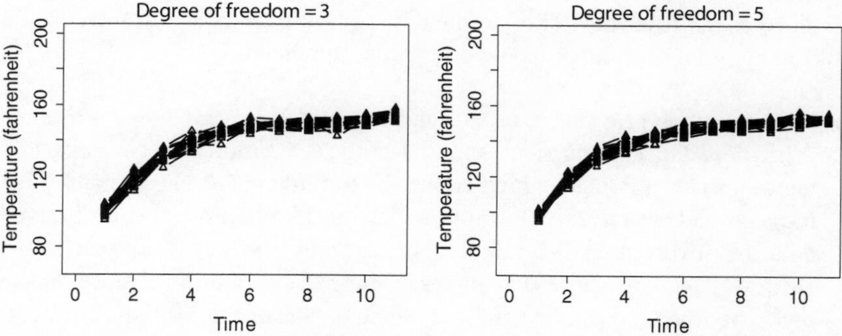

Figure 10.6 Simulated temperature data from the fitted linear mixed effects model using B-spline with cubic degree

random effects $\mathbf{b}_i$ estimated from historical data as the prior distribution of $\mathbf{b}_p$. Then we can use the Bayesian approach to compute the posterior distribution of $\mathbf{b}_p$ according to the observed data $\mathbf{y}_p^*$ from the unit p. In this way, both the information contained in the historical data and the unit specific information contained in the observations up to time t^* can be integrated together. In more detail, we have the posterior distribution of $\mathbf{b}_p$ as

$$f(\mathbf{b}_p \mid \mathbf{y}_p^*) \propto f(\mathbf{y}_p^* \mid \mathbf{b}_p)\pi(\mathbf{b}_p), \tag{10.3}$$

where $\pi(\mathbf{b}_p)$ is the prior distribution of $\mathbf{b}_p$ and $f(\mathbf{y}_p^* \mid \mathbf{b}_p)$ is the likelihood function of $\mathbf{y}_p^*$ given $\mathbf{b}_p$. We assign the estimated distribution of the random effects as the prior distribution so we have $\pi(\mathbf{b}_p) = f_N(\mathbf{b}_p; \hat{\mathbf{b}}, \hat{\sigma}_\epsilon^2 \hat{\mathbf{D}})$, where $f_N(\cdot; \boldsymbol{\mu}, \boldsymbol{\Sigma})$ represents the multivariate normal density function with mean $\boldsymbol{\mu}$ and covariance matrix $\boldsymbol{\Sigma}$, the hat means estimated values from historical data. Given a fixed value of $\mathbf{b}_p$, $\mathbf{y}_p$ follows a normal distribution $\mathcal{N}(\mathbf{Z}_p\mathbf{b}_p, \hat{\sigma}_\epsilon^2 \mathbf{I}_{n_p})$ according to the relationship in (10.1). Thus, the likelihood term is

$$f(\mathbf{y}_p^* \mid \mathbf{b}_p) = \prod_{j=1}^{n_p} \frac{1}{\sqrt{2\pi\hat{\sigma}_\epsilon^2}} \exp\{-\frac{[y_p(t_{pj}) - \mathbf{z}^T(t_{pj})\mathbf{b}_p]^2}{2\hat{\sigma}_\epsilon^2}\}. \tag{10.4}$$

With (10.3) and (10.4) and the specified prior distribution $\pi(\mathbf{b}_p)$, we can show that the posterior distribution $f(\mathbf{b}_p \mid \mathbf{y}_p^*)$ is a normal distribution $\mathcal{N}(\hat{\mathbf{b}}_p^*, \hat{\boldsymbol{\Sigma}}_{\mathbf{b}_p^*})$, where

$$\begin{aligned} \hat{\mathbf{b}}_p^* &= \hat{\boldsymbol{\Sigma}}_{\mathbf{b}_p^*} \cdot [\mathbf{Z}_p^{*T} \mathbf{y}_p^* / \hat{\sigma}_\epsilon^2 + (\hat{\sigma}_\epsilon^2 \hat{\mathbf{D}})^{-1} \hat{\mathbf{b}}] \\ \hat{\boldsymbol{\Sigma}}_{\mathbf{b}_p^*} &= [(\hat{\sigma}_\epsilon^2 \hat{\mathbf{D}})^{-1} + \mathbf{Z}_p^{*T} \mathbf{Z}_p^* / \hat{\sigma}_\epsilon^2]^{-1} \end{aligned} . \tag{10.5}$$

The proof of this result can be found in the appendix of this chapter. We can treat $\hat{\mathbf{b}}_p^*$ as the true value of $\mathbf{b}_p$ to predict the future value of $y_p(t)$. We would like to mention several points regarding this approach:

- The estimation provided in (10.5) can be viewed as an online updating step. With more data coming in, we can update the estimation $\hat{\mathbf{b}}_p^*$ accordingly and hence provide an updated prediction. The formula in (10.5) is a "batch" formula instead of a recursive formula, which means all the previously observed data are involved in the calculation, not only the newly observed one. However, we can show that under a normal distribution assumption, a recursive updating can be achieved. Some details of the recursive formulation can be found in Section 10.5 and Son et al. [2016].

- The results in (10.5) is obtained through Bayesian updating assuming a fixed prior distribution. The results are slightly different from that obtained from the frequentist approach, e.g., [Jones, 1993; Laird and Ware, 1982; Harville, 1976]. The estimation of the mean $\hat{\mathbf{b}}_p^*$ in both the Bayesian and the frequentist approach are identical but the estimation of variance $\hat{\Sigma}_{\mathbf{b}_p^*}$ is different. This difference is due to the fact that in the Bayesian approach, we assume the parameters of the prior distribution are constant while in the frequentist approach, the population mean and variance are estimated values with variability. Thus, the variance estimation provided by the Bayesian method tends to be smaller than that in the frequentist approach. However, if we have a very large set of historical data, the variability in the population parameter estimation will be small and the difference in these two approaches will be small as well.

In the following, we will use two examples to illustrate the prediction approach.

Example 10.4 Consider the data and the mixed effects model in Example 10.1. Assume we want to predict the resistance value for a new unit. We can simulate the resistance values from the new unit with underlying $\mathbf{b}_p$ values (4.86 −0.05 0.01) and we get the observed resistance values up to $t^* = 6$ as $\mathbf{y}_p^* = (4.85\,4.88\,4.90\,4.89\,4.96\,5.04)^T$. We want to predict the future value of the resistance through the quadratic mixed effects model. Based on the Bayesian updating strategy, we can use the following R code to compute the updated mean and variance of $\mathbf{b}_p$.

```
Sigma.bp <- solve(solve(Sigma.b)+(t(Zp))%*%
         (Zp/sigma.epsilon.square))
Mu.bp <- Sigma.bp %*%(t(Zp)%*% yp/sigma.epsilon.square
         +solve(Sigma.b)%*% Mu.b)
```

where `solve` function returns the inverse of a matrix and `yp,Zp` are the observed values from unit p up to t^* and the corresponding design matrix, respectively. With the updated estimation, we can compute the future values of resistance as shown in the following figure.

Figure 10.7 illustrates the Bayesian updating steps at $t^* = 2, 4, 6$, respectively. We can see clearly that with more data available, the updated result is closer to the underlying truth. Table 10.3 shows the comparison of the prior value, updated (posterior) value, and the true value of the model coefficient $\mathbf{b}_p$.

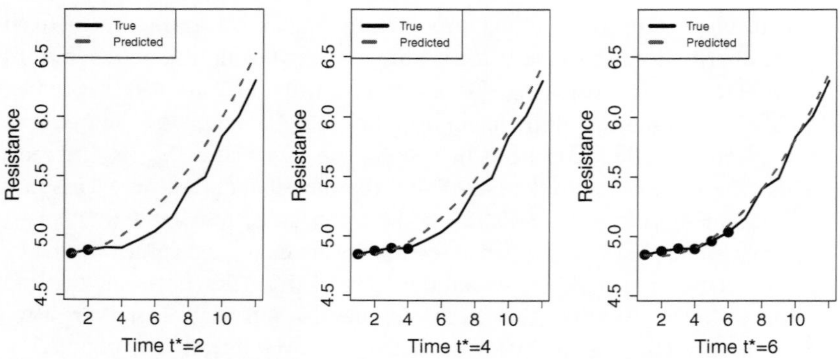

Figure 10.7 Prediction of battery resistance using the Bayesian approach

Table 10.3 Updated parameters at different time instances

Parameter	true	prior	posterior $t^* = 2$	posterior $t^* = 4$	posterior $t^* = 6$
b_0	4.86	4.81	4.84	4.85	4.88
b_1	−0.05	−0.01	−0.01	−0.03	−0.05
b_2	0.01	0.01	0.01	0.01	0.01

Example 10.5 Consider the data and the mixed effects model in Example 10.3. We can demonstrate the flexibility in using B-spline functions as the regressors. Figure 10.8 shows the prediction of the temperature within a specific ice making cycle using the mixed effects model approach introduced in Sections 10.2 and 10.3. We have one, five, and seven observations from the cycle we want to predict, respectively, represented by the black dots in the three panels in Figure 10.8. The dotted line, dashed line, and solid line are the prior, posterior, and true temperature cure. The prior curve indicates the average behavior of the temperature profile. We can see clearly that this specific temperature curve deviates from the average behavior between time unit five and eight. In the mixed effects modeling, we use eight B-spline bases with polynomial degrees of four. With more observations from the specific cycle becoming available, the predicted curve (i.e., the posterior curve) starts to deviate from the prior curve and follow the true curve. It is particularly notable that the posterior curve can reflect the subtle "local" features of the true curve. This demonstrates that B-spline bases are quite flexible in a linear mixed effects model and they can be used to describe and predict complex data.

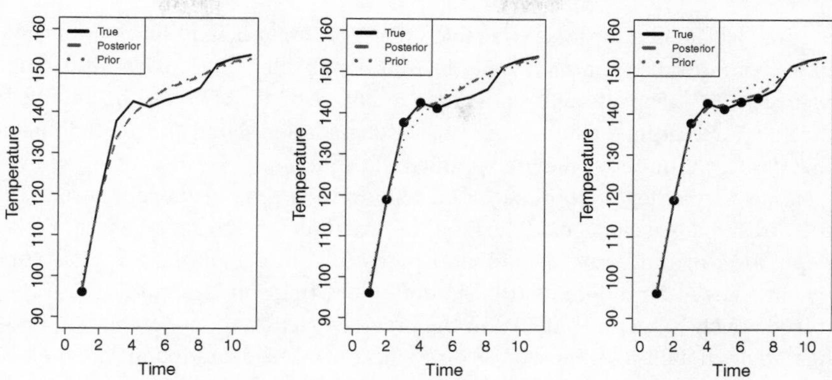

Figure 10.8 Prediction of temperature using Bayesian approach

10.3 Estimation of Time-to-Failure Distribution

One of the important methods for predicting the failure time of an in-service unit is based on the forecasting of the degradation signal collected from the unit. The basic idea is illustrated in Figure 10.9. In this method, the failure event is defined as the event when the degradation signal reaches a pre-specified threshold. For example, if a battery's capacity is below 60% of its initial value, we claim the battery fails although it may be still functioning. Another example is to use the 60% decreasing of light output from a light bulb to define its failure. This type of failure is often called "soft" failure. The concept of "soft" failure was

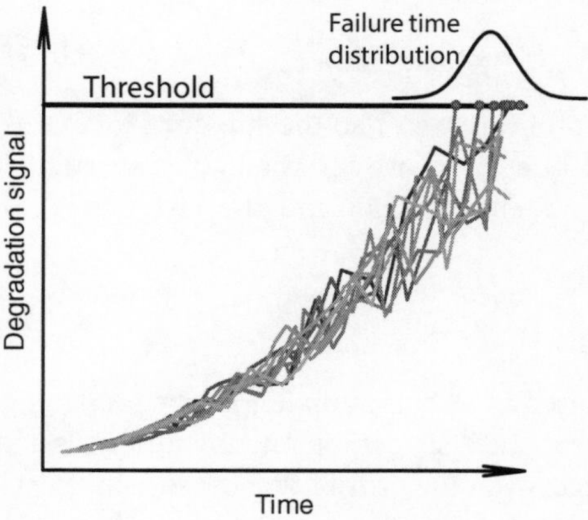

Figure 10.9 Predicting failure time based on degradation signal observations

originally proposed for highly reliable units, because it is hard to get sufficient time-to-failure data for such units in traditional reliability analyses [Lu and Meeker, 1993]. For soft failure, the time to failure distribution can be obtained through forecasting the future value of the degradation signal and predicting its first passage time above the pre-specified threshold.

Based on the degradation path of a unit and the time instance it surpasses the threshold, we can decide the (soft) failure time for that specific unit. For a population of units, the degradation paths for different units are different, which leads to a random distribution of failure times. It is natural to use the mixed effects model introduced in the previous section to model the randomness in degradation signals across different units. The degradation signal path for the ith unit can be expressed as a general mixed effect model:

$$y_i(t) = \eta(t; \mathbf{b}_i) + \epsilon_i(t), i = 1, 2, \ldots, m, \tag{10.6}$$

where $\epsilon_i(t)$ is the error term. $\eta(t; \mathbf{b}_i)$ is the path of the degradation signal for the ith unit at time t with unknown parameters of $\mathbf{b}_i$, which is a vector of random-effects for the ith unit, varies from unit to unit. It is obvious that the linear mixed effects model in (10.1) is a special case of (10.6). For degradation signal path modeling, $\mathbf{b}_i$ is often assumed to follow a multivariate normal distribution with unknown parameters that need to be estimated; however, any general multivariate distribution can be used as well. The error term $\epsilon_i(t)$ is often assumed to follow a normal distribution with mean zero and variance σ_ϵ^2; and $\epsilon_i(t)$ is independent with $\epsilon_i(t')$ for $t \neq t'$; but it can also be assumed to be generated by other stochastic processes such as the Brownian motion. If we let T denote the failure time, and set D as the threshold when the failure occurs, the Cumulative Distribution Function (CDF) of T is

$$\Pr(T \leq t) = F_T(t) = \Pr(y(t) \geq D). \tag{10.7}$$

This definition is applicable to an arbitrary unit i so the subscript index i is ignored in (10.7). Knowing the CDF of T, $F_T(t)$, we can obtain the expectation of T based on the well known property of CDF and the fact that T is non-negative,

$$E[T] = \int_0^\infty [1 - F_T(t)]\mathrm{d}t. \tag{10.8}$$

The expectation value $E[T]$ can be used as a point estimator for the remaining useful life of the unit. In general cases, the distribution function of T does not have a closed form due to the complexity of the distribution of random effects $\mathbf{b}_i$ and $\eta(\cdot)$. However, under certain simple function η and a relatively simple parametric probability density function for $\mathbf{b}_i$, the closed form can be obtained.

Let us consider the case that the degradation paths are modeled by the linear mixed effects modeled in (10.1).

Example 10.6 Assume the degradation signal is modeled by a simple linear function $\eta(t) = b_1 + b_2 \cdot t$ and b_1 is fixed and $b_2 \sim \mathcal{N}(\mu_{b_2}, \sigma_{b_2}^2)$. To ensure we have a positive failure time, the degradation signal $\eta(t)$ should be monotonic so we need $b_1 > 0$. To achieve this, we further assume $\sigma_{b_2} << \mu_{b_2}$ and thus $\Pr(b_2 \le 0)$ is negligible. Considering the degradation signal function $\eta(t) = b_1 + b_2 \cdot t$ we have $\eta(t) \sim \mathcal{N}(b_1 + \mu_{b_2} t, \sigma_{b_2}^2 t^2)$. Then $F_T(t) = \Pr(T \le t) = \Pr(\eta(t) \ge D) = \Pr(-\eta(t) \le -D)$. We know $-\eta(t) \sim \mathcal{N}(-b_1 - \mu_{b_2} t, \sigma_{b_2}^2 t^2)$, thus $\Pr(-\eta(t) \le -D) \approx \Phi(\frac{-D-(-b_1-\mu_{b_2} t)}{\sigma_{b_2} t}) = \Phi(\frac{t-(D-b_1)/\mu_{b_2}}{\sigma_{b_2} t/\mu_{b_2}})$, where $\Phi(\cdot)$ is the CDF of the standard normal distribution. The approximation is due to the negligible $\Pr(b_2 \le 0)$. As a result, we have

$$F_T(t) \approx \Phi(\frac{t-(D-b_1)/\mu_{b_2}}{\sigma_{b_2} t/\mu_{b_2}}) \tag{10.9}$$

for the simple linear mixed degradation model $\eta(t)$. Plugging (10.9) into (10.8), we can obtain the point estimate of the RUL. Generally, numerical integration is needed to compute (10.8).

The results in the above example can be extended into more complex cases. For example, we can assume both the intercept and the slope of a simple linear degradation signal model are random effects, i.e., $\eta(t) = b_1 + b_2 t$, where $b_1 \sim \mathcal{N}(\mu_{b_1}, \sigma_{b_1}^2)$ and $b_2 \sim \mathcal{N}(\mu_{b_2}, \sigma_{b_2}^2)$, and b_1 is independent of b_2. We have $\eta(t) \sim \mathcal{N}(\mu_{b_1} + \mu_{b_2} t, \sigma_{b_1}^2 + \sigma_{b_2}^2 t^2)$. Again, we ignore the probability that b_1, b_2 take the values that make $\eta(t)$ decreasing, we can get

$$F_T(t) \approx \Phi(\frac{t-(D-\mu_{b_1})/\mu_{b_2}}{\sqrt{(\sigma_{b_1}^2 + \sigma_{b_2}^2 t^2)/\mu_{b_2}^2}}). \tag{10.10}$$

In these two examples, we assume the random effects follow normal distribution. Other distributions (e.g., Weibull distribution, lognormal distribution) can be used as well. Interested readers can find such examples in Lu and Meeker [1993]. It needs to be pointed out that in the above example, the failure is defined as the underlying degradation signal $\eta(t)$ reaching the threshold D. The error term is not considered. However, in reality, $\eta(t)$ is not directly observable and we only observe $y_i(t)$ directly. If we use the value of $y_i(t)$ to determine the occurrence of failure, then the error term should be considered. It is actually fairly straightforward to consider the error term, noting that the mean of $y_i(t)$ is identical to that of $\eta(t)$, while the variance of $y_i(t)$ is larger than that of $\eta(t)$ with the addition of the variance of the error term σ_ϵ^2. We can replace the variance of $\eta(t)$ in (10.9) and (10.10) by the variance of $y_i(t)$, then we get the failure time distribution based on the observations of $y_i(t)$.

Now let us consider the problem of predicting the RUL at time instance t^* for an individual unit p discussed in Section 10.2. Noting that at t^*, no failure has occurred and we want to predict the future RUL after t^*, we have

$$F_T(t|t^*) = \Pr(T - t^* \leq t | T \geq t^*) = \Pr(y_p(t^* + t) \geq D | T \geq t^*).$$

From (10.2), we have

$$y_p(t + t^*) = \mathbf{z}_p(t + t^*)\mathbf{b}_p + \epsilon_p(t + t^*), \tag{10.11}$$

where $\mathbf{z}_p(t + t^*)$ is the row design vector at $t + t^*$ and $\mathbf{b}_p$ contains the random effects. Please note that the mean ($\hat{\mathbf{b}}_p^*$) and covariance ($\hat{\Sigma}_{\mathbf{b}_p^*}$) of $\mathbf{b}_p$ are updated based on the observations from unit p up to time t^* and are given in (10.5). It is straightforward to get that $y_p(t + t^*)$ follows normal distribution and the predicted mean and variance are given as

$$\hat{y}_p(t + t^*) = \mathbf{z}_p(t + t^*)\hat{\mathbf{b}}_p^*$$

$$\hat{\sigma}^2_{y_p(t+t^*)} = \mathbf{z}_p(t + t^*)\hat{\Sigma}_{\mathbf{b}_p^*}\mathbf{z}_p^T(t + t^*) + \sigma_\epsilon^2.$$

The first equation gives the predicted mean of $y_p(t + t^*)$ and the second equation gives the predicted variance. Noting that the mean of the error term is zero, we can easily obtain the mean of $y_p(t + t^*)$ based on (10.11). To get the variance of $y_p(t + t^*)$, we used the fact that the random effects $\mathbf{b}_p$ and the error term ϵ_p are independent of each other and the covariance formula $\text{cov}(\mathbf{Au}, \mathbf{Bu}) = \mathbf{A\Sigma B}^T$, where $\mathbf{u}$ is a random vector with covariance matrix $\mathbf{\Sigma}$ and $\mathbf{A}$ and $\mathbf{B}$ are two constant matrices. In the above derivation, $\mathbf{u}$ is $\mathbf{b}_p^*$ and $\mathbf{A}$ and $\mathbf{B}$ are the same as $\mathbf{z}_p(t + t^*)$.

Knowing the distribution of $y_p(t + t^*)$, we can follow a similar derivation in Example 10.6 to get the failure time cumulative distribution as

$$F_T(t + t^*) = \Pr(y_p(t + t^*) \geq D) = \Pr(-y_p(t + t^*) \leq -D) \approx \Phi(\hat{y}_p^0(t + t^*)), \tag{10.12}$$

where $\hat{y}_p^0(t + t^*)$ is a standardized value of $\hat{y}_p(t + t^*)$, defined as $\frac{\hat{y}_p(t+t^*) - D}{\sqrt{\hat{\sigma}^2_{y_p(t+t^*)}}}$.

This equation gives us the expression of the marginal failure time cumulative distribution. To obtain the conditional distribution $F_T(t|t^*)$, we can employ the conditional probability formula as follows.

$$
\begin{aligned}
F_T(t \mid t^*) &= \Pr(T \le t + t^* \mid T \ge t^*) \\
&= \Pr(y_p(t+t^*) \ge D \mid y_p(t^*) \le D) \\
&= \frac{\Pr((y_p(t+t^*) \ge D) \text{ and } (y_p(t^*) \le D))}{\Pr(y_p(t^*) \le D)} \\
&\approx \frac{\Phi(\hat{y}_p^0(t+t^*)) - \Phi(\hat{y}_p^0(t^*))}{1 - \Phi(\hat{y}_p^0(t^*))}.
\end{aligned} \tag{10.13}
$$

The last step of the above derivation used the definition of $\hat{y}_p^0(\cdot)$ given in (10.12) and the identity for a continuous random variable u, $\Pr(u \ge D) = 1 - \Pr(u \le D)$.

Equation (10.13) provides an estimation of failure time cumulative distribution based on the degradation signal that is modeled as a mixed effects longitudinal model and updated using the Bayesian approach as described in Section 10.2. Now we use an example to illustrate the steps of RUL prediction using this approach.

Example 10.7 Let us consider Example 10.4. Assume we set the failure threshold as eight and we want to predict the RUL at $t^* = 2$, $t^* = 4$, $t^* = 6$, respectively. We can use the results in (10.13) and (10.8). This can be achieved through several lines of R code as follows. In this code, `Mu.bp` and `Sigma.bp` are the updated $\hat{\mathbf{b}}_p^*$ and $\hat{\Sigma}_{\mathbf{b}_p^*}$, respectively, based on the observations of y_p up to t^*. The function `failuredist` computes the failure time CDF $F_T(t)$ as given in (10.13). The last two lines of the code evaluate the values $F_T(t|t^*)$ between `tstar` and `MaxT`, where `MaxT` is selected as a relatively large time instance for the horizon we want to predict on. Here we set `MaxT` as 18.

```
xt=function(ti) matrix(c(1,ti,ti^2),3,1)
pred=Vectorize(function(ti) t(xt(ti))%*% Mu.bp-D)
Zp = function(ti) pred(ti) /
      sqrt(t(xt(ti))%*% Sigma.bp %*% xt(ti)
      +sigma.epsilon.square)
failuredist = function(ti)
              (pnorm(Zp(ti))-pnorm(Zp(tstar))) /
              (1-pnorm(Zp(tstar)))
ti<-seq(tstar,MaxT,length.out =100)
res<-lapply(ti,failuredist)
```

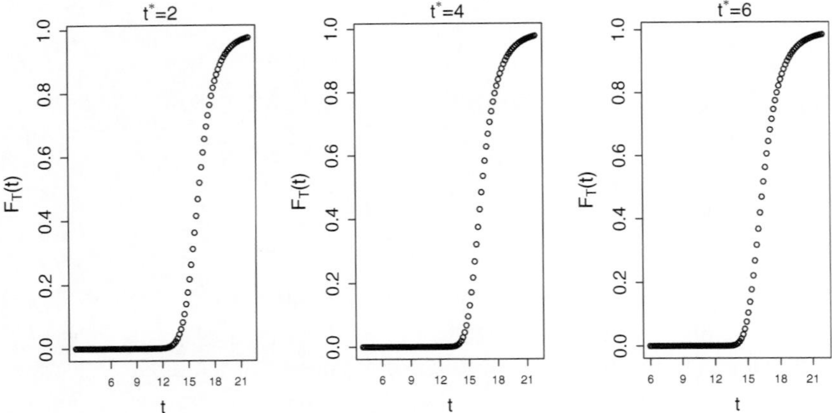

Figure 10.10 Predicted failure time distribution based on degradation signal observations

After running the code, we can plot $F_T(t|t^*)$ from `tstar` to `MaxT` as shown in Figure 10.10. It can be seen that $F_T(t|t^*)$ always starts at t^*. At the beginning, the failure probability is low and goes up as time passes.

Using the value of $F_T(t|t^*)$, we can compute the mean failure time as a point estimation of the RUL. This can be realized by the following `R` codes, where `trapz` computes the *trapezoidal integration* of a function.

```
ti<-seq(tstar,MaxT,length.out=1000)
trapz(ti, 1-unlist(lapply(ti,failuredist)))
```

With the above code, we can obtain the expected RUL at $t^* = 2, 4, 6$ to be 14.53, 12.73, and 10.73, respectively, which gives us similar overall life expectancy of 16.53, 16.73, and 16.73, respectively.

10.4 Mixed Effects Model with Mixture Prior Distribution

The most common choice for the distribution of the random effects in the mixed effects model is the normal distribution as introduced in Section 10.1 due to its simplicity and tractability. However, in many cases, a more flexible distribution may be needed to capture some specific features of the data. Use the application of failure prediction as an example. Due to technological advancements, most modern engineering systems are reliable and thus premature failures are quite rare. Consequently, the historical data used for mixed effects modeling are often imbalanced, i.e., most data are from reliable units with long life and very few are from early failed units. It is well known that the data imbalance issue will cause detrimental effects in statistical modeling and analysis [He and Garcia, 2009]. Unsurprisingly, it also causes trouble for the degradation signal based failure prediction. As shown in previous sections, the mixed effects model is used to describe the progression of the degradation signals. In such models, a prior distribution that accounts for variations among a population of similar units is fitted from the historical data. Then a Bayesian update strategy is used to integrate the prior with online observations to obtain a unit specific posterior distribution from which the unit-specific RUL is predicted. The prior information of the population is an important component in the Bayesian framework. It allows the integration of historical data with newly observed data. Precise and representative priors can improve the accuracy of a degradation model. If we use the regular normal distribution as the prior distribution of random effects and when the historical data is imbalanced in which only a very small proportion of units failed early, the prior distribution established from historical data will be mostly representative of the majority group with a long useful life. Such a prior distribution will make the updating process inefficient for the minority group, i.e., early failed units. This phenomenon is illustrated in Figure 10.11, which shows the results of the predicted degradation signal using a mixed effects model. Figure 10.11(a) and (b) are for the early failed units where the signal increases rapidly whereas Figure 10.11(c) and (d) show degradation signals for the regular units.

In Figure 10.11, the solid and the dashed lines represent the true and the predicted signal trajectory, respectively. We can clearly see from Figure 10.11 (a) and (b) that, because the unit is from the minority group, there is a noticeable discrepancy between the true and the predicted signals even when the number of observations is large. On the other hand, the Bayesian updating works satisfactorily for the units in the majority group with a long useful life as shown in Figure 10.11(c) and (d). This example illustrates that, if a conventional mixed effects model is used, we will not be able to achieve satisfactory accuracy of RUL prediction for the early failed units.

To address this issue, we can extend the choices of the prior distribution in the mixed effects model for longitudinal data to capture the data imbalance. Specifically, we can use a mixture prior distribution to capture the characteristics of different groups, i.e., failed and survived units. Of course, the corresponding model fitting and the Bayesian updating procedures need to be modified accordingly. This method was first reported in Kontar et al. [2017a].

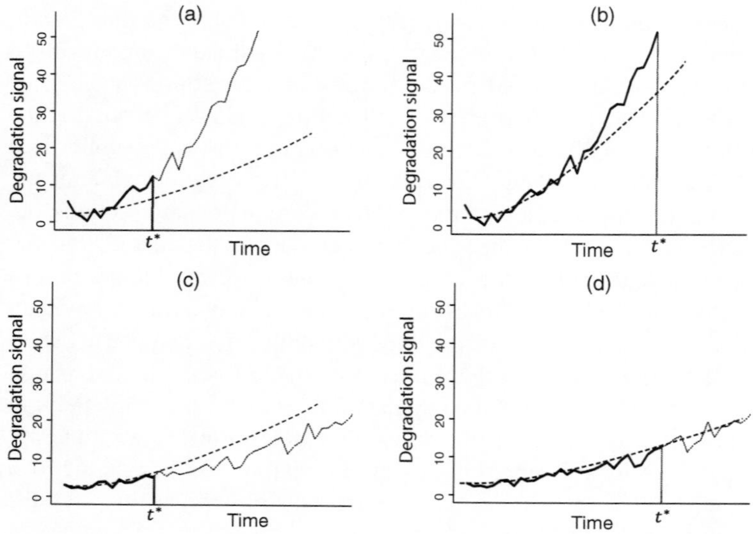

Figure 10.11 Predicting degradation signal using normal distribution as prior distribution. t^* is the time of prediction and the solid and dashed lines are the true and predicted degradation signal, respectively. The signals in panels (a) and (b) are from the minority group and that in panels (c) and (d) are from the majority group

10.4.1 Mixture Distribution

As the name implies, a random mixture distribution is a mixture of two or more probability distributions. Or more specifically, a mixture distribution is formed from the weighted combination of two or more component distributions. The component distributions can be either univariate or multivariate. The component distributions do not need to have the same form but they should have the same dimensionality. Intuitively, a mixture distribution can be viewed as the probability distribution of a random variable, which is selected by chance from a collection of random variables. For example, let $Y = \delta \cdot Y_1 + (1-\delta) \cdot Y_2$, where Y_1 and Y_2 are normally distributed random variables with different mean and/or variance, and δ is a random variable following Bernoulli distribution, i.e., $\Pr(\delta = 0) = \lambda$ and $\Pr(\delta = 1) = 1 - \lambda$. Then the random variable Y is a random variable following mixture distribution with two components corresponding to the distribution of Y_1 and Y_2. Please do not confuse the variable Y with $Y' = Y_1 + Y_2$, which is just a regular normally distributed random variable.

Consider a more general case where the mixture distribution has L components. Assume $Y = \sum_{j=1}^{L} I(\delta = j) \cdot Y_j$, where δ is a categorical random variable $\delta \in \{1, 2, \ldots, L\}$ and $I(\cdot)$ is the indicator function. Then using the law of total probability, we can get

$$
\begin{aligned}
F(y) &= \Pr(Y \le y) = \Pr(Y \le y, \delta \in \{1, 2, \ldots, L\}) = \sum_{j=1}^{L} \Pr(Y \le y, \delta = j) \\
&= \sum_{j=1}^{L} \Pr(\delta = j) \cdot \Pr(Y \le y \mid \delta = j) = \sum_{j=1}^{L} \lambda_j \cdot F_j(y).
\end{aligned}
\tag{10.14}
$$

The term λ_j is called *weights* and it is obvious that $\sum_{j=1}^{L} \lambda_j = 1$. Equation (10.14) provides the fundamental relationship between the mixture distribution $F(y)$ and the individual component distribution $F_j(\cdot)$, $j = 1, \ldots, L$. It provides a way of obtaining the CDF for a mixture random variable from the distribution of its components. The same relations hold for probability density functions. Figure 10.12 illustrates a simple example of mixture distribution with three components $\mathcal{N}(0,1)$, $\mathcal{N}(3,1)$, $\mathcal{N}(10,1)$ and corresponding weights of 0.5, 0.3, and 0.2, respectively.

Attributed to its flexibility, mixture distribution has been used widely in constructing complex shaped distribution functions and in clustering techniques, see [McLachlan and Peel, 2004; Fraley and Raftery, 1998].

The first step in utilizing the mixture distribution is to fit the model with given data. If the class information of the data, i.e., the number of components and which component the data belong to, is known, then the model fitting is straightforward. The weights will simply be the proportion of each component in the dataset and the parameters of each component distribution can be estimated using the data from that component. If the class information is unknown,

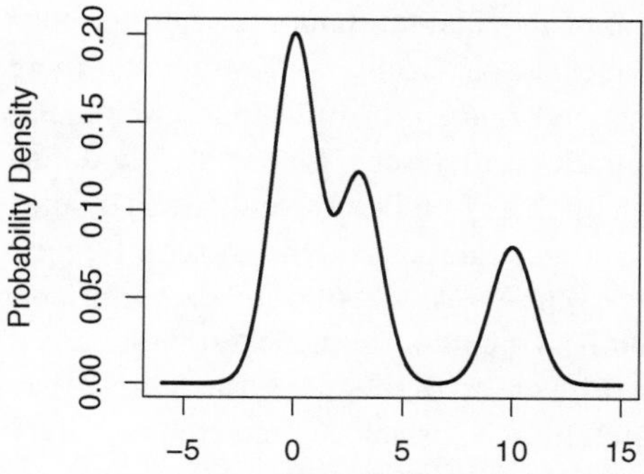

Figure 10.12 Example of a mixture distribution with three components

the mixture distribution estimation becomes a challenging problem. The maximum likelihood estimation (MLE) method can be used to estimate the model parameters. However, if the class information is unknown, a large number of parameters include the weights and the parameters for all the components need to be estimated simultaneously in the MLE procedure; this will be a very challenging high dimensional nonlinear optimization problem. Fortunately, a successive approximation procedure can be applied to the mixture model estimation problem to address this challenge. In such a procedure, we can separate the estimation of the weights and the estimation of the component distribution parameters:

1. Start from an initial guess of the weights and component parameters
2. Update the estimation of the weights given the component parameters
3. Update the estimation of the component parameters given the weights
4. Repeat steps 2 and 3 until convergence

This procedure is actually a special case of the famous expectation maximization (EM) algorithm, which is a general way of maximizing the likelihood when some variables are unobserved. In the mixture distribution estimation problem, if the class information is unknown, then we only have observations on Y but the variable δ in the model (10.14) is unobservable. Thus, the EM algorithm fits very well with the mixture distribution fitting problem. Limited by the scope, we will not discuss the EM algorithm further here. The interested reader can refer to McLachlan and Krishnan [2007] for more information.

10.4.2 Mixed Effects Model with Mixture Prior for Longitudinal Data

The basic idea in addressing the data imbalance issue in degradation signal prediction is to assume the population of degradation signals is heterogeneous and use a mixture distribution to model it. Specifically, we assume that there are two sub-populations (i.e., components), one for the units that have regular useful life and one for the units that fail prematurely. We can simply define these two groups as one that survived up to a given time T_s and one that failed before T_s. We denote those two sub-populations as the survived group $\mathcal{S}=\{i:\delta_i=0\}$ and the failed group $\mathcal{F}=\{i:\delta_i=1\}$, δ_i is an indicator variable which shows if the ith unit fails before T_s. Under most practical situations, $\mathcal{S}$ is the majority group and $\mathcal{F}$ is the minority group in the historical data. We can put more emphasis on $\mathcal{F}$ while maintaining satisfactory prediction accuracy for the units in group $\mathcal{S}$ to address the data imbalance issue.

Without loss of generality, we assume the proportion of $\mathcal{S}$ in the population is λ_0 where $0 \leq \lambda_0 \leq 1$. The model for the degradation signal path can now be extended as

$$y_i(t) = I(\delta_i = 0)\{\mathbf{z}^T(t)\mathbf{b}_{i,0} + \epsilon_{i,0}\} + I(\delta_i = 1)\{\mathbf{z}^T(t)\mathbf{b}_{i,1} + \epsilon_{i,1}\}, \quad (10.15)$$

where $I(\cdot)$ is an indicator function. $\mathbf{b}_{i,0}$ and $\mathbf{b}_{i,1}$ are the random effects corresponding to group $\mathcal{S}$ and $\mathcal{F}$, respectively. They can be assumed to follow normal distribution. With slight abuse of notation, we denote $\mathbf{b}_{i,0} \sim \mathcal{N}(\boldsymbol{\mu}_0, \boldsymbol{\Sigma}_0)$ and $\mathbf{b}_{i,1} \sim \mathcal{N}(\boldsymbol{\mu}_1, \boldsymbol{\Sigma}_1)$. The error term is assumed to follow zero mean normal distribution as $\epsilon_{i,0} \sim \mathcal{N}(0, \sigma^2_{\epsilon 0})$ and $\epsilon_{i,1} \sim \mathcal{N}(0, \sigma^2_{\epsilon 1})$. In addition, we assume δ_i is a Bernoulli random variable with its probability mass function defined as $\Pr(\delta_i = 0) = \lambda_0$ and $\Pr(\delta_i = 1) = 1 - \lambda_0$.

It is clear from (10.15) that the degradation signal $y_i(t)$ is defined by $\mathbf{b}_{i,0}$ and $\epsilon_{i,0}$ if $i \in \mathcal{S}$; and by $\mathbf{b}_{i,1}$ and $\epsilon_{i,1}$ if $i \in \mathcal{F}$. From the property of mixture distribution, it is straightforward to get the probability density function for $y_i(t)$ as

$$\begin{aligned} f(y_i(t)) = {} & \lambda_0 f_N(y_i(t); \mathbf{z}^T(t)\boldsymbol{\mu}_0, \mathbf{z}^T(t)\boldsymbol{\Sigma}_0\mathbf{z}(t) + \sigma^2_{\epsilon 0}) \\ & + (1 - \lambda_0) f_N(y_i(t); \mathbf{z}^T(t)\boldsymbol{\mu}_1, \mathbf{z}^T(t)\boldsymbol{\Sigma}_1\mathbf{z}(t) + \sigma^2_{\epsilon 1}) \end{aligned}, \quad (10.16)$$

where $f_N(y; \mu, \sigma)$ is the normal density function with mean and variance as μ and σ, respectively. Clearly, $y_i(t)$ follows a mixture distribution corresponding to the two groups. For $i \in \mathcal{S}$, $y_i(t) \sim \mathcal{N}(\mathbf{z}^T(t)\boldsymbol{\mu}_0, \mathbf{z}^T(t)\boldsymbol{\Sigma}_0\mathbf{z}(t) + \sigma^2_{\epsilon 0})$; otherwise, $y_i(t) \sim \mathcal{N}(\mathbf{z}^T(t)\boldsymbol{\mu}_1, \mathbf{z}^T(t)\boldsymbol{\Sigma}_1\mathbf{z}(t) + \sigma^2_{\epsilon 1})$. The mean and variance of $y_i(t)$ can be easily obtained based on the linear relationship between $y_i(t)$ and the random effects in (10.15) within each group.

The unknown parameters in the model (10.16) can be collectively denoted as $\boldsymbol{\Psi} = \{\lambda_0, \boldsymbol{\Psi}_0, \boldsymbol{\Psi}_1\}$, where $\boldsymbol{\Psi}_0 = \{\boldsymbol{\mu}_0, \boldsymbol{\Sigma}_0, \sigma^2_{\epsilon 0}\}$, and $\boldsymbol{\Psi}_1 = \{\boldsymbol{\mu}_1, \boldsymbol{\Sigma}_1, \sigma^2_{\epsilon 1}\}$. These parameters can be estimated using historical data. In practice, we will typically know if a unit is in the survived or the failed group. So the class information for each historical data point is known. As a result, the estimation of $\boldsymbol{\Psi}$ is straightforward. Specifically, assume we have a historical dataset collected from m units. For each unit i, we observe its degradation signal $\mathbf{y}_i$ and the indicator of survival δ_i. Then the weight parameter λ_0 can be easily estimated as

$$\hat{\lambda}_0 = \sum_{i=1}^{m} (1 - \delta_i)/m,$$

which is the conventional maximum likelihood estimator for Bernoulli distribution. To estimate $\boldsymbol{\Psi}_0$ and $\boldsymbol{\Psi}_1$, we can use the restricted maximum likelihood approach for mixed effects model based on the data from corresponding groups. In other words, we can estimate $\boldsymbol{\Psi}_0$ using $\{\mathbf{y}_i, i \in \mathcal{S}\}$ and estimate

$\boldsymbol{\Psi}_1$ using $\{\mathbf{y}_i, i \in \mathcal{F}\}$. Collectively, we can denote the estimated parameters as $\widehat{\boldsymbol{\Psi}} = \{\hat{\lambda}_0, \widehat{\boldsymbol{\Psi}}_0, \widehat{\boldsymbol{\Psi}}_1\}$.

Similar to the treatment in Section 10.2, we can use a Bayesian approach to make a prediction for a new unit if some observations from the new unit is available. We can use the estimated $\widehat{\boldsymbol{\Psi}}$ as the prior information. Using the same notation as that used in (10.2), we can obtain the predicted distribution of $y_p(t)$ for a new unit p given some observations of $y_p(t)$ up to t^*, $t^* \le t$, denoted as $\mathbf{y}_p^*$

$$\begin{aligned} f(y_p(t) \mid \mathbf{y}_p^*) = {} & \hat{\lambda}_0^* f_N(y_i(t); \mathbf{z}^T(t)\hat{\boldsymbol{\mu}}_{0,p}^*, \mathbf{z}^T(t)\widehat{\boldsymbol{\Sigma}}_{0,p}^*\mathbf{z}(t) + \hat{\sigma}_{\epsilon 0}^2) \\ & + (1-\hat{\lambda}_0^*) f_N(y_i(t); \mathbf{z}^T(t)\hat{\boldsymbol{\mu}}_{1,p}^*, \mathbf{z}^T(t)\widehat{\boldsymbol{\Sigma}}_{1,p}^*\mathbf{z}(t) + \hat{\sigma}_{\epsilon 1}^2), \end{aligned} \quad (10.17)$$

where $\hat{\lambda}_0^*$, $\hat{\boldsymbol{\mu}}_{0,p}^*$, $\widehat{\boldsymbol{\Sigma}}_{0,p}^*$, $\hat{\boldsymbol{\mu}}_{1,p}^*$, $\widehat{\boldsymbol{\Sigma}}_{1,p}^*$ are the updated weight parameter and the component distribution parameters. If we define $\mathbf{a}_0 = \mathbf{Z}_p^* \hat{\boldsymbol{\mu}}_0$, $\mathbf{b}_0 = \mathbf{Z}_p^* \widehat{\boldsymbol{\Sigma}}_0 \mathbf{Z}_p^{*T} + \hat{\sigma}_{\epsilon 0}^2 \mathbf{I}_{n_p}$, $\mathbf{a}_1 = \mathbf{Z}_p^* \hat{\boldsymbol{\mu}}_1$, $\mathbf{b}_1 = \mathbf{Z}_p^* \widehat{\boldsymbol{\Sigma}}_1 \mathbf{Z}_p^{*T} + \hat{\sigma}_{\epsilon 1}^2 \mathbf{I}_{n_p}$, where $\mathbf{Z}_p^*$ is defined in (10.2), we have

$$\begin{aligned} \hat{\lambda}_{0,p}^* &= \frac{\hat{\lambda}_0^* f_N(\mathbf{y}_p^*; \mathbf{a}_0, \mathbf{b}_0)}{\hat{\lambda}_0^* f_N(\mathbf{y}_p^*; \mathbf{a}_0, \mathbf{b}_0) + (1-\hat{\lambda}_0^*) f_N(\mathbf{y}_p^*; \mathbf{a}_1, \mathbf{b}_1)} \\ \hat{\boldsymbol{\mu}}_{0,p}^* &= \widehat{\boldsymbol{\Sigma}}_{0,p}^* \left[\mathbf{Z}_p^{*T} \mathbf{y}_p^* / \hat{\sigma}_{\epsilon 0}^2 + \widehat{\boldsymbol{\Sigma}}_0^{-1} \hat{\boldsymbol{\mu}}_0 \right] \\ \widehat{\boldsymbol{\Sigma}}_{0,p}^* &= \left[\widehat{\boldsymbol{\Sigma}}_0^{-1} + \mathbf{Z}_p^{*T} \mathbf{Z}_p^* / \hat{\sigma}_{\epsilon 0}^2 \right]^{-1} \\ \hat{\boldsymbol{\mu}}_{1,p}^* &= \widehat{\boldsymbol{\Sigma}}_{1,p}^* \left[\mathbf{Z}_p^{*T} \mathbf{y}_p^* / \hat{\sigma}_{\epsilon 1}^2 + \widehat{\boldsymbol{\Sigma}}_1^{-1} \hat{\boldsymbol{\mu}}_1 \right] \\ \widehat{\boldsymbol{\Sigma}}_{1,p}^* &= \left[\widehat{\boldsymbol{\Sigma}}_1^{-1} + \mathbf{Z}_p^{*T} \mathbf{Z}_p^* / \hat{\sigma}_{\epsilon 1}^2 \right]^{-1}. \end{aligned} \quad (10.18)$$

The proof of this result can be found in the Appendix of this chapter. With the updated parameters, we can predict the future values of y_p beyond t^* and similar to the treatment in Section 10.3, we can use the prediction to estimate the RUL of the unit. Specifically, for a new unit p, we can predict its RUL at t^* using the updated parameters in (10.18). Let D be the failure threshold and T_p be the failure time of unit p after t^*, we have $T_p = \min\{t | y_p(t) \ge D\} - t^*$. Then the CDF of T_p, $F_p(t \mid t^*)$ can be derived as

$$F_p(t|t^*) = \frac{\hat{\lambda}_{0,p}^* \Phi(\hat{y}_{0,p}^0(t+t^*)) + (1-\hat{\lambda}_{0,p}^*)\Phi(\hat{y}_{1,p}^0(t+t^*)) - \omega(t^*)}{1-\omega(t^*)}, \quad (10.19)$$

where $\Phi(\cdot)$ is the CDF of a standard normal random variable and

$$\hat{y}^0_{0,p}(t) = (\mathbf{z}^T(t)\hat{\boldsymbol{\mu}}^*_{0,p} - D) / \sqrt{\mathbf{z}^T(t)\boldsymbol{\Sigma}^*_{0,p}\mathbf{z}(t) + \hat{\sigma}^2_{\epsilon 0}}$$
$$\hat{y}^0_{1,p}(t) = (\mathbf{z}^T(t)\hat{\boldsymbol{\mu}}^*_{1,p} - D) / \sqrt{\mathbf{z}^T(t)\boldsymbol{\Sigma}^*_{1,p}\mathbf{z}(t) + \hat{\sigma}^2_{\epsilon 1}}$$
$$\omega(t) = \hat{\lambda}^*_{0,p}\Phi(\hat{y}^0_{0,p}(t)) + (1 - \hat{\lambda}^*_{0,p})\Phi(\hat{y}^0_{1,p}(t)).$$

From the above CDF, the pdf can be obtained through a derivative operation. The proof of the above result can be found in the Appendix of this chapter. It needs to be pointed out that, although the expressions look complex, everything can be computed based on closed form expressions. It is also interesting to note that the result in Section 10.3 is a special case of the above result achieved by setting λ_0 either to 0 or 1.

Based on (10.19), the survival function for the unit p can be easily obtained as $S_p(t|t^*) = 1 - F_p(t|t^*)$. The estimator for the RUL could either be the mean of T_p or the median of T_p. Both are used in practice. It is generally agreed that for a severely skewed distribution, the median is a more robust estimator than the mean [Liu et al., 2013]. If we use the mean of T_p as the RUL, we can compute $\text{RUL} = \int_0^\infty (1 - F_p(t \mid t^*))dt$. If we use the median as the estimator of the RUL, then $\text{RUL} = F_p^{-1}(0.5 \mid t^*)$. We can compute RUL by numerically solving the equation $F_p(t|t^*) = 0.5$ with respect to $t \geq t^*$. In the following, we use an example to demonstrate and provide more insight into this method.

Example 10.8 Consider the data shown in Figure 10.13 below. This dataset simulates the degradation signal from 1000 units. The dataset clearly exhibits grouping structure: there is a small number of units that degrade faster than the majority group and will fail early (i.e., reaching the failure threshold early, which is set as 40 in the figure).

If we ignore the grouping structure and use a single normal distribution as the distribution of the random effects, we can use the techniques introduced in Sections 10.1, 10.2, and 10.3 by adopting a mixed effects model to describe the data, estimate the parameters for a new unit, and finally obtain the prediction of failure time distribution. If we take the grouping structure into consideration, then we can use a mixture distribution as the distribution of the random effects and use the techniques introduced in this section to model the data and predict the failure distribution. In this example, we will compare the performance of these two methods to illustrate the importance of considering the grouping structure. The dataset in Figure 10.13 consists of the degradation signal from 1000 units, among which 100 are units with fast degradation and premature failure. We can use the data to fit a quadratic polynomial mixed effects model as shown in (10.1) with single normal distribution as the prior distribution,

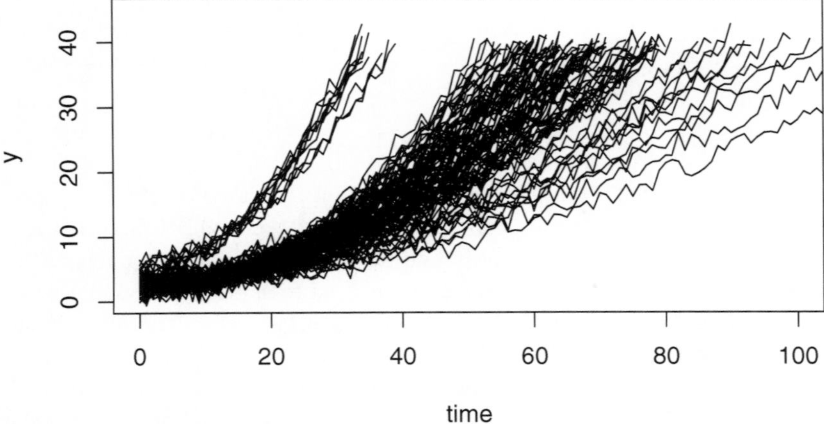

Figure 10.13 Degradation signals with grouping structure

and a similar mixed effects model but with mixture normal distribution as the prior distribution as shown in (10.15). The estimated parameters are presented in Table 10.4.

Using the prior distributions estimated from the available data, we can obtain the parameters for the degradation signal for a new unit p using the Bayesian updating approach introduced in Section 10.2 and this section, respectively. Figure 10.14 illustrates the prediction performance for a fast degrading unit using both a single and a normal mixture distribution as prior distribution for the random effects. Figure 10.15 illustrates the corresponding prediction performance using both methods for a unit in the majority group.

Based on Figures 10.14 and 10.15 we can obtain many important insights. First, the model with mixture prior updates its parameters and gets close to the true degradation curve significantly faster than the regular model due to its flexible model structure. In other words, the predicted function based on mixture prior is more sensitive to the newly collected degradation signals. Second, we can see that the regular model severely overestimates the predicted failure for the fast degrading unit. This is because the offline parameter estimates for the regular model strongly reflects the behavior of the majority group. On the other hand, the model with mixture prior quickly adjusts itself to accommodate the potential risk of fast degradation. This feature is very desirable in practice because it enables early detection for the premature failure in an efficient manner despite the imbalanced historical data.

With the prediction of the degradation signals, the failure time distribution (or the survival function) of the unit can be obtained. The predicted survival

Table 10.4 Estimated parameters using the same dataset under different models

	Model with regular normal prior	Model with mixture prior (failed component)	Model with mixture prior (survived component)
$\hat{\mathbf{b}}$	$\begin{pmatrix}2.75 & 0.02 & 0.009\end{pmatrix}^T$	$\begin{pmatrix}4.93 & 0.10 & 0.018\end{pmatrix}^T$	$\begin{pmatrix}2.51 & 0.01 & 0.008\end{pmatrix}^T$
$\hat{\boldsymbol{\Sigma}}_{\mathbf{b}}$	$\begin{pmatrix}0.73 & 0.017 & 0.0022 \\ 0.017 & 5.1\times10^{-4} & 7.5\times10^{-5} \\ 0.0022 & 7.5\times10^{-5} & 1.5\times10^{-5}\end{pmatrix}$	$\begin{pmatrix}0.15 & -0.0020 & 1.97\times10^{-4} \\ -0.0020 & 0.000055 & 7.1\times10^{-6} \\ 1.97\times10^{-4} & 7.1\times10^{-6} & 6.3\times10^{-6}\end{pmatrix}$	$\begin{pmatrix}0.15 & 7.8\times10^{-4} & -9.7\times10^{-5} \\ 7.8\times10^{-4} & 2.6\times10^{-5} & 5.0\times10^{-6} \\ -9.7\times10^{-5} & 5.0\times10^{-6} & 6.1\times10^{-6}\end{pmatrix}$
$\hat{\sigma}^2$	0.998	0.998	1.008
$\hat{\lambda}_0$	-	0.1	0.9

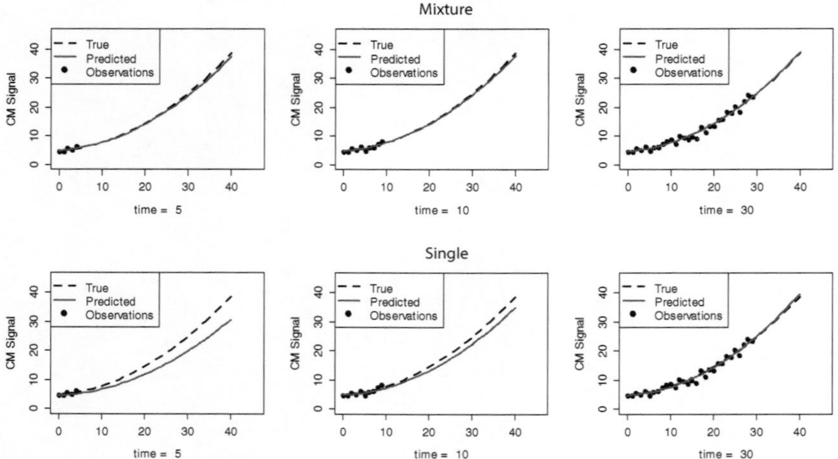

Figure 10.14 Prediction of degradation for a fast degrading unit. The upper row uses mixture prior distribution while the lower row uses a regular normal prior distribution.

function for these two units under these two different models are shown in Figures 10.16 and 10.17, respectively. The true underlying failure time is shown by the solid vertical line while the median point estimator from both methods

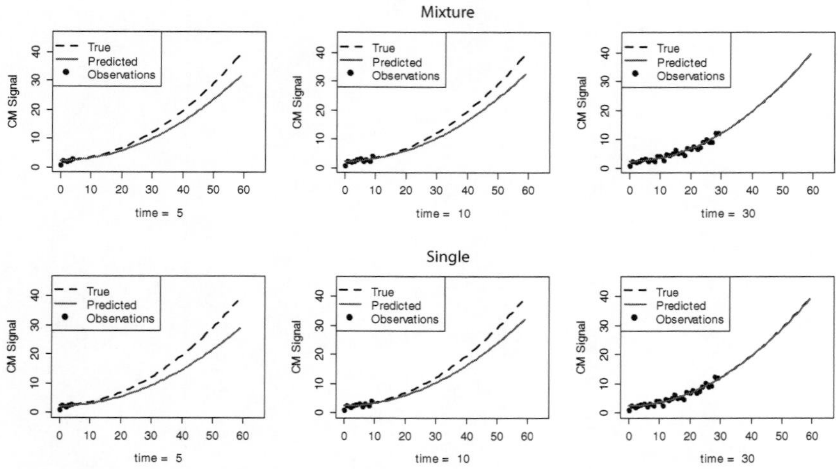

Figure 10.15 Prediction of degradation for a unit in the majority group. The upper row uses mixture prior distribution while the lower row uses a regular normal prior distribution.

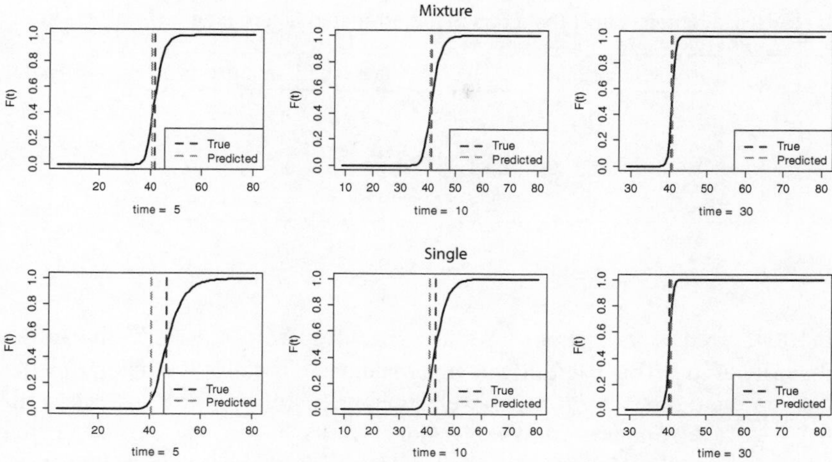

Figure 10.16 Prediction of failure distribution for a fast degrading unit. The upper row uses mixture prior distribution while the lower row uses a regular normal prior distribution.

is shown by the dash vertical line. Once again the survival curves confirm the flexibility of the mixture based model and its ability to quickly adjust its model parameters for fast degradation which might lead to premature failure. We note that, the enhanced prediction accuracy obtained by using mixture prior in

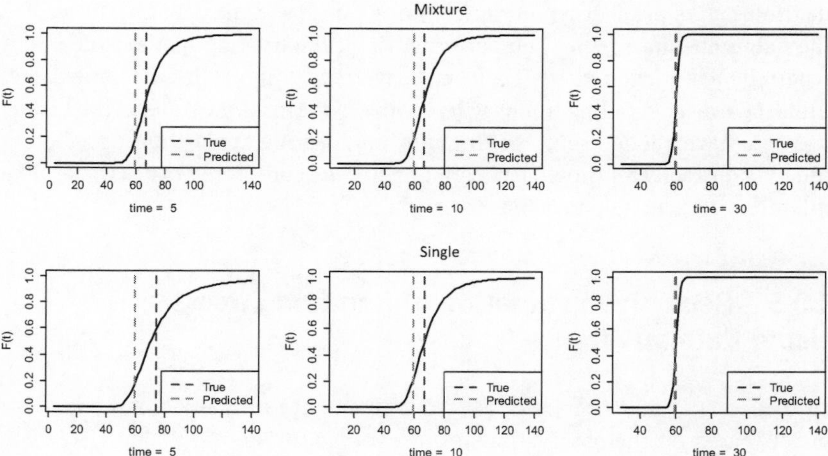

Figure 10.17 Prediction of failure distribution for a unit in the majority group. The upper row uses mixture prior distribution while the lower row uses a regular normal prior distribution.

Table 10.5 Comparison of the errors in the estimated failure time

	Regular model			Model with mixture prior		
	t_1^*	t_2^*	t_3^*	t_1^*	t_2^*	t_3^*
Unit 1	6.12	2.60	0.60	0.85	0.54	0.24
Unit 2	10.81	6.95	0.35	7.55	6.52	0.15

the RUL prediction is larger for a fast degrading unit than for a unit in the majority group. This is intuitively understandable because the regular model can represent the majority group well enough. Therefore, if the in-service unit behaves in a similar way to the majority units in the historical data, the performance gain may not be noticeable. However, for the fast degrading units, the model with mixture prior takes less time to adjust the model to accommodate the unexpected fast increasing trend in the degradation signals.

With the predicted failure time distribution, we can obtain the point estimation of the unit failure time using the median of the failure time distribution. Table 10.5 shows the error of the point estimation of the failure time of two units with these two models, respectively. We can see that the model with mixture prior outperforms the regular model for both failed and survived units specifically at earlier stages when few data are available.

The unique feature of the model with mixture prior is that it uses a mixture distribution as prior distribution for the online Bayesian updating to resolve the data imbalance issue. The performance of the existing methods degrades when the historical data suffers from a severe data imbalance, i.e., only a few units have fast degrading trends whereas the majority of the units in the historical data have a long useful life. In many engineering applications, predicting the RUL correctly for those fast degrading units is considered to be crucial. The mixture prior can help to achieve this goal.

10.5 Recursive Estimation of Random Effects Using Kalman Filter

In Section 10.2, equation (10.5) is used to update the estimation of the random effects based on the observations from a unit p. It is obvious that (10.5) is a "batch" formula, which means that all the observations $\mathbf{y}_p^*$ appear in the formula. With the accumulation of observations, the complexity of the calculation in (10.5) will increase. In fact, the updates of random effects can be achieved recursively through the Kalman filtering framework. In this section, we will first introduce the basics of the Kalman filter, then we will introduce how to realize (10.5) recursively.

10.5.1 Introduction to the Kalman Filter

The Kalman filter is a powerful mathematical tool for system state tracking and prediction [Kalman, 1960], and it is the optimal estimator for a large class of problems. The Kalman filter can be derived under different frameworks but here we provide a Bayesian approach, which will be easy to follow and compare with the procedure we presented in Section 10.2.

The standard Kalman filter formulation concerns the estimation of the unobservable system states $\mathbf{q}_k$ from the system observations $\mathbf{y}_k$ of the following linear system:

$$\begin{aligned} \mathbf{q}_k &= \mathbf{A}_{k-1}\mathbf{q}_{k-1} + \mathbf{w}_k \\ \mathbf{y}_k &= \mathbf{H}_k\mathbf{q}_k + \mathbf{v}_k \end{aligned}, \tag{10.20}$$

where $\mathbf{q}_k$ is the state vector, $\mathbf{y}_k$ is the observation vector, $\mathbf{A}_k$ and $\mathbf{H}_k$ are coefficient matrices, and the subscript k is the discrete time index. $\mathbf{w}_k$ and $\mathbf{v}_k$ represent process error and observation error, respectively. The error terms are independent and identically distributed, i.e., $\mathbf{w}_i$ is independent of $\mathbf{w}_j$ for any $i \neq j$ and similarly for $\mathbf{v}_i$ and $\mathbf{v}_j$. This model is called the linear state space model. The first equation in (10.20) is the state transition equation and the second equation in (10.20) is called the observation equation.

From the model in (10.20), it is straightforward to get the following *Markovian* property,

$$\begin{aligned} f(\mathbf{q}_k \mid \mathbf{q}_{1:k\text{-}1}, \mathbf{y}_{1:k\text{-}1}) &= f(\mathbf{q}_k \mid \mathbf{q}_{k\text{-}1}) \\ f(\mathbf{q}_{k\text{-}1} \mid \mathbf{q}_{k:T}, \mathbf{y}_{k:T}) &= f(\mathbf{q}_{k-1} \mid \mathbf{q}_k) \end{aligned}. \tag{10.21}$$

The notation $(\cdot)_{i:j}$ represents the aggregated data from time instance i to j, where $i \leq j$. The above property essentially says that the system state $\mathbf{q}_k$ is only dependent on the immediate past $\mathbf{q}_{k-1}$ and the immediate future state $\mathbf{q}_{k+1}$, and is not dependent on other states. From the observation equation in (10.20), we can also have

$$f(\mathbf{y}_k \mid \mathbf{q}_{1:k}, \mathbf{y}_{1:k-1}) = f(\mathbf{y}_k \mid \mathbf{q}_k). \tag{10.22}$$

The basic problem we want to solve is to estimate the current unobservable state $\mathbf{q}_k$ based on the observations up to time instance k, $\mathbf{y}_{1:k}$. Using the Bayesian approach, we need to obtain the posterior distribution $f(\mathbf{q}_k \mid \mathbf{y}_{1:k})$ and the mean value of the distribution can then be used as a point estimation of $\mathbf{q}_k$. To compute $f(\mathbf{q}_k \mid \mathbf{y}_{1:k})$, we have two steps:

(1) The prediction step. First, we note that

$$f(\mathbf{q}_k|\mathbf{y}_{1:k-1})=\int f(\mathbf{q}_k,\mathbf{q}_{k-1}|\mathbf{y}_{1:k-1})d\mathbf{q}_{k-1}.$$

However, noting the chain rule of probability $f(A,B)=f(A|B)\cdot f(B)$ and the Markovian property, we have

$$\begin{aligned}f(\mathbf{q}_k,\mathbf{q}_{k-1}|\mathbf{y}_{1:k-1})&=f(\mathbf{q}_k|\mathbf{q}_{k-1},\mathbf{y}_{1:k-1})f(\mathbf{q}_{k-1}|\mathbf{y}_{1:k-1})\\&=f(\mathbf{q}_k|\mathbf{q}_{k-1})f(\mathbf{q}_{k-1}|\mathbf{y}_{1:k-1}).\end{aligned}$$

Combining the above two equations, we can get the prediction expression, also called Chapman–Kolmogorov equation

$$f(\mathbf{q}_k|\mathbf{y}_{1:k-1})=\int f(\mathbf{q}_k|\mathbf{q}_{k-1})f(\mathbf{q}_{k-1}|\mathbf{y}_{1:k-1})d\mathbf{q}_{k-1}. \tag{10.23}$$

(2) The update step. Using the Bayesian rule and noting that the result that is true for unconditional probability remains true if everything is conditioned on the same event, we have

$$\begin{aligned}f(\mathbf{q}_k|\mathbf{y}_{1:k})=f(\mathbf{q}_k|\mathbf{y}_k,\mathbf{y}_{1:k-1})&=\frac{f(\mathbf{y}_k|\mathbf{q}_k,\mathbf{y}_{1:k-1})f(\mathbf{q}_k|\mathbf{y}_{1:k-1})}{f(\mathbf{y}_k|\mathbf{y}_{1:k-1})}\\&=\frac{f(\mathbf{y}_k|\mathbf{q}_k)f(\mathbf{q}_k|\mathbf{y}_{1:k-1})}{f(\mathbf{y}_k|\mathbf{y}_{1:k-1})},\end{aligned} \tag{10.24}$$

where $f(\mathbf{y}_k|\mathbf{q}_k)$ is the likelihood of the observations $\mathbf{y}_k$ given the state $\mathbf{q}_k$ and the denominator is

$$f(\mathbf{y}_k|\mathbf{y}_{1:k-1})=\int f(\mathbf{y}_k,\mathbf{q}_k|\mathbf{y}_{1:k-1})d\mathbf{q}_k=\int f(\mathbf{y}_k|\mathbf{q}_k)f(\mathbf{q}_k|\mathbf{y}_{1:k-1})d\mathbf{q}_k.$$

The prediction step (10.23) and the update step (10.24) provide a recursive Bayesian filtering scheme. The recursion starts from a pre-specified prior distribution $f(\mathbf{q}_0)$ and for $k=1$, we define $\mathbf{y}_{1:k-1}$ as the null set and thus $f(\cdot\,|\,\mathbf{y}_{1:k-1})=f(\cdot)$. Then we can invoke (10.23) and (10.24) iteratively to compute $f(\mathbf{q}_k\,|\,\mathbf{y}_{1:k})$ as the observation $\mathbf{y}_k$ becoming available sequentially.

The prediction step (10.23) and the update step (10.24) are generally true as long as the properties in (10.21) and (10.22) hold. Under specific cases, closed form expressions without integral can be obtained. In the standard Kalman filter framework, we assume the error terms $\mathbf{w}_k$, $\mathbf{v}_k$, and the system starting state $\mathbf{q}_0$ all follow normal distribution. Specifically, we assume $\mathbf{w}_k\sim\mathcal{N}(\mathbf{0},\boldsymbol{\Sigma}_{\mathbf{w}_k})$ and $\mathbf{v}_k\sim\mathcal{N}(\mathbf{0},\boldsymbol{\Sigma}_{\mathbf{v}_k})$. With these assumptions, the state space model in (10.20) is also called the *Gauss–Markov model*. From the state transition equation and the observation equation, we can immediately get

$$f(\mathbf{q}_k|\mathbf{q}_{k-1}) = f_{\mathcal{N}}(\mathbf{q}_k; \mathbf{A}_{k-1}\mathbf{q}_{k-1}, \boldsymbol{\Sigma}_{\mathbf{w}_k})$$
$$f(\mathbf{y}_k|\mathbf{q}_k) = f_{\mathcal{N}}(\mathbf{y}_k; \mathbf{H}_k\mathbf{q}_k, \boldsymbol{\Sigma}_{\mathbf{v}_k}) \quad .$$

Also, with the distribution assumptions on $\mathbf{q}_0$, $\mathbf{w}_k$, and $\mathbf{v}_k$, we can see that $\mathbf{q}_k$ and $\mathbf{y}_k$ are linear combinations of normally distributed random variables and thus $\mathbf{q}_k$ and $\mathbf{y}_k$ will also follow normal distributions.

Based on the properties of normal distribution and through some algebraic manipulation, we can get the prediction step as

$$f(\mathbf{q}_k \mid \mathbf{y}_{1:k-1}) = f_{\mathcal{N}}(\mathbf{q}_k; \mathbf{A}_{k-1}\boldsymbol{\mu}_{\mathbf{q}_{k-1}}, \mathbf{A}_{k-1}\boldsymbol{\Sigma}_{\mathbf{q}_{k-1}}\mathbf{A}_{k-1}^T + \boldsymbol{\Sigma}_{\mathbf{w}_k}). \tag{10.25}$$

The updating step can be obtained as

$$\begin{aligned} \boldsymbol{\mu}_{\mathbf{q}_k} &= \boldsymbol{\mu}_{\mathbf{q}_k}^- + \mathbf{K}_k(\mathbf{y}_k - \mathbf{H}_k\boldsymbol{\mu}_{\mathbf{q}_k}^-) \\ \boldsymbol{\Sigma}_{\mathbf{q}_k} &= (\mathbf{I} - \mathbf{K}_k\mathbf{H}_k)\boldsymbol{\Sigma}_{\mathbf{q}_k}^- \end{aligned} \quad , \tag{10.26}$$

where $\boldsymbol{\mu}_{\mathbf{q}_k}^-$ is defined as $\mathbf{A}_{k-1}\boldsymbol{\mu}_{\mathbf{q}_{k-1}}$, $\boldsymbol{\Sigma}_{\mathbf{q}_k}^-$ is defined as $\mathbf{A}_{k-1}\boldsymbol{\Sigma}_{\mathbf{q}_{k-1}}\mathbf{A}_{k-1}^T + \boldsymbol{\Sigma}_{\mathbf{w}_k}$, $(\mathbf{y}_k - \mathbf{H}_k\boldsymbol{\mu}_{\mathbf{q}_k}^-)$ is called the *measurement residue*, $\mathbf{K}_k$, called the *Kalman gain*, defined as

$$\mathbf{K}_k = \boldsymbol{\Sigma}_{\mathbf{q}_k}^- \mathbf{H}_k^T(\mathbf{H}_k\boldsymbol{\Sigma}_{\mathbf{q}_k}^-\mathbf{H}_k^T + \boldsymbol{\Sigma}_{\mathbf{v}_k})^{-1},$$

and $\mathbf{I}$ is the identity matrix with the appropriate dimension.

Combining prediction step (10.25) and the update step (10.26), we can obtain the following recursive Kalman filtering scheme:

1. Initialization: set $\mathbf{q}_0 \sim \mathcal{N}(\boldsymbol{\mu}_{\mathbf{q}0}, \boldsymbol{\Sigma}_{\mathbf{q}0})$
2. Prediction:

$$\begin{aligned} \boldsymbol{\mu}_{\mathbf{q}_k}^- &= \mathbf{A}_{k-1}\,\boldsymbol{\mu}_{\mathbf{q}_{k-1}} \\ \boldsymbol{\Sigma}_{\mathbf{q}_k}^- &= \mathbf{A}_{k-1}\,\boldsymbol{\Sigma}_{\mathbf{q}_{k-1}}\mathbf{A}_{k-1}^T + \boldsymbol{\Sigma}_{\mathbf{w}_k}. \end{aligned}$$

3. Update:

$$\begin{aligned} \boldsymbol{\mu}_{\mathbf{q}_k} &= \boldsymbol{\mu}_{\mathbf{q}_k}^- + \mathbf{K}_k(\mathbf{y}_k - \mathbf{H}_k\boldsymbol{\mu}_{\mathbf{q}_k}^-) \\ \boldsymbol{\Sigma}_{\mathbf{q}_k} &= (\mathbf{I} - \mathbf{K}_k\mathbf{H}_k)\boldsymbol{\Sigma}_{\mathbf{q}_k}^-. \end{aligned}$$

With this algorithm, we can simply take $\boldsymbol{\mu}_{\mathbf{q}_k}$ as an estimation of underlying state $\hat{\mathbf{q}}_k$. The Kalman filter has several nice properties:

- In the above derivation, all the terms have subscript time index, which means that these terms could be time variant. If the model is time-invariant, then the Kalman filter procedure can be further simplified. Indeed, if $\mathbf{A}_k$, $\mathbf{H}_k$, $\mathbf{\Sigma}_{\mathbf{w}_k}$, and $\mathbf{\Sigma}_{\mathbf{v}_k}$ are all fixed and do not change with k, then the Kalman gain will converge to a constant matrix, i.e., $\mathbf{K}_k \to \mathbf{K}$ and the filter becomes $\boldsymbol{\mu}_{\mathbf{q}_k} = (\mathbf{A} - \mathbf{KHA})\boldsymbol{\mu}_{\mathbf{q}_{k-1}} + \mathbf{K}\mathbf{y}_k$.
- It is noteworthy that the covariance update (the second equation in the update step) does not depend on the observation $\mathbf{y}_k$ and thus can be computed offline and stored. This feature can certainly speed up the online update step and make the Kalman filter fit well with real time applications.
- To implement the Kalman filter, we will need to specify system parameters $\mathbf{A}_k$, $\mathbf{H}_k$, the error term covariance matrices $\mathbf{\Sigma}_{\mathbf{w}_k}$ and $\mathbf{\Sigma}_{\mathbf{v}_k}$, and the initial state $\mathbf{q}_0$. If the specified values are the correct ones, then Kalman filtering is the optimal linear estimator of the underlying state for the Gauss–Markov model (10.20) in the senses of minimum mean squared error, maximum likelihood, and maximum a posteriori.
- When we implement the Kalman filter, we often do not know the values of $\mathbf{\Sigma}_{\mathbf{w}_k}$ and $\mathbf{\Sigma}_{\mathbf{v}_k}$, and the initial state $\mathbf{q}_0$ distribution. These values need to be tuned to get the best performance from the filter. Typical to most Bayesian data analysis procedures, the influence of the values of the distribution of the initial state $\mathbf{q}_0$ on the estimation $\hat{\mathbf{q}}_k$ will diminish when k gets larger. $\mathbf{\Sigma}_{\mathbf{w}_k}$ reflects the modeling error and $\mathbf{\Sigma}_{\mathbf{v}_k}$ reflects the observation error. In some specific cases, $\mathbf{\Sigma}_{\mathbf{w}_k}$ can be set to zero but $\mathbf{\Sigma}_{\mathbf{v}_k}$ should always be non-zero.

10.5.2 Random Effects Estimation Using the Kalman Filter

The estimation of random effects for an individual unit (i.e., (10.5)) can be achieved through the recursive Kalman filter. To see this clearly, we can first re-write the model in (10.2) into a state space form as

$$\begin{aligned} \mathbf{b}_k &= \mathbf{b}_{k-1} \\ y_k &= \mathbf{z}_k^T \mathbf{b}_k + \epsilon_k. \end{aligned} \tag{10.27}$$

For the sake of notation simplicity, we omit the subscript p and the superscript * in (10.27). Also, please note that the observation equation in (10.27) is the scalar expression corresponding to the vector expression (10.2), where y_k and ϵ_k are the kth element of $\mathbf{y}_p^*$ and ϵ_p^*, respectively, and $\mathbf{z}_k^T$ is the k th row of $\mathbf{Z}_p^*$ and $\mathbf{b}_i$, $i = 1 \dots k$, are simply $\mathbf{b}_p$.

It is obvious that the model (10.27) is a degenerated state space model where $\mathbf{A}_k$ is an identity matrix $\mathbf{I}$, $\mathbf{\Sigma}_{\mathbf{w}_k}$ is a zero matrix, and the observation is a scalar. Under this special case, the prediction step in the Kalman filter becomes trivial as $\boldsymbol{\mu}_{\mathbf{b}_k}^- = \boldsymbol{\mu}_{\mathbf{b}_{k-1}}$ and $\mathbf{\Sigma}_{\mathbf{b}_k}^- = \mathbf{\Sigma}_{\mathbf{b}_{k-1}}$.

If we set the distribution of the initial state $\mathbf{b}_0$ as the prior distribution we used in (10.2), we have the following simplified Kalman filter procedure for the model of (10.27).

1. Initialization: set $\mathbf{b}_0 \sim \mathcal{N}(\hat{\mathbf{b}}, \hat{\sigma}^2\hat{\mathbf{D}})$
2. Update:

$$\begin{aligned}
\mathbf{K}_k &= \widehat{\boldsymbol{\Sigma}}_{k-1}\mathbf{z}_k(\mathbf{z}_k^T\widehat{\boldsymbol{\Sigma}}_{k-1}\mathbf{z}_k + \sigma^2)^{-1} \\
\hat{\mathbf{b}}_k &= \hat{\mathbf{b}}_{k-1} + \mathbf{K}_k(y_k - \mathbf{z}_k^T\hat{\mathbf{b}}_{k-1}) \\
\widehat{\boldsymbol{\Sigma}}_k &= (\mathbf{I} - \mathbf{K}_k\mathbf{z}_k^T)\widehat{\boldsymbol{\Sigma}}_{k-1}
\end{aligned} \quad . \tag{10.28}$$

We can show that the results based on the recursive update in (10.28) is identical to the results obtained in (10.5).

Lemma 10.1 Assume from unit p, we have k observations up to the time instance t^*. Then the results obtained in (10.5) is identical to that obtained in (10.28), i.e., $\hat{\mathbf{b}}_p^* = \hat{\mathbf{b}}_k$ and $\widehat{\boldsymbol{\Sigma}}_{\mathbf{b}_p^*} = \widehat{\boldsymbol{\Sigma}}_k$.

The proof of this result is outlined in the Appendix of this chapter. The recursive updating procedure obviously has advantages in computation over the batch update procedure. Another advantage of the recursive procedure is that we can add certain constraints in each step of the updating step. For example, in Son et al. [2016], a constrained Kalman filtering approach is developed to model the degradation signal progression path. The Kalman filtering method enforces a set of inequality constraints on the parameter space so that the updated model parameters (i.e., $\mathbf{b}_p$) can stay in the specified region which can be justified by either physical knowledge of system degradation or empirical domain knowledge from experts. By having the additional information (constraints) borrowed from physical or empirical knowledge, the model updating method can be significantly improved. More details about the constrained Kalman filter approach can be found in Pizzinga [2012].

In `R` language, the Kalman filter algorithm is implemented in several different packages. For detailed descriptions and comparisons of these implementations, please refer to Tusell et al. [2011]. Here we adopt the `dlm` package, which has a nice interface and is easy to use. The `dlm` function in the dlm package can specify the dynamic system in (10.20) or the degenerated system in (10.27), while dimFilter function can conduct the "filtering", i.e., estimating the state based on the given data. The syntax of these two functions are quite straightforward. The inputs to dlm are FF, V, GG, W, m0, and C0, which correspond to $\mathbf{H}_k$, $\mathbf{\Sigma}_\mathbf{v}$, $\mathbf{A}_k$, $\mathbf{\Sigma}_\mathbf{w}$, and the initial state mean and variance, respectively, in (10.20). The output of this function is a dlm model object. The main inputs to dlmFilter function are the observations and a dlm model object. This function will filter the data and provide the estimate of the states.

We can use the Kalman filter and the degenerated state space model in (10.27) to recursively estimate the random effects for a mixed effects model. Compared to the general state space model, we can easily see that for the state space model in (10.27), we have $\mathbf{A}_k$ the identity matrix, $\mathbf{\Sigma}_\mathbf{w}$ a zero matrix, $\mathbf{H}_k$ the design matrix $\mathbf{z}_k^T$, and $\mathbf{\Sigma}_\mathbf{v}$ the variance of observation error. Furthermore, we can set the initial state mean m0 and variance C0 as the estimated mean and variance of the random effects $\mathbf{b}_i$ from the historical data. Below we will illustrate a simple example of using the Kalman filter for recursive estimation of random effects.

Example 10.9 Consider the data and the mixed effects model in Example 10.3. The random effects $\mathbf{b}_i$ is a 6×1 vector. In Example 10.3, we estimated the distribution parameters of the random effects. Assume the estimated mean and variance of $\mathbf{b}_i$ from the historical data are represented by the variables MUBf and SIGMABf, and sigma2f is the estimated error term variance from the historical data, then we can use the following R code to realize the Kalman filtering for recursive estimation of the random effects.

```
state_est <- matrix(0,nrow = 10,ncol = 6)
MUB <- MUBf
SIGMAB <- SIGMABf
for(i in 1:10)
{
dlmfit<-dlm(FF=t(Zp[i,]),V=sigma2f,GG=diag(6),
        W=0*diag(6),m0=MUB,C0=t(SIGMAB))
dlmfil<-dlmFilter(testy[[i]],dlmfit)
MUB <- as.vector(dropFirst(dlmfil$m))
SIGMAB <- dlmfil$U.C[[2]] %*%
          diag(dlmfil$D.C[2,]^2)%*% t(dlmfil$U.C[[2]])
state_est[i,]<-MUB
}
```

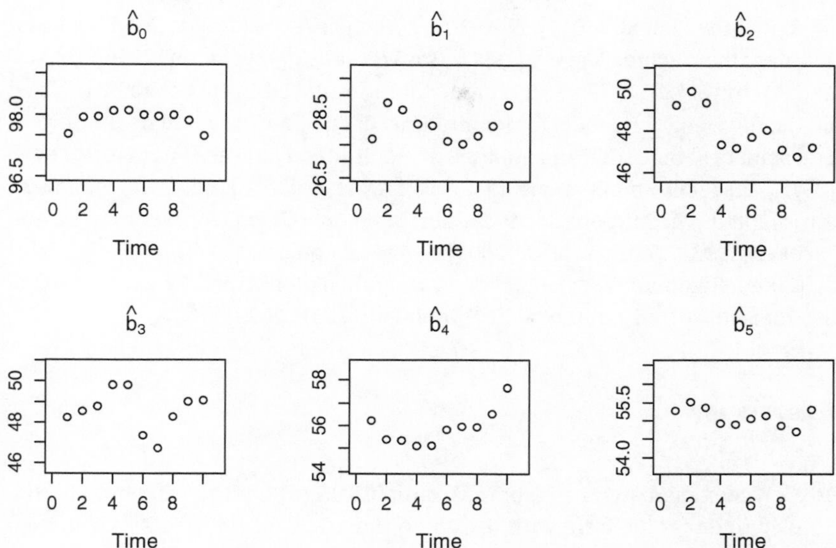

Figure 10.18 Random effects estimation using the Kalman filtering approach

In the above code, `testy` contains the observations from a new unit/cycle, `Zp` is the design matrix and each row of `Zp` contains 1 and the value of the five B-spline bases evaluated at the corresponding time instance, and i is the time index. In this code, for each time instance, we use `dlm` to set up a state space model for a given time instance and then use `dlmFilter` to conduct Kalman estimation of the state value. The line `MUB=...` and `SIGMAB=...` provide the updated mean and variance of the random effects after each observation. The estimated state values are stored in `state_est`. In Figure 10.18, we show the estimated random effects at each time instance obtained using the Kalman filter.

By comparing the results in Figure 10.18 with the "batch" estimation results using (10.5), we can see that the estimates from iterative Kalman filter are identical to that obtained through batch Bayesian updating.

Biographical Notes

Remaining useful life (or failure event) prediction is a well studied area. Extensive literature exists in the fields of system degradation analysis and failure prognosis using continuous condition monitoring/degradation signal. Excellent reviews can be found in Jardine et al. [2006], Si et al. [2011], Peng et al. [2010], Tsui et al. [2015], among others. Engineering books that cover this topic include Si et al. [2017], Niu [2017], Vachtsevanos [2006]. Good books on Kalman filter include Harvey [1990], Brown et al. [1992], Simon [2006], Grewal and Andrews [2014]. The RUL prediction introduced in this chapter is based on the mixed effects model for longitudinal data. Some excellent books on general statistical modeling and analysis of longitudinal data are Fitzmaurice et al. [2012], Hedeker and Gibbons [2006], Singer et al. [2003], Diggle et al. [2002], Taris [2000]. The interested readers can find more related techniques in Lu and Meeker [1993], Gebraeel et al. [2005], Gebraeel and Pan [2008], Chakraborty et al. [2009], Rizopoulos et al. [2014] and the publications from the author [Son et al., 2015a,b, 2016; Kontar et al., 2017a; Jahani et al., 2020].

Exercises

1. Consider the battery resistance Example 10.1 where the response is modeled in quadratic form with respect to time. The mixed effects model fitted on the historical data gives the following estimates:

$$\widehat{\mathbf{b}} = (-0.1\ 0.2\ 0.7)^T, \hat{\sigma}_\epsilon^2\,\widehat{\mathbf{D}} = \begin{pmatrix} 0.016 & -0.007 & 0.006 \\ -0.007 & 0.006 & -0.005 \\ 0.006 & -0.005 & 0.005 \end{pmatrix}, \hat{\sigma}_\epsilon^2 = 0.2.$$

   We get observations from the in-service unit p at time points $\mathbf{t} = (1\ 2)^T$ where $\mathbf{y}_p = (1\ 7)^T$. The failure threshold is set to 50.
   (a) Update the distribution of $\mathbf{b}$ at $t^* = 2$ using the Bayesian framework.
   (b) Find the conditional probability that the unit will have an RUL of three time units, knowing that it has survived until $t^* = 2$. In other words, find $F_T(T = 3\,|t^* = 2)$.
2. Assume we use a linear mixed effects model with mixture prior to model a population of degradation signals where we have two sub-populations. From historical data we get the following estimates:

$$\hat{\lambda}_0 = 0.25, \hat{\boldsymbol{\mu}}_0 = (1\ 5)^T,\ \widehat{\boldsymbol{\Sigma}}_0 = \begin{pmatrix} 2 & 1 \\ 1 & 2 \end{pmatrix}, \hat{\boldsymbol{\mu}}_1 = (2\ 3)^T,$$

$$\widehat{\boldsymbol{\Sigma}}_1 = \begin{pmatrix} 3 & 1 \\ 1 & 3 \end{pmatrix}, \hat{\sigma}_{\epsilon 0}^2 = \hat{\sigma}_{\epsilon 1}^2 = 0.2.$$

   From the new in-service unit p we get the observations of $\mathbf{t} = (1\ 2)^T$ and $\mathbf{y} = (2\ 6)^T$.
   (a) Get the updated distribution estimates using the Bayesian approach at $t^* = 2$.

(b) Find the conditional probability that the in-service unit will have an RUL of three time units, knowing that it has survived until $t^* = 2$ and the failure threshold is set to 40. In other words, find $F_T(T = 5 \mid t^* = 2)$.

3. Consider the temperature data collected during the ice-making cycles of an ice machine. This data set is balanced using an interpolation technique. "`train.csv`" and "`test.csv`" contain the training and testing data, respectively. The "`ID`" column in the two files represents the signal number, "`X`" represents the time of observation and "`Y`" shows the measured temperature. We want to model the evolution of temperature signals over time using a B-spline based mixed effects framework. Considering B-spline basis functions of degree three with seven degrees of freedom, fit a mixed effects model on the training data in `R`. Conduct Bayesian updating using the mixed effects model estimated and the testing data to get the posterior distribution of random coefficients in `R`.

4. In this exercise we want to investigate the effect of using the wrong parametric form on RUL prediction in a mixed effects model framework. First, we simulate 15 signals in `R` where each signal is generated according to $y_{i,j}(t) = wt_{i,j}^2 + \epsilon_{i,j}$ where $i = 1, 2, \ldots, 15$, $j = 1, 2, \ldots, 30$ and $t \in [0, 10]$. For the ith output, $w \sim \mathcal{N}(1, 0.25^2)$. The number of observations per signal is 30 evenly spaced points and the observation standard deviation is set to $\sigma_\epsilon = 0.5$ for all outputs. The failure threshold is set to 50. Consider two parametric forms for the mixed effects model: (1) a quadratic form where the design matrix is set to $\mathbf{z}_{i,j} = (1, t_{i,j}, t_{i,j}^2)$ (2) an exponential form where the design matrix is set to $\mathbf{z}_{i,j} = (\exp(t_{i,j}))$.

   (a) Use `R` to simulate 15 signals as explained above. Consider the first 14 signals as the training set and the remaining one as the testing signal. For the testing signal consider three scenarios where we only have 25%, 50% and 75% of its observations. Conduct Bayesian updating at each of the testing signal life percentiles (t^*) to get the posterior distribution of random coefficients for each of the parametric mixed effects models explained above.

   (b) Assume that after each Bayesian updating the signal propagates according to the new updated coefficients. Estimate the failure time of testing signal considering the mean values of these coefficients at each updating stage for each parametric form. (Hint: you can define a function to get the difference between the projected signal propagation and failure threshold and use it in the `uniroot.all` function of `rootSolve` package in `R` to find the estimated failure time.)

   (c) Perform the above simulation in parts (a) and (b) 1000 times for each parametric form at each testing signal life percentile (t^*) and estimate its failure time. In each simulation, calculate the absolute error between the true failure time T_p and the estimated failure time $\widehat{T}_p(t^*)$ as follows:

$$AE(t^*) = |\widehat{T}_p(t^*) - T_p|.$$

Draw box plots comparing the absolute error predictions of two parametric forms at each observation percentile. What do you conclude from comparing the performance of the two parametric forms as visualized in these box plots? (Hint: The true failure time is the time when the generated testing signal in part (1) actually hits the failure threshold.)

Appendix

1. Proof of the results in (10.5)

We can write the likelihood function (10.4) in a matrix form as

$$f(\mathbf{y}_p^* | \mathbf{b}_p) = (2\pi\hat{\sigma}_\epsilon^2)^{-n_p/2} \exp[-(\mathbf{y}_p^* - \mathbf{Z}_p^*\mathbf{b}_p)^T(\mathbf{y}_p^* - \mathbf{Z}_p^*\mathbf{b}_p)/(2\hat{\sigma}_\epsilon^2)].$$

Then we have

$$f(\mathbf{b}_p | \mathbf{y}_p^*) \propto f(\mathbf{y}_p^* | \mathbf{b}_p)\pi(\mathbf{b}_p)$$

$$\propto \exp\left\{-\frac{1}{2\hat{\sigma}_\epsilon^2}\left[(\mathbf{y}_p^* - \mathbf{Z}_p^*\mathbf{b}_p)^T(\mathbf{y}_p^* - \mathbf{Z}_p^*\mathbf{b}_p)\right]\right\}\exp\left\{-\frac{1}{2}\left[(\mathbf{b}_p - \hat{\mathbf{b}})^T(\hat{\sigma}_\epsilon^2\widehat{\mathbf{D}})^{-1}(\mathbf{b}_p - \hat{\mathbf{b}})\right]\right\}$$

$$\propto \exp\left\{-\frac{1}{2}\left[\mathbf{b}_p^T\left(\frac{\mathbf{Z}_p^{*T}\mathbf{Z}_p^*}{\hat{\sigma}_\epsilon^2} + (\hat{\sigma}_\epsilon^2\widehat{\mathbf{D}})^{-1}\right)\mathbf{b}_p - \mathbf{b}_p^T\left(\frac{\mathbf{Z}_p^{*T}\mathbf{y}_p^*}{\hat{\sigma}_\epsilon^2} + (\hat{\sigma}_\epsilon^2\widehat{\mathbf{D}})^{-1}\hat{\mathbf{b}}_p\right) - \left(\frac{\mathbf{y}_p^{*T}\mathbf{Z}_p^*}{\hat{\sigma}_\epsilon^2} + \hat{\mathbf{b}}^T(\hat{\sigma}_\epsilon^2\widehat{\mathbf{D}})^{-1}\right)\mathbf{b} + C_1\right]\right\}$$

$$\propto \exp\left\{-\frac{1}{2}\left[\mathbf{v}^T\left(\frac{\mathbf{Z}_p^{*T}\mathbf{Z}_p^*}{\hat{\sigma}_\epsilon^2} + (\hat{\sigma}_\epsilon^2\widehat{\mathbf{D}})^{-1}\right)\mathbf{v} + C_2\right]\right\}.$$

where C_1 and C_2 are constants not involving $\mathbf{b}_p$, and $\mathbf{v}$ is defined as

$$\mathbf{v} = \mathbf{b}_p - \left(\frac{\mathbf{Z}_p^{*T}\mathbf{Z}_p^*}{\hat{\sigma}_\epsilon^2} + (\hat{\sigma}_\epsilon^2\widehat{\mathbf{D}})^{-1}\right)^{-1}\left(\frac{\mathbf{Z}_p^{*T}\mathbf{y}_p^*}{\hat{\sigma}_\epsilon^2} + (\hat{\sigma}_\epsilon^2\widehat{\mathbf{D}})^{-1}\hat{\mathbf{b}}\right).$$

Then we can define $\hat{\mathbf{b}}_p^*$ and $\widehat{\mathbf{\Sigma}}_{\mathbf{b}_p^*}$ as that in equation (10.5) and from above derivation, we have

$$f(\mathbf{b}_p | \mathbf{y}_p^*) \propto \exp\left\{-\frac{1}{2}[(\mathbf{b}_p - \hat{\mathbf{b}}_p^*)^T\widehat{\mathbf{\Sigma}}_{\mathbf{b}_p^*}^{-1}(\mathbf{b}_p - \hat{\mathbf{b}}_p^*)]\right\}.$$

This density defines a multivariate normal distribution $\mathcal{N}(\hat{\mathbf{b}}_p^*, \widehat{\mathbf{\Sigma}}_{\mathbf{b}_p^*})$.

2. Proof of the result in (10.17)

From the law of total probability, we have that

$$P(\mathbf{b}_p|\mathbf{y}_p^*)=P(\mathbf{b}_{p,0}|\delta_p=0,\mathbf{y}_p^*)P(\delta_p=0|\mathbf{y}_p^*)+P(\mathbf{b}_{p,1}|\delta_p=1,\mathbf{y}_p^*)P(\delta_p=1|\mathbf{y}_p^*),$$

where $P(\delta_p=0|\mathbf{y}_p^*)=\hat{\lambda}_p^{0*}$ and $P(\delta_p=1|\mathbf{y}_p^*)=1-\hat{\lambda}_p^{0*}$ represent the probabilities that the newly observed data belong to the survived or failed group, respectively.

Following Bayes theorem, $\hat{\lambda}_p^{0*}$ can be expressed as

$$\hat{\lambda}_p^{0*}=\frac{P(\delta_p=0,\mathbf{y}_p^*)}{P(\mathbf{y}_p^*)}=\frac{P(\delta_p=0)P(\mathbf{y}_p^*\mid\delta_p=0)}{P(\delta_p=0)P(\mathbf{y}_p^*\mid\delta_p=0)+P(\delta_p=1)P(\mathbf{y}_p^*\mid\delta_p=1)}=$$

$$\frac{\hat{\lambda}^0\, f_N\left(\mathbf{y}_p^*\mid\mathbf{Z}_p^*\hat{\mathbf{b}}_0,\mathbf{Z}_p^*\hat{\Sigma}_0(\mathbf{Z}_p^*)^t+\hat{\sigma}_{\epsilon 0}^2\mathbf{I}\right)}{\hat{\lambda}^0 f_N(\mathbf{y}_p^*\mid\mathbf{Z}_p^*\hat{\mathbf{b}}_0,\mathbf{Z}_p^*\hat{\Sigma}_0(\mathbf{Z}_p^*)^T+\hat{\sigma}_{\epsilon 0}^2\mathbf{I})+(1-\hat{\lambda}^0)\, f_N(\mathbf{y}_p^*\mid\mathbf{Z}_p^*\hat{\mathbf{b}}_1,\mathbf{Z}_p^*\hat{\Sigma}_1(\mathbf{Z}_p^*)^T+\hat{\sigma}_{\epsilon 1}^2\mathbf{I})},$$

where $P(\delta_p=0)=\hat{\lambda}^0$ and $P(\delta_p=1)=1-\hat{\lambda}^0$ represents the prior probabilities of belonging to the survived or failed group respectively, which partly proves the result for $\hat{\lambda}_p^{0*}$.

For $P(\mathbf{b}_{p,0}\mid\delta_p=0,\mathbf{y}_p^*)$ we have that $P(\mathbf{y}_p^*\mid\delta_p=0,\mathbf{b}_{p,0})=f_N(\mathbf{y}_p^*\mid\mathbf{Z}_p^*\mathbf{b}_{p,0},\hat{\sigma}_{\epsilon 0}^2\mathbf{I})$, therefore,

$$\begin{aligned}P(\mathbf{b}_{p,0}|\delta_p=0,\mathbf{y}_p^*)&\propto P(\mathbf{y}_p^*|\delta_p=0,\mathbf{b}_{p,0})\pi(\mathbf{b}_{p,0})\\&\propto f_N(\mathbf{y}_p^*|\mathbf{Z}_p^*\mathbf{b}_{p,0},\hat{\sigma}_{\epsilon 0}^2\mathbf{I})f_N(\mathbf{b}_{p,0}|\hat{\mathbf{b}}_0,\hat{\Sigma}_0)\\&\propto C\cdot f_N(\hat{\mathbf{b}}_{0,p}^*,\hat{\Sigma}_{0,p}^*),\end{aligned}$$

where C is a normalizing constant and $\pi(\mathbf{b}_{p,0})$ is the estimated prior for a unit in the failed group. Therefore $P(\mathbf{b}_{p,0}\mid\delta_p=0,\mathbf{y}_p^*)=f_N(\hat{\mathbf{b}}_{0,p}^*,\hat{\Sigma}_{0,p}^*)$, where

$$\begin{cases}\hat{\mathbf{b}}_{0,p}^*=\hat{\Sigma}_{0,p}^*[(\mathbf{Z}_p^*)^T\mathbf{y}_p^*/\hat{\sigma}_{\epsilon 0}^2+\hat{\Sigma}_0^{-1}\hat{\boldsymbol{\mu}}_0],\\ \hat{\Sigma}_{0,p}^*=[\hat{\Sigma}_0^{-1}+(\mathbf{Z}_p^*)^T\mathbf{Z}_p^*/\hat{\sigma}_{\epsilon 0}^2]^{-1}.\end{cases}$$

Similarly, we can obtain the posterior distribution for the failed group, i.e., $P(\mathbf{b}_{p,1}\mid\delta_p=1,\mathbf{y}_p^*)$, as $f_N(\hat{\mathbf{b}}_{1,p}^*,\hat{\Sigma}_{1,p}^*)$.

3. Proof of the result in (10.19)

The conditional CDF of T_p of unit p given the observations up to t^*, denoted as $\mathbf{y}_p^*$ can be expressed as

$$P(T_p\le t\mid\mathbf{y}_p^*)=P(y_p(t+t^*)\ge D\mid\mathbf{y}_p^*)=1-P(y_p(t+t^*)\le D\mid\mathbf{y}_p^*).$$

Considering the distribution of $y_p(t)$ given in (10.17), we have

$$P(T_p \le t \mid \mathbf{y}_p^*) = \hat{\lambda}_{0,p}^* \Phi\left(\frac{\mathbf{z}^T(t+t^*)\hat{\boldsymbol{\mu}}_{0,p}^* - D}{\sqrt{\mathbf{z}^T(t+t^*)\hat{\boldsymbol{\Sigma}}_{0,p}^*\mathbf{z}(t+t^*) + \hat{\sigma}_{\epsilon 0}^2}}\right)$$
$$+ (1-\hat{\lambda}_{0,p}^*)\Phi\left(\frac{\mathbf{z}^T(t+t^*)\hat{\boldsymbol{\mu}}_{1,p}^* - D}{\sqrt{\mathbf{z}^T(t+t^*)\hat{\boldsymbol{\Sigma}}_{1,p}^*\mathbf{z}(t+t^*) + \hat{\sigma}_{\epsilon 1}^2}}\right)$$
$$= \hat{\lambda}_{0,p}^*\Phi(z_{0,p}(t+t^*)) + (1-\hat{\lambda}_{0,p}^*)\Phi(z_{1,p}(t+t^*)).$$

However, when we make a prediction at t^*, we know $T_p \ge t^*$. Thus, the above CDF needs to be truncated as

$$F_p(t \mid t^*) = P(T_p \le t \mid \mathbf{y}_p^*, T_p \ge t^*) = \frac{P(t^* \le T_p \le t \mid \mathbf{y}_p^*)}{P(T_p \ge t^* \mid \mathbf{y}_p^*)}$$
$$= \frac{\hat{\lambda}_{0,p}^*\Phi\{z_{0,p}(t+t^*)\} + (1-\hat{\lambda}_{0,p}^*)\Phi\{z_{1,p}(t+t^*)\} - \omega(t^*)}{1-\omega(t^*)},$$

where $\omega(\cdot)$ is defined in (10.19). This completes the proof.

4. Proof of Lemma 10.1

First, we can re-write (10.5) into a "recursive" form. To simplify the notation, we can omit the subscript p and superscript $*$ in (10.5) and at the kth step of updating, we have k observations up to now and

$$\hat{\mathbf{b}}_k = \hat{\boldsymbol{\Sigma}}_k \cdot [\mathbf{Z}_k^T \mathbf{y}_{1:k} / \hat{\sigma}^2 + (\hat{\sigma}^2 \hat{\mathbf{D}})^{-1}\hat{\mathbf{b}}]$$
$$\hat{\boldsymbol{\Sigma}}_k = [(\hat{\sigma}^2\hat{\mathbf{D}})^{-1} + \mathbf{Z}_k^T\mathbf{Z}_k / \hat{\sigma}^2]^{-1}.$$

Noting $\mathbf{y}_{1:k} = \begin{bmatrix} \mathbf{y}_{1:k-1} \\ y_k \end{bmatrix}$, $\mathbf{Z}_k = \begin{bmatrix} \mathbf{Z}_{k-1} \\ \mathbf{z}_k^T \end{bmatrix}$, where $\mathbf{z}_k^T$ is the kth row of $\mathbf{Z}_k$ that corresponds to the kth observation, $\mathbf{Z}_k^T\mathbf{Z}_k = \mathbf{Z}_{k-1}^T\mathbf{Z}_{k-1} + \mathbf{z}_k\mathbf{z}_k^T$ and adding a superscript "B" indicating the Bayesian batch updating, we have

$$\hat{\mathbf{b}}_k^B = \hat{\boldsymbol{\Sigma}}_k[\mathbf{z}_k \cdot y_k/\hat{\sigma}^2 + \hat{\boldsymbol{\Sigma}}_{k-1}^B \hat{\mathbf{b}}_{k-1}]$$
$$\hat{\boldsymbol{\Sigma}}_k^B = [(\hat{\boldsymbol{\Sigma}}_{k-1}^B)^{-1} + \mathbf{z}_k\mathbf{z}_k^T / \hat{\sigma}^2]^{-1}. \tag{10.29}$$

Noting that the distribution of $\mathbf{b}_0$ is set the same in both Bayesian updating and Kalman filer in (10.28), thus we have $\hat{\mathbf{b}}_0 = \hat{\mathbf{b}}_0^B$ and $\hat{\boldsymbol{\Sigma}}_0 = \hat{\boldsymbol{\Sigma}}_0^B$. Without loss of generality, we can assume $\hat{\mathbf{b}}_{k-1} = \hat{\mathbf{b}}_{k-1}^B$ and $\hat{\boldsymbol{\Sigma}}_{k-1} = \hat{\boldsymbol{\Sigma}}_{k-1}^B$. We can show $\hat{\boldsymbol{\Sigma}}_k$ in (10.28) is the same as $\hat{\boldsymbol{\Sigma}}_k^B$ in (10.29).

$$\begin{aligned}
&\widehat{\boldsymbol{\Sigma}}_k \times (\widehat{\boldsymbol{\Sigma}}_k^B)^{-1} \\
&= [\widehat{\boldsymbol{\Sigma}}_{k-1} - \widehat{\boldsymbol{\Sigma}}_{k-1}\mathbf{z}_k(\mathbf{z}_k^T\widehat{\boldsymbol{\Sigma}}_{k-1}\mathbf{z}_k + \widehat{\sigma}^2)^{-1}\mathbf{z}_k^T\widehat{\boldsymbol{\Sigma}}_{k-1}] \times [\widehat{\boldsymbol{\Sigma}}_{k-1}^{-1} + \mathbf{z}_k\mathbf{z}_k^T\widehat{\sigma}^{-2}] \\
&= \mathbf{I} - \widehat{\boldsymbol{\Sigma}}_{k-1}\mathbf{z}_k[(\mathbf{z}_k^T\widehat{\boldsymbol{\Sigma}}_{k-1}\mathbf{z}_k + \widehat{\sigma}^2)^{-1} - \widehat{\sigma}^{-2} + (\mathbf{z}_k^T\widehat{\boldsymbol{\Sigma}}_{k-1}\mathbf{z}_k + \widehat{\sigma}^2)^{-1}\mathbf{z}_k^T\widehat{\boldsymbol{\Sigma}}_{k-1}\mathbf{z}_k\widehat{\sigma}^{-2}]\mathbf{z}_k^T \\
&= \mathbf{I} - \widehat{\boldsymbol{\Sigma}}_{k-1}\mathbf{z}_k[(\mathbf{z}_k^T\widehat{\boldsymbol{\Sigma}}_{k-1}\mathbf{z}_k + \widehat{\sigma}^2)^{-1}(\mathrm{I} + \mathbf{z}_k^T\widehat{\boldsymbol{\Sigma}}_{k-1}\mathbf{z}_k\widehat{\sigma}^{-2}) - \widehat{\sigma}^{-2}]\mathbf{z}_k^T \\
&= \mathbf{I} - \widehat{\boldsymbol{\Sigma}}_{k-1}\mathbf{z}_k(\widehat{\sigma}^{-2} - \widehat{\sigma}^{-2})\mathbf{z}_k^T \\
&= \mathbf{I}.
\end{aligned}$$

Thus, we have $\widehat{\boldsymbol{\Sigma}}_k = \widehat{\boldsymbol{\Sigma}}_k^B$. We can simplify the expression for the Kalman gain by multiplying $\widehat{\boldsymbol{\Sigma}}_k\widehat{\boldsymbol{\Sigma}}_k^{-1}$ and $\widehat{\sigma}^2\widehat{\sigma}^{-2}$ as

$$\begin{aligned}
\mathbf{K}_k &= \widehat{\boldsymbol{\Sigma}}_k\widehat{\boldsymbol{\Sigma}}_k^{-1}\widehat{\boldsymbol{\Sigma}}_{k-1}\mathbf{z}_k\widehat{\sigma}^2\widehat{\sigma}^{-2}(\mathbf{z}_k^T\widehat{\boldsymbol{\Sigma}}_{k-1}\mathbf{z}_k + \widehat{\sigma}^2)^{-1} = \widehat{\boldsymbol{\Sigma}}_k\widehat{\boldsymbol{\Sigma}}_k^{-1}\widehat{\boldsymbol{\Sigma}}_{k-1}\mathbf{z}_k\widehat{\sigma}^{-2}(\mathbf{z}_k^T\widehat{\boldsymbol{\Sigma}}_{k-1}\mathbf{z}_k\widehat{\sigma}^{-2} + \mathbf{I})^{-1} \\
&= \widehat{\boldsymbol{\Sigma}}_k(\mathbf{I} + \mathbf{z}_k^T\widehat{\boldsymbol{\Sigma}}_{k-1}\mathbf{z}_k\widehat{\sigma}^{-2})\mathbf{z}_k\widehat{\sigma}^{-2}(\mathbf{z}_k^T\widehat{\boldsymbol{\Sigma}}_{k-1}\mathbf{z}_k\widehat{\sigma}^{-2} + \mathbf{I})^{-1} \\
&= \widehat{\boldsymbol{\Sigma}}_k\mathbf{z}_k\widehat{\sigma}^{-2}.
\end{aligned}$$

Then, $\widehat{\mathbf{b}}_k^B$ can be expressed as

$$\begin{aligned}
\widehat{\mathbf{b}}_k^B &= \widehat{\boldsymbol{\Sigma}}_k(\mathbf{z}_k y_k\widehat{\sigma}^{-2} + \widehat{\boldsymbol{\Sigma}}_{k-1}^{-1}\widehat{\mathbf{b}}_{k-1}) = \mathbf{K}_k y_k + (\mathbf{I} - \mathbf{K}_k\mathbf{z}_k^T)\widehat{\boldsymbol{\Sigma}}_{k-1}\widehat{\boldsymbol{\Sigma}}_{k-1}^{-1}\widehat{\mathbf{b}}_{k-1} \\
&= \widehat{\mathbf{b}}_{k-1} + \mathbf{K}_k(y_k - \mathbf{z}_k^T\widehat{\mathbf{b}}_{k-1}) \\
&= \widehat{\mathbf{b}}_k.
\end{aligned}$$

This concludes the proof.

11

Prognosis Using Gaussian Process Model

In Chapter 10, we discussed the methods of using the mixed effects model for the forecasting of longitudinal signals. The mixed effects model can extract information from historical longitudinal data and then through a Bayesian updating method, the historical information can be integrated with the information from individual units to make an individualized prediction. The mixed effects models introduced in Chapter 10 are parametric models that assume all units have the same functional form. In other words, units behave according to a similar trend (e.g., quadratic polynomial) but at different rates (i.e., different parameter values). The B-spline regression based mixed effects model introduced in Example 10.2 provides more flexibility but its flexibility is still limited by the knot value selection and the degree of the basis.

One limitation of the parametric mixed effects model is that all the historical data are restricted to the same functional form and the predictions for the new unit are restricted to the same functional form as well. This assumption may not hold in many real-life applications. For instance, units may not be from the same generation or are operated under different environmental conditions, such as different levels of speed, load, temperature, and humidity [Zhang et al., 2015]. Under these conditions, data might exhibit different evolution trends and rates. The mathematical form of the evolution is unknown or even no closed expression exists. If the specified form in the parametric mixed effects model is far from the truth, the modeling and prognosis results will be misleading.

To address this limitation, in this chapter, we introduce a nonparametric approach for longitudinal data modeling and prediction. This approach is based on the Gaussian process (GP) model. GP is a stochastic process and can be viewed as a distribution over functions with a continuous domain, e.g. time or

Industrial Data Analytics for Diagnosis and Prognosis: A Random Effects Modelling Approach, First Edition. Shiyu Zhou and Yong Chen.

space. Thus, it can be viewed as an extension to the mixed effects model. In the conventional mixed effects model, the mixed effects are random variables, while for the GP, the mixed effects are random functions over time or space.

This chapter is organized as follows. Section 11.1 introduces the structure of the GP model. It will be compared with other models. Section 11.2 will discuss the estimation and prediction methods for the GP model. The pairwise GP model is introduced in Section 11.3. The extension of a single output GP model to the general Multiple Output Gaussian Process (MOGP) model, which plays a critical role in longitudinal data prediction, is presented in Section 11.4. The time-to-failure distribution based on the MOGP model is also discussed in Section 11.4.

11.1 Introduction to Gaussian Process Model

A Gaussian process is a stochastic process, which is a collection of random variables indexed by time or space. It is called the "Gaussian process" because every finite collection of random variables from it has a multivariate normal distribution. Thus, the distribution of a GP is the joint distribution of all those (infinitely many) random variables.

Without loss of generality, assume we have a dataset $\mathcal{D}$ of n observations, $\mathcal{D} = \{(\mathbf{x}_i, y_i) \mid i = 1, \ldots, n\}$. In this dataset, y_i is the univariate response/output, and $\mathbf{x}_i$ is the multivariate input/predictor. The GP model for this dataset can be written as

$$y_i = \mu(\mathbf{x}_i) + g(\mathbf{x}_i),$$

where $\mu(\mathbf{x}_i)$ is the deterministic mean function and $g(\cdot)$ is a GP with zero mean. Please note that $g(\mathbf{x}_i)$ is a normally distributed random variable which is indexed by $\mathbf{x}_i$. Then the joint multivariate distribution of the responses is

$$\begin{pmatrix} y_1 \\ \vdots \\ y_n \end{pmatrix} \sim \mathcal{N}(\begin{pmatrix} \mu(\mathbf{x}_1) \\ \vdots \\ \mu(\mathbf{x}_n) \end{pmatrix}, \begin{pmatrix} C(\mathbf{x}_1, \mathbf{x}_1) & \cdots & C(\mathbf{x}_1, \mathbf{x}_n) \\ \vdots & \ddots & \vdots \\ C(\mathbf{x}_n, \mathbf{x}_1) & \cdots & C(\mathbf{x}_n, \mathbf{x}_n) \end{pmatrix}), \tag{11.1}$$

where $\mu(\mathbf{x}_i)$ is the mean of y_i and $C(\mathbf{x}_i, \mathbf{x}_j)$ is the covariance between y_i and y_j that is determined by the GP $g(\cdot)$. If we let $\mathbf{y} = (y_1 \cdots y_n)^T$ and $\mathbf{X} = (\mathbf{x}_1 \cdots \mathbf{x}_n)^T$, we can rewrite (11.1) into a vector form as

$$\mathbf{y} \sim \mathcal{N}(\boldsymbol{\mu}(\mathbf{X}), \mathbf{C}(\mathbf{X})). \tag{11.2}$$

It is well known that Gaussian probability distribution is completely defined by the first and the second order statistics. Thus, (11.2) completely defines the multivariate distribution of $\mathbf{y}$, which is

$$f(\mathbf{y};\mathbf{X}) = \frac{1}{(2\pi)^{n/2}\left|\mathbf{C}(\mathbf{X})\right|^{1/2}} \times \exp\left(-\frac{1}{2}(\mathbf{y}-\boldsymbol{\mu}(\mathbf{X}))^T \mathbf{C}(\mathbf{X})^{-1}(\mathbf{y}-\boldsymbol{\mu}(\mathbf{X}))\right). \tag{11.3}$$

To establish a GP model in (11.2) for the dataset $\mathcal{D}$, we need to parameterize the mean $\boldsymbol{\mu}(\mathbf{X})$ and the covariance $\mathbf{C}(\mathbf{X})$. Otherwise, there will be too many parameters in the model and we will not be able to estimate them from the data. For example, the covariance matrix $\mathbf{C}(\mathbf{X})$ alone will have $n+n(n-1)/2$ parameters if no parameterization is adopted.

In principle, we can pick any function mapping the input $\mathbf{x}$ to the real number as the mean function, e.g., a linear regression function $\mu(\mathbf{x}) = \beta_0 + \mathbf{x}^T\boldsymbol{\beta}$. However, it is known that treating the mean as a constant is often sufficient for GP model, particularly when we have dense data in the input space. The intuition is that the GP model interpolates the existing data when it is used for prediction. Thus, the mean function will not have much effect when we have observations that are close in the input space to the input point at which we want to make prediction. We will further illustrate this point later. Thus, the commonly used GP model in practice becomes

$$y_i = \mu + g(\mathbf{x}_i).$$

Different from mean parameterization, the parameterization of the covariance $\mathbf{C}(\mathbf{X})$ plays a critical role in GP modeling. It in fact determines the behavior of the model. To parameterize $\mathbf{C}(\mathbf{X})$, we need to define a covariance function that maps two inputs $\mathbf{x}_i$ and $\mathbf{x}_j$ to the covariance of corresponding responses y_i and y_j. With slight amendment to the notation, we denote the covariance function as $C(\mathbf{x}_i, \mathbf{x}_j; \boldsymbol{\theta})$, where $\boldsymbol{\theta}$ is the parameter. Because the covariance matrix $\mathbf{C}(\mathbf{X})$ should be a positive semi-definite symmetric matrix, an arbitrary function may not produce a valid covariance matrix.

In the literature, various ways of specifying the covariance function have been proposed. Table 11.1 summarizes a number of commonly used covariance functions.

Please note that for the periodic covariance function, the input should be univariate. The length-scale parameter l that appears in a few covariance function determines "how close" two input points $\mathbf{x}_i$ and $\mathbf{x}_j$ have to be to influence the corresponding responses significantly.

Table 11.1 Commonly used covariance functions for GP model

Name	Expression $C(\mathbf{x}_i, \mathbf{x}_j; \theta)$	Parameters θ
Constant	σ^2	σ
Linear	$\mathbf{x}_i^T \mathbf{D} \mathbf{x}_j$	$\mathbf{D}$: diagonal matrix with positive elements
Gaussian noise	σ^2; if $\mathbf{x}_i = \mathbf{x}_j$ 0; otherwise	σ
Squared exponential	$\exp(-\frac{\lvert \mathbf{x}_i - \mathbf{x}_j \rvert^2}{2l^2})$	l : length-scale
Ornstein–Uhlenbeck	$\exp(-\frac{\lvert \mathbf{x}_i - \mathbf{x}_j \rvert}{l})$	l : length-scale
Matérn*	$\frac{2^{1-\upsilon}}{\Gamma(\upsilon)}(\frac{\sqrt{2\upsilon}\lvert \mathbf{x}_i - \mathbf{x}_j \rvert}{l})^{\upsilon} K_{\upsilon}(\frac{\sqrt{2\upsilon}\lvert \mathbf{x}_i - \mathbf{x}_j \rvert}{l})$	l: length-scale υ: degree
Periodic	$\exp(-\frac{2\sin^2((\mathbf{x}_i - \mathbf{x}_j)/2)}{l^2})$	l : length-scale

*In the Matérn covariance function, $\Gamma(\upsilon)$ is the gamma function and K_{υ} is the modified Bessel function of order υ.

There are several important characterization properties for covariance functions. If the covariance only depends on the difference between the inputs, the covariance function is called stationary, i.e., $C(\mathbf{x}_i, \mathbf{x}_j; \theta) = C(\mathbf{x}_i - \mathbf{x}_j; \theta)$. If the covariance function only depends on the distance $\lvert \mathbf{x}_i - \mathbf{x}_i \rvert$, the covariance function is called isotropic. Obviously the isotropic condition is a stronger condition than the stationary condition. For a detailed description of the covariance functions and the methods of creating covariance functions, please refer to Chapter 4 of Rasmussen and Williams [2006].

With the covariance function, we can obtain the covariance between any two given inputs. In Figure 11.1, we illustrate several simulated GPs with different covariance functions. Please note that the input x in these simulations is an univariate variable. We can see that the behavior of the data that can be expressed by a GP with different covariance functions is quite different. In fact, even with the same covariance function form, by varying the parameters of the function, we can get quite different behaviors as well. Figure 11.2 shows a simulated GP with a squared exponential covariance function with different length-scale parameters.

Clearly, a small length-scale parameter leads to a wiggly curve while a large length-scale parameter results in smoother and slow changing curves.

With the parameterization of the mean function and covariance function, we can estimate the parameters using the existing data and then predict the response value at a new input location using the estimated GP.

11.2 GP Parameter Estimation and GP Based Prediction

The parameters of a specified GP are often estimated through the MLE method. For a dataset $\mathcal{D} = \{(\mathbf{x}_i, y_i) \mid i = 1, \dots, n\}$, we can get the negative log-likelihood function according to the probability density function (11.3) as

$$-l(\boldsymbol{\theta}, \boldsymbol{\beta}) \propto \ln \left|\mathbf{C}(\mathbf{X}; \boldsymbol{\theta})\right| + (\mathbf{y} - \boldsymbol{\mu}(\mathbf{X}; \boldsymbol{\beta}))^T \mathbf{C}(\mathbf{X}; \boldsymbol{\theta})^{-1}(\mathbf{y} - \boldsymbol{\mu}(\mathbf{X}; \boldsymbol{\beta})), \quad (11.4)$$

where $\boldsymbol{\theta}$ is the collection of parameters of the specified covariance function and $\boldsymbol{\beta}$ is the collection of parameters of the mean function. As mentioned previously, we can often let the mean function be a constant. In that case, $\boldsymbol{\mu}(\mathbf{X}; \boldsymbol{\beta}) = \mu_0$ and $\boldsymbol{\beta}$ becomes a simple number μ_0. The maximum likelihood estimation of the parameters $\boldsymbol{\theta}, \boldsymbol{\beta}$ is

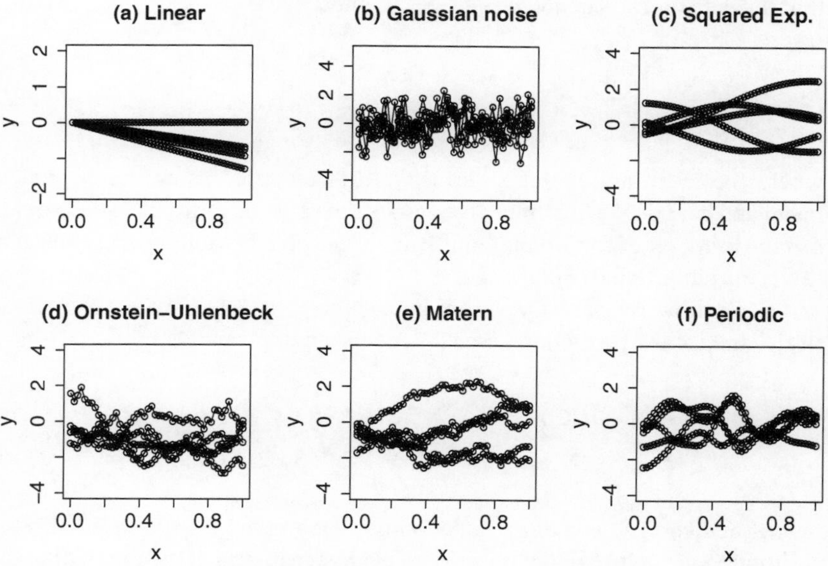

Figure 11.1 Simulated GPs with different covariance functions

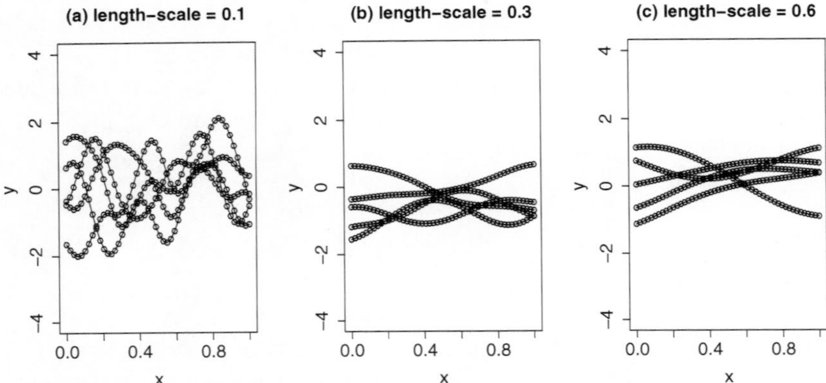

Figure 11.2 Simulated GPs with squared exponential covariance functions

$$(\hat{\theta}, \hat{\beta}) = \underset{\theta,\beta}{\operatorname{argmin}} - l(\theta, \beta). \tag{11.5}$$

With the estimated parameters, we can obtain the mean function and covariance function, which can be used to predict the response values at new input locations. Assume we want to make prediction at one or more new input locations denoted as $\mathbf{X}^*$ and the corresponding responses are denoted as $\mathbf{y}^*$. Noting that the data in $\mathcal{D}$ is denoted as $\{\mathbf{X}, \mathbf{y}\}$, then we have

$$\begin{pmatrix} \mathbf{y} \\ \mathbf{y}^* \end{pmatrix} \sim \mathcal{N}\left(\begin{pmatrix} \mu(\mathbf{X}) \\ \mu(\mathbf{X}^*) \end{pmatrix}, \begin{pmatrix} \mathbf{C}(\mathbf{X}, \mathbf{X}) & \mathbf{C}(\mathbf{X}, \mathbf{X}^*) \\ \mathbf{C}^T(\mathbf{X}, \mathbf{X}^*) & \mathbf{C}(\mathbf{X}^*, \mathbf{X}^*) \end{pmatrix} \right), \tag{11.6}$$

where the mean and variance terms can be computed using the estimated mean and covariance function. To estimate the value of $\mathbf{y}^*$ from (11.6), we can use the property of conditional multivariate normal distribution in (3.11) and (3.12) introduced in Chapter 3.

It is clear that we can let $\mathbf{y}^*$ and $\mathbf{y}$ be $\mathbf{x}_1$ and $\mathbf{x}_2$ in (3.11) and (3.12), respectively, and get the prediction based GP model as

$$\begin{aligned} \hat{\mathbf{y}}^* &= \mu(\mathbf{X}^*) + \mathbf{C}^T(\mathbf{X}, \mathbf{X}^*)\mathbf{C}^{-1}(\mathbf{X}, \mathbf{X})(\mathbf{y} - \mu(\mathbf{X})) \\ \widehat{\Sigma}_{\mathbf{y}^*} &= \mathbf{C}(\mathbf{X}^*, \mathbf{X}^*) - \mathbf{C}^T(\mathbf{X}, \mathbf{X}^*)\,\mathbf{C}^{-1}(\mathbf{X}, \mathbf{X})\,\mathbf{C}(\mathbf{X}, \mathbf{X}^*) \end{aligned}. \tag{11.7}$$

The prediction $\hat{\mathbf{y}}^*$ is a linear function of the observed responses $\mathbf{y}$. Also, in addition to the point prediction $\hat{\mathbf{y}}^*$, the covariance matrix of the prediction $\widehat{\Sigma}_{\mathbf{y}^*}$ is also given, which can give us confidence in the prediction.

Example 11.1 In this example, we will first simulate data from a GP with zero mean. Then these data will be used to estimate the model parameters and finally, we will make a prediction using the fitted GP model.

A GP can be viewed as a very high dimensional multivariate Gaussian distribution. Thus, the simulation of a GP can be achieved through the simulation of a multivariate Gaussian random distribution, where the covariance of the Gaussian distribution is obtained through the covariance function. In `R`, this can be achieved through the `mvrnorm` function in the `MASS` package. Assume the variable `Sigma` contains the covariance of n responses corresponding to n inputs. The n inputs are represented by the variable `x`. Then the simulated GP can be obtained as

```
data.frame("x" = x, "y" = mvrnorm(mu=rep(0,times=n),
    Sigma=Sigma))
```

The resulting data frame contains the simulated GP $\mathbf{x}$ versus $\mathbf{y}$. The covariance matrix can be obtained through evaluating the covariance function (e.g., the functions in Table 11.1) for the n inputs. A few `R` packages such as `mlegp`, `GPfit`, and `GPFDA` provide the capability of estimating the parameters of a GP. Most of these packages assume the covariance function of the GP is in the form of an exponential, of which the "Squared exponential" in Table 11.1 is a special case. The `GPfit` is also able to handle the Matérn function. It also needs to be mentioned that in these packages, the correlation function $\text{corr}(\mathbf{x}_i, \mathbf{x}_j)$, instead of the covariance function, is used to describe the GP. The diagonal elements of a correlation matrix are 1s and the off-diagonal elements are between -1 and 1. When the GP is stationary, the covariance function and the correlation function differ by a scaling parameter

$$\mathbf{C}(\mathbf{X}) = \sigma_{GP}^2 \text{corr}(\mathbf{X}). \tag{11.8}$$

If a correlation function is used, the scaling parameter σ_{GP}^2 and the parameters of the correlation functions will be estimated. Figure 11.3 shows the estimated GP and the corresponding prediction. In this figure, the circles are the observed data from a simulated GP, which has mean zero and correlation function as $\text{corr}(x_i, x_j) = \exp(-50(x_i - x_j)^2)$. Thus, the underlying true parameter values are $\sigma_{\text{GP}} = 1$ and $\theta = 50$. Assume the data frame `gps` contains the simulated GP data (i.e., the circles in the figure), we can use the following code to estimate the GP

```
gp_res<-mlegp(gps$x,gps$y)
```

The function `mlegp` estimates the GP model, and it returns a GP object. Using `summary(gp_res)`, we can get the estimated parameters as $\hat{\sigma}_{GP}^2 = 1.41$ and $\hat{\theta} = 47.32$.

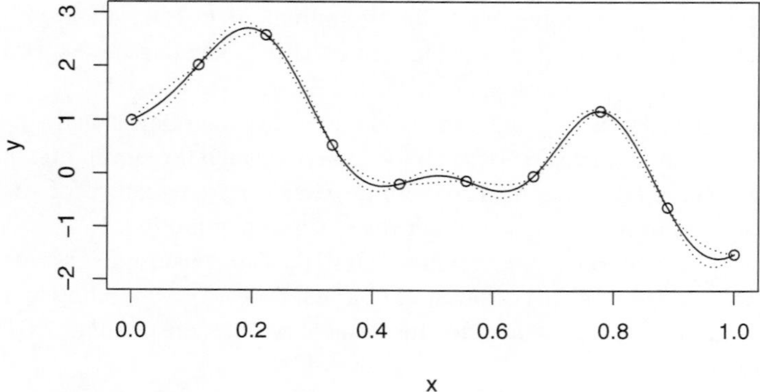

Figure 11.3 Estimated Gaussian processes: ———, predicted values; ......, confidence interval

With the estimated parameters, we can predict the response at other input locations. As shown in Figure 11.3, the solid line is the predicted value for a given input x and the dotted line shows the confidence interval of the prediction. A clear feature is that the prediction curve passes through the observed values, and there is no variation in the prediction at those input locations. This feature makes the GP model a popular tool in surrogate modeling for computer experiments [Santner et al., 2013]. In computer experiments, the response is deterministic for a given fixed input and as a result, repeated measures at the same location are not needed. Thus, the prediction for computer experiments needs to pass through the observed values.

Figure 11.3 can be easily obtained through the following `R` code. The first line generates 1000 new inputs in addition to the observed input locations and the second line makes prediction from the fitted `gp.res`, where `se.fit = TRUE` requires the prediction to report the standard error of the prediction that is obtained through (11.7). The next two lines sort the data based on the input values and the rest of the code plots the values.

```
newData <-c(gps$x,runif(1000,min=0,max=1))
pdata<-predict(gp.res,newData=matrix(newData),
  se.fit=TRUE)
full.pdata<-data.frame(x=newData,y = pdata$fit[,1],
                              se = pdata$se.fit[,1])
full.pdata<-full.pdata[order(full.pdata$x),]
#plot observed values
plot(gps$x,gps$y,xlab="x",ylab="y",ylim=c(-2,3))
```

```
#plot predicted values
lines(full.pdata$x,full.pdata$y)
# plot the 1-sigma confidence bands
lines(full.pdata$x,full.pdata$y+full.pdata$se,
  lty = "dotted")
lines(full.pdata$x, full.pdata$y-full.pdata$se,
  lty = "dotted")
```

Example 11.2 In this example, we illustrate that in a GP model specification, a constant mean is often sufficient. We first simulate a GP with covariance function $C(x_1, x_2) = \exp(-5.56(x_1 - x_2)^2)$ which gives the true underlying parameters as $\sigma_{GP}^2 = 1$ and $\theta = 5.56$. Then we add a trend $2 + 2x$. Thus the underlying function actually has a linear mean with a slope of 2 instead of a constant mean. We use a GP with constant mean and linear mean to fit the data and then plot the predictions, respectively. The GP with linear trend can be easily fitted using `mlegp` by setting the input parameter `constantMean=0`. The result is shown in Figure 11.4. It is clear that the fitting of a GP with linear mean is not better than that of a GP with constant mean. In fact, the confidence band is wider, which is due to the uncertainty in the estimation of the trend parameters. As a result, it is suggested that in most cases, for the GP model specification, we can simply set the mean as a constant, particularly when the data are limited.

In the previous two examples, the GP is only sampled once at a given input. No measurement noise is considered. This sampling scheme is consistent with

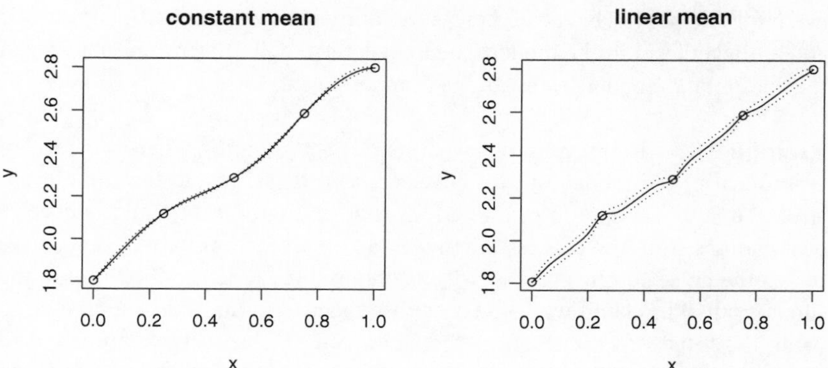

Figure 11.4 Estimated Gaussian processes with constant mean and linear mean function: ———, predicted values; . . . . . ., confidence interval

computer simulations, where the response is deterministic for a given input site. Indeed, one important application of a GP model is to build "surrogate" model, i.e., an approximated model, for computationally expensive simulations. However, in physical experiments, to reduce the influence of the measurement noise, repeated experiments at the same input sites are often conducted. In such cases, an error term should be added to the GP model as follows

$$y_i = \mu + g(\mathbf{x}_i) + \epsilon_i, \tag{11.9}$$

where ϵ_i is the error term due to measurement noise and is often assumed to be a simple white noise, i.e., $\epsilon_i \sim \mathcal{N}(0, \sigma_\epsilon^2)$ and ϵ_i is independent with $g(\cdot)$ and ϵ_j for $i \neq j$. In this model, the GP $g(\cdot)$ is used to model the smooth trend in the signal and the sudden changes in the signal are modeled by the error term ϵ_i. In the GP literature, the error term is often called "nugget" term. With this term, the covariance of the responses $\mathbf{y}$

$$\mathbf{\Sigma}_\mathbf{y} = \sigma_{GP}^2 \text{corr}(\mathbf{X}) + \sigma_\epsilon^2 \mathbf{I}, \tag{11.10}$$

where $\mathbf{I}$ is the identity matrix with the appropriate dimension. The additional parameter σ_ϵ^2 can be estimated using the maximum likelihood estimation method as in (11.5). Please note that the negative log-likelihood function in (11.4) will be adjusted to include the additional parameter: the covariance term $\mathbf{C}(\cdot)$ that only contains GP parameters will be replaced by $\mathbf{\Sigma}_\mathbf{y}$ as in (11.10) that includes both the GP parameters and σ_ϵ^2. Similarly, the prediction formula in (11.7) will also be adjusted by replacing $\mathbf{C}(\cdot)$ with the corresponding covariance matrix with the error term included. One feature of the prediction with nugget term is that the prediction will be smoother and typically not pass through the observations. It also needs to be mentioned that in many cases, even if we do not have repeated measures at the same input sites, people often add a nugget term in the model. The nugget term will improve the stability of the computation and also encourage a smoother fit.

Example 11.3 In this example, we illustrate the situation where we have an error term in the model and we have repeated measures on the same input sites. We first simulate the same GP as that in Example 11.1 and collect 15 observations from the process. Then we add an error term with σ_ϵ^2 as 0.2. We repeat the noise adding process twice to obtain three repeated measures at the same input sites. Then we fit a GP with nugget term and make a prediction using the fitted process. Assume the simulated 15 observations are saved in `gps0`, `gps1`, and `gps2`. We can use `gps<-rbind(gps0,gps1,gps2)` to combine the data and give it to `mlegp` to estimate the model parameters. The function `mlegp` can automatically detect the existence of repeated measurements and add the nugget term in the model accordingly. The estimated GP process is shown in Figure 11.5. Clearly, the predicted curve does not pass through the observed values in this example.

The examples above illustrate that GP is a flexible model for descriptive and predictive analytics. The prediction equation in (11.7) shows that the prediction based on GP model, $\hat{\mathbf{y}}^*$, is a linear function with respect to the observed data $\mathbf{y}$. In fact, there is a close relationship between the GP based prediction and linear regression. An excellent reference discussing the relationship between the GP model and other predictive models is Rasmussen and Williams [2006]. Here we only provide a brief discussion.

To see the relationship between the GP model and linear regression, let us consider a very simple Bayesian regression example. Assume we want to conduct a linear regression between a univariate exploratory variable x and the response variable y without the intercept term, i.e., we want to fit a model $y = \beta x + \epsilon$, where β is the unknown regression parameter and ϵ is the error term and we assume $\epsilon \sim \mathcal{N}(0, \sigma_\epsilon^2)$. We employ the Bayesian approach to estimate the parameter β and we specify the prior distribution of β as $\beta \sim \mathcal{N}(0, \Sigma_\beta)$. We assume we have a dataset of n observations $\mathcal{D} = \{(x_i, y_i) \mid i = 1, \ldots, n\}$ and we want to predict the value of y^* at a new input x^*. Given $\mathbf{x} = (x_1 x_2 \ldots x_n)^T$ and $\mathbf{y} = (y_1\ y_2 \cdots y_n)^T$, it is known that the estimation of β using a Bayesian approach is [Rasmussen and Williams, 2006]:

$$\hat{\beta} = E(\beta \mid \mathbf{y}, \mathbf{x}) = \frac{1}{\sigma_\epsilon^2}\left(\frac{1}{\sigma_\epsilon^2}\mathbf{x}^T\mathbf{x} + \Sigma_\beta^{-1}\right)^{-1}\mathbf{x}^T\mathbf{y}.$$

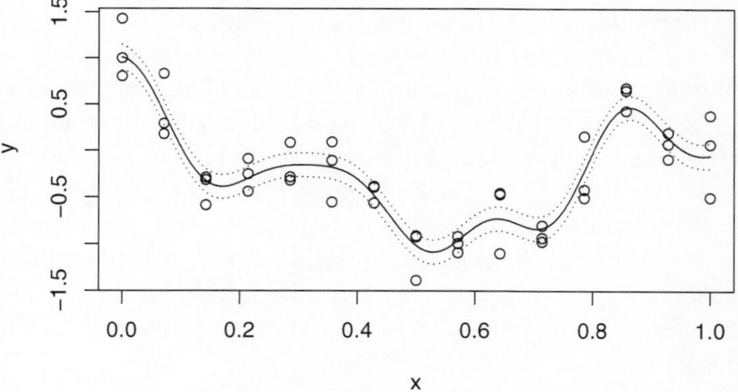

Figure 11.5 Estimated Gaussian processes with repeated measurements: ———, predicted values; . . . . . ., confidence interval

Then the prediction of the response $\hat{y}^*$ at a new input x^* is

$$\hat{y}^*_{lr} = x^*(\mathbf{x}^T\mathbf{x} + \sigma_\epsilon^2\Sigma_\beta^{-1})^{-1}\mathbf{x}^T\mathbf{y}. \tag{11.11}$$

Please note that we add subscript lr to indicate that it is a linear regression prediction.

Now let us use a zero mean GP to model the data and make the prediction. We use the model $y = g(x) + \epsilon$ where $g(x)$ is a zero mean GP and we specify a linear covariance function as $C(x_i, x_j) = \sigma^2 x_i x_j$ for $g(x)$. Also, we have the same assumption for $\epsilon \sim \mathcal{N}(0, \sigma_\epsilon^2)$. Thus, under this case, Equation (11.6) becomes

$$\begin{pmatrix} \mathbf{y} \\ y^* \end{pmatrix} \sim \mathcal{N}\left(0, \sigma^2 \begin{pmatrix} \mathbf{x}\mathbf{x}^T + \sigma_\epsilon^2 & x^*\mathbf{x} \\ x^*\mathbf{x}^T & x^*x^* + \sigma_\epsilon^2 \end{pmatrix}\right). \tag{11.12}$$

Plugging in the covariance function in (11.12) into the prediction formula (11.7), we have

$$y^*_{gp} = x^*\mathbf{x}^T(\mathbf{x}\mathbf{x}^T + \sigma_\epsilon^2)^{-1}\mathbf{y}. \tag{11.13}$$

If we select Σ_β as one (or a simple identity matrix if the dimension of β is more than 1), then the two expressions in (11.11) and (11.13) are very similar. If fact, if we use the matrix inverse lemma (also called the Woodbury formula)

$$(\mathbf{A} + \mathbf{UCV})^{-1} = \mathbf{A}^{-1} - \mathbf{A}^{-1}\mathbf{U}(\mathbf{C}^{-1} + \mathbf{V}\mathbf{A}^{-1}\mathbf{U})^{-1}\mathbf{V}\mathbf{A}^{-1},$$

where $\mathbf{A}$, $\mathbf{U}$, $\mathbf{C}$, $\mathbf{V}$ are all matrices with appropriate dimensions, we can prove that these two expressions are identical given that we set Σ_β as the identity matrix. The detailed algebraic manipulation for the proof is omitted here.

The above simple illustration shows that the GP based prediction using a linear covariance function is identical to the prediction using Bayesian linear regression with appropriate prior distribution. What will happen if we select another covariance function for the GP, e.g., the popular squared exponential function? It turns out that the relationship between GP model and the Bayesian regression still holds, but the regressors will become projected "features" from the input. Projecting the initial inputs into feature space is a commonly used idea to make the model more flexible and expressive. In the previous example, we set the model structure as $y = \beta x + \epsilon$, which limits the relationship between y and x to be a simple linear relationship. However, we can make a projection of the scalar x into the space of powers of x as

$$x \to \phi(x) = (1\ x\ x^2 \ldots)^T,$$

and then set the model structure as

$$y = \phi(x)^T \beta + \epsilon. \tag{11.14}$$

Clearly, the model structure in (11.14) allows any polynomial relationship between y and x, although the relationship between y and the model parameters β is still linear. In the feature space projection, the terms in $\phi(x)$ are also called the basis functions of the feature space.

It can be shown the GP based prediction is equivalent to the prediction using Bayesian regression where the regressors are the projected features of the input, which are determined by the covariance function of the GP. For example, it can be shown that the popular squared exponential covariance function $\exp\left(-\frac{(x_i - x_j)^2}{2l^2}\right)$ corresponds to an infinite number of basis functions as $\phi_c(x) = \exp\left(-\frac{(x-c)^2}{2l'^2}\right)$ where l' is proportional to l and c is an arbitrary real number. In other words, the GP based prediction is equivalent to the prediction of a Bayesian regression where y is regressed on an infinite number of basis functions $\phi_c(x)$. It is somewhat hard to wrap one's head around this equivalence because it is hard to understand how to conduct a regression with an infinite number of covariates. This is achieved through a so-called "kernel trick", through which we can compute the inner product of an infinite number of basis functions through the evaluation of a kernel function. Here, a covariance function can be viewed as a kernel function. The intuition is that, in many linear regression predictions, we do not need to find the unknown coefficients first and then conduct the prediction. Instead, we only need to compute the inner product (or weighted inner product) of the regressors through the kernel function, even if we have infinite number of regressors.

The "kernel trick" is widely used in machine learning to fight the curse of dimensionality and to improve the model flexibility. Limited by the scope of this book, we will not discuss the kernel trick further. Interested readers can refer to Scholkopf and Smola [2001].

11.3 Pairwise Gaussian Process Model

In previous sections, we introduced the basic structure of a GP and showed that the GP is a very flexible way to describe a complex response. It can provide a prediction of the response at a new input site with a confidence interval. It should be pointed out that the GP based prediction can be viewed as a type of "interpolation". So the new input site should not be too far from the sites we have observations. In all the previous discussions, we assume the response is univariate. In other words, one GP only models one response signal. However, as discussed in Chapter 10 and shown in Figure 10.1, we often have longitudinal data such as the degradation signal collected from *multiple* units. In Chapter 10, we introduced a mixed effects model approach to model the set of longitudinal data together. Such an approach is a parametric approach: we assume the signals follow the same polynomial form and they only differ in the values of the parameters.

Can we use the GP model to describe multiple longitudinal signals? It turns out that this can be done through a Multi-Output Gaussian Process (MOGP), in which one output corresponds to one longitudinal signal. In fact this is a quite effective approach. Indeed, the GP model is a nonparametric and generally more flexible model compared with the mixed effects model. Further, in the MOGP model, the prediction conducted at a new input site for a specific signal draws information not only from the same signal but also from other "similar" signals. This will bring tremendous benefit comparing with the single output GP approach where the information can only be from the same signal. For example, if we want to make a prediction in the far future for a temporally evolving signal, the single output GP approach will not be effective because the new input site will be far from any observed sites. However, in MOGP, if we have observed other signals at the time close to the time we want to make a prediction, we will have close observed sites in the model and thus improve the prediction significantly.

In this section, we will introduce the basic structure of MOGP and discuss the MOGP based prediction method. The material in this chapter is based on Kontar et al. [2018].

11.3.1 Introduction to Multi-output Gaussian Process

To simplify the presentation, we only consider the GP model with constant mean. The MOGP model for N signals/outputs is an extension of the single output model in (11.9) as

$$y_{ik} = \mu_i + g_i(\mathbf{x}_{ik}) + \epsilon_{ik}, \tag{11.15}$$

where i, $1 \leq i \leq N$ is the signal index and k is the sample index. If for the ith signal, we have n_i observations and let $\mathbf{y}_i = (y_{i1} \cdots y_{in_i})^T$ and $\mathbf{X}_i = (\mathbf{x}_{i1}, \ldots, \mathbf{x}_{in_i})$ be the responses and the corresponding input sites for the ith signal, the observed dataset for N signals can be denoted as $\mathcal{D} = \{(\mathbf{y}_1, \mathbf{X}_1), \ldots, (\mathbf{y}_i, \mathbf{X}_i), \ldots, (\mathbf{y}_N, \mathbf{X}_N)\}$.

As discussed previously, the specification of the covariance function is of critical importance for a GP model. Unsurprisingly, such specification for MOGP is more complicated compared with that for a single output GP model. Specifically, we need to specify the covariance function between the ith output and the jth output, $C_{ij}(\mathbf{x}, \mathbf{x}')$, which include both the within-output covariance when $i = j$ and the between-output covariance when $i \neq j$. Specifying a flexible covariance function but without over-parameterization (i.e., using too many parameters) is challenging. In many cases, signals are either modeled independently or jointly with a separable covariance structure. If the signals are modeled independently, then we can set $C_{ij}(\mathbf{x}, \mathbf{x}') = 0$ for all $i \neq j$. However, this will challenge our goal of learning information from other signals. A separable covariance structure models the within-output covariance and the between-output covariance separately. A popular choice for separable covariance is proposed in Qian et al. [2008], Zhou et al. [2011], which is given in the form of a correlation function as follows

$$\text{corr}_{ij}(\mathbf{x}, \mathbf{x}') = \tau_{i,j} \exp\left(-(\mathbf{x} - \mathbf{x}')^T \mathbf{D}\, (\mathbf{x} - \mathbf{x}')\right), \tag{11.16}$$

where $\mathbf{D}$ is a diagonal matrix and $\tau_{i,j}$ is a scalar. In this correlation matrix, the exponential part is often called the Gaussian correlation function, which is an extension of the squared exponential covariance function. This part is only related to the input sites and it is the same for all signals. The parameter $\tau_{i,j}$ is only related to the output index, representing the correlation between two outputs. The parameters of this model include the diagonal elements of $\mathbf{D}$, $\tau_{i,j}$s, and the scaling parameter σ^2_{GP} (see (11.8)). One challenging issue for estimating this model is that we need to make sure the estimated parameters make the final correlation matrix a valid one. In Qian et al. [2008], the authors proved that the matrix $\{\tau_{i,j}\}$, $i, j = 1, \ldots, N$, should be a positive definite matrix with diagonal elements being 1. This constraint is relatively hard to reinforce during the parameter estimation procedure. In Zhou et al. [2011], the authors proposed a re-parameterization of $\{\tau_{i,j}\}$, $i, j = 1, \ldots, N$, and changed the constraint into a simple box constraint on the re-defined parameters, which makes the parameter estimation simpler.

The key advantage of the separable covariance function is that the number of parameters in the model is relatively small compared with the non-separable models. However, this also unavoidably causes inflexibility in the model. For example, we can easily see that all the signals share the same parameters in $\mathbf{D}$. In other words, we use the same GP to model different signals. This will sometimes become a critical limitation. A different way of specifying the covariance function is to use a non-separable covariance function through the convolution process. This approach is based on the idea that a GP $g(\mathbf{x})$ can be constructed by convolving a Gaussian white noise process $W(\mathbf{x})$ with a smoothing kernel function $K(\mathbf{x})$ as

$$g(\mathbf{x}) = \int_{\mathcal{S}} K(\mathbf{x} - \mathbf{u}) W(\mathbf{u}) d\mathbf{u}, \tag{11.17}$$

where $\mathcal{S}$ is the input space, $\text{cov}(W(\mathbf{x}), W(\mathbf{x}')) = \delta(\mathbf{x} - \mathbf{x}')$ and δ is the Dirac delta function. The Dirac delta function is defined as $\delta(\mathbf{x} - \mathbf{x}') = +\infty$ if $\mathbf{x} = \mathbf{x}'$ and is zero otherwise and is constrained to satisfy $\int_{\mathcal{S}} \delta(\mathbf{u}) d\mathbf{u} = 1$. The kernel function $K(\cdot)$ needs to be square or absolutely integrable. The resulting covariance function of $g(\mathbf{x})$ is

$$C(\mathbf{x}, \mathbf{x}') = \int_{\mathcal{S}} K(\mathbf{x} - \mathbf{u}) K(\mathbf{x}' - \mathbf{u}) d\mathbf{u}.$$

The details of this construction method can be found in Matérn [2013], Higdon [2002].

With this method, we can directly parameterize the covariance function through parameters in the smoothing kernel. In other words, once we know the kernel function, we will know the corresponding covariance function. For example, if we select the popular square exponential function as the kernel function, we will have the corresponding covariance function as

$$K(\mathbf{x}) \propto \exp(-\frac{1}{2} ||\mathbf{x}||^2) \rightarrow C(\mathbf{x}, \mathbf{x}') \propto \exp(-\frac{1}{4} ||\mathbf{x} - \mathbf{x}'||^2).$$

The symbol $\propto$ means the left term differs from the right term by a constant multiplicative term.

Equation (11.17) shows a way of constructing a single output GP. How can we use this approach to build a joint covariance function among multiple Gaussian processes that correspond to multiple signals? The basic idea is to build multiple GPs where all the GPs $g_i(\mathbf{x})$ depend on some common latent Gaussian white noise process $W(\mathbf{x})$. Through the common latent process, the covariance between different signals can then be modeled. Let us consider a pairwise example first.

11.3.2 Pairwise GP Modeling Through Convolution Process

We consider constructing the covariance functions for two GPs, g_i and g_j, as shown in Figure 11.6. We want to illustrate how the covariance is captured in the model through the convolution process. In the figure, W_i, W_j, and W_0 are independent Gaussian white noise processes and K_{ii}, K_{0i}, K_{0j}, and K_{jj} are kernel functions. The symbol $*$ represents convolution operation. In this construction, g_i is a summation of $W_i * K_{ii}$ and $W_0 * K_{0i}$, while g_j is a summation of $W_j * K_{jj}$ and $W_0 * K_{0j}$. Clearly W_0 is the shared latent process between g_i and g_j.

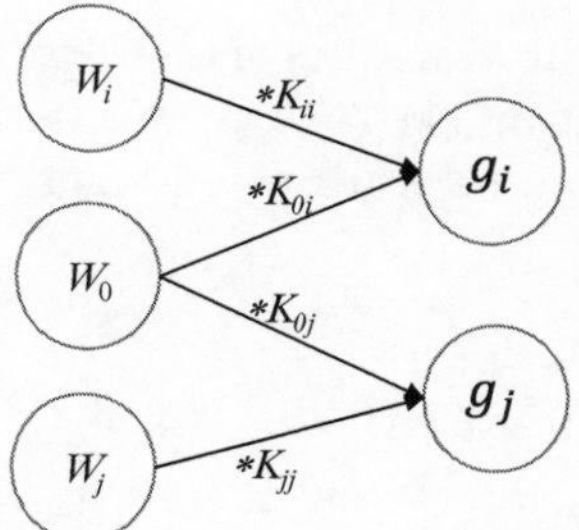

Figure 11.6 Construct MOGP model with two outputs

From the construction scheme, we have

$$\begin{aligned} g_i(\mathbf{x}) &= W_0(\mathbf{x}) * K_{0i}(\mathbf{x}) + W_i(\mathbf{x}) * K_{ii}(\mathbf{x}) \\ &= \int_S K_{0i}(\mathbf{u}) W_0(\mathbf{x}-\mathbf{u}) \mathrm{d}\mathbf{u} + \int_S K_{ii}(\mathbf{u}) W_i(\mathbf{x}-\mathbf{u}) \mathrm{d}\mathbf{u}. \end{aligned}$$

The expression of g_j is identical except that the index i is replaced by index j in the above equation. Now noting the Gaussian processes that we are considering have mean zero, we have

$$C_{ij}(\mathbf{x}, \mathbf{x}') = E(g_i(\mathbf{x}) g_j(\mathbf{x}')),$$

which is

$$\begin{aligned} &E(g_i(\mathbf{x}) g_j(\mathbf{x}')) = \\ &E\Big(\sum_{q=\{0,i\}} \int_S K_{qi}(\mathbf{u}) W_q(\mathbf{x}-\mathbf{u}) \mathrm{d}\mathbf{u} \times \sum_{q'=\{0,j\}} \int_S K_{q'j}(\mathbf{u}') W_{q'}(\mathbf{x}'-\mathbf{u}') \mathrm{d}\mathbf{u}' \Big). \end{aligned}$$

Noting only $W_q(\cdot)$ and $W_{q'}(\cdot)$ are random variables, we have

$$\begin{aligned} &E\Big(\int_S K_{qi}(\mathbf{u}) W_q(\mathbf{x}-\mathbf{u}) \mathrm{d}\mathbf{u} \int_S K_{q'j}(\mathbf{u}') W_{q'}(\mathbf{x}'-\mathbf{u}') \mathrm{d}\mathbf{u}' \Big) \\ &= \int_S \int_S K_{qi}(\mathbf{u}) K_{q'j}(\mathbf{u}) E(W_q(\mathbf{x}-\mathbf{u}) W_{q'}(\mathbf{x}'-\mathbf{u}')) \mathrm{d}\mathbf{u}\mathrm{d}\mathbf{u}'. \end{aligned} \tag{11.18}$$

However, W_q and $W_{q'}$ are white noise processes and the covariance of them are zero for $q \neq q'$. Thus, (11.18) is non-zero only if $q = q'$. In such a case, we have

$$\begin{aligned} &\int_S \int_S K_{qi}(\mathbf{u}) K_{qj}(\mathbf{u}) E(W_q(\mathbf{x}-\mathbf{u}) W_q(\mathbf{x}'-\mathbf{u}')) \mathrm{d}\mathbf{u}\mathrm{d}\mathbf{u}' \\ &= \int_S \int_S K_{qi}(\mathbf{u}) K_{qj}(\mathbf{u}) \delta(\mathbf{x}-\mathbf{u}-\mathbf{x}'+\mathbf{u}') \mathrm{d}\mathbf{u}\mathrm{d}\mathbf{u}' \\ &= \int_S K_{qi}(\mathbf{u}) K_{qj}(\mathbf{u}-\mathbf{d}) \mathrm{d}\mathbf{u}, \end{aligned}$$

where $\mathbf{d}=\mathbf{x}-\mathbf{x}'$. In the above derivation, we use the properties of the Dirac function $\int f(u)\delta(u-x)\mathrm{d}u = f(x)$. Using the results in (11.18), we have the covariance function of g_i and g_j constructed using the structure in Figure 11.6 as

$$\begin{cases} C_{ii}(\mathbf{x},\mathbf{x}')=\sum_{q=\{0,i\}}\int_{\mathcal{S}} K_{qi}(\mathbf{u})K_{qi}(\mathbf{u}-\mathbf{d})\mathrm{d}\mathbf{u} \\ C_{jj}(\mathbf{x},\mathbf{x}')=\sum_{q=\{0,j\}}\int_{\mathcal{S}} K_{qj}(\mathbf{u})K_{qj}(\mathbf{u}-\mathbf{d})\mathrm{d}\mathbf{u} \\ C_{ij}(\mathbf{x},\mathbf{x}')=\int_{\mathcal{S}} K_{0i}(\mathbf{u})K_{0j}(\mathbf{u}-\mathbf{d})\mathrm{d}\mathbf{u} \text{ for } i\neq j \end{cases} \cdot \tag{11.19}$$

The result in (11.19) is general and can be applied to an arbitrary kernel function. The Gaussian kernel is a common choice, as it can model various features and can provide a large degree of flexibility with a small number of parameters. Specifically, let us consider a one-dimensional input space and define the Gaussian kernel as

$$K_{qi}(x)=\frac{\alpha_{qi}\pi^{-1/4}}{\sqrt{|\lambda_{qi}|}}\exp\left(-\frac{x^2}{2\lambda_{qi}^2}\right), \tag{11.20}$$

where $q\in\{0,i\}$ and λ_{qi} and α_{qi} are the model parameters. Plugging the kernel definition in (11.20) into (11.19) and noting $\mathcal{S}=[-\infty,+\infty]$, we can obtain the following corresponding covariance functions

$$\begin{cases} C_{ii}(x,x')=\alpha_{0i}^2\exp\left(-\frac{d^2}{4\lambda_{0i}^2}\right)+\alpha_{ii}^2\exp\left(-\frac{d^2}{4\lambda_{ii}^2}\right) \\ C_{jj}(x,x')=\alpha_{0j}^2\exp\left(-\frac{d^2}{4\lambda_{0j}^2}\right)+\alpha_{jj}^2\exp\left(-\frac{d^2}{4\lambda_{jj}^2}\right) \\ C_{ij}(x,x')=\alpha_{0i}\alpha_{0j}\sqrt{\frac{2|\lambda_{0i}\lambda_{0j}|}{\lambda_{0i}^2+\lambda_{0j}^2}}\exp\left(-\frac{d^2}{2(\lambda_{0i}^2+\lambda_{0j}^2)}\right) \quad \text{for } i\neq j \end{cases}, \tag{11.21}$$

where $d=x-x'$. Equation (11.21) can be derived from (11.19) by using the integral formula. A more general derivation dealing with multi-dimensional inputs can be found in the supplementary materials of Kontar et al. [2018].

Please note that the covariance between g_i and g_j for $i\neq j$ is proportional to α_{0i} and α_{0j}. If either α_{0i} or α_{0j} is zero, then the covariance will be zero and g_i and g_j will become two separate standard single output GPs with squared exponential covariance function as shown in the first two equations of (11.21).

Equation (11.21) provides the expression of a non-separable covariance function of an MOGP model with two outputs. It is interesting to note that in this model, we have a total of eight unknown parameters. If we use a separable covariance function, we will only have three free parameters (two for the within-signal covariance and one for the between-signal covariance). Clearly, the parameter number is much larger for the non-separable case.

We can use the MLE method to estimate the parameter. Specifically, considering the model in (11.15) with only two outputs

$$\begin{cases} y_{s1} = \mu_1 + g_1(x_{s1}) + \epsilon_{s1} \\ y_{s2} = \mu_2 + g_2(x_{s2}) + \epsilon_{s2} \end{cases}. \tag{11.22}$$

Assume we have n_1 observations for the first output, which is denoted as $\mathbf{y}_1 = (y_{11} \cdots y_{1n_1})^T$, $\mathbf{X}_1 = (x_{11} \dots x_{1n_1})^T$ and have n_2 observations for the second output, which is denoted as $\mathbf{y}_2 = (y_{21} \cdots y_{2n_2})^T$, $\mathbf{X}_2 = (x_{21} \dots x_{2n_2})^T$, we have

$$\text{cov}\left(\begin{pmatrix} \mathbf{y_1} \\ \mathbf{y_2} \end{pmatrix}\right) = \begin{pmatrix} \mathbf{C}_{11}(\mathbf{X}_1, \mathbf{X}_1) & \mathbf{C}_{12}(\mathbf{X}_1, \mathbf{X}_2) \\ \mathbf{C}_{12}(\mathbf{X}_1, \mathbf{X}_2)^T & \mathbf{C}_{22}(\mathbf{X}_2, \mathbf{X}_2) \end{pmatrix} + \begin{pmatrix} \sigma_1^2 \mathbf{I}_{n_1} & 0 \\ 0 & \sigma_2^2 \mathbf{I}_{n_2} \end{pmatrix}, \tag{11.23}$$

where we assume the error term is independent and identically distributed for each output. The elements of the first matrix on the right-hand side can be obtained using (11.21).

With the covariance in (11.23) and noting the means for $\mathbf{y}_1$ and $\mathbf{y}_2$ are μ_1 and μ_2, respectively, we can write out the likelihood function similar to (11.4). Through the maximization, we can get the estimates for the parameters in the model, namely, $\boldsymbol{\theta} = \{\alpha_{01}, \alpha_{02}, \alpha_{11}, \alpha_{22}, \lambda_{01}, \lambda_{02}, \lambda_{11}, \lambda_{22}, \sigma_1^2, \sigma_2^2\}$. With the estimated parameters, we can compute the covariance term at a new input location. Similar to the single output GP prediction, the prediction in MOGP can be achieved. Without loss of generality, assume we want to predict the value of the second output at a new input site x^*, then similar to (11.6), we can write the joint distribution of $\mathbf{y} = (\mathbf{y}_1^T \ \mathbf{y}_2^T)^T$ and the corresponding y_2^* as

$$\begin{pmatrix} \mathbf{y_1} \\ \mathbf{y_2} \\ y_2^* \end{pmatrix} \sim \mathcal{N}\left(\begin{pmatrix} \mu_1 \cdot \mathbf{1}_{n_1} \\ \mu_2 \cdot \mathbf{1}_{n_2} \\ \mu_2 \end{pmatrix}, \boldsymbol{\Sigma}_{MOGP} + \boldsymbol{\Sigma}_\epsilon\right), \tag{11.24}$$

where $\boldsymbol{\Sigma}_{MOGP}$ is the covariance of the underlying GPs and $\boldsymbol{\Sigma}_\epsilon$ is the covariance due to the error term, which are given as

$$\boldsymbol{\Sigma}_{MOGP} = \begin{pmatrix} \mathbf{C}_{11}(\mathbf{X}_1, \mathbf{X}_1) & \mathbf{C}_{12}(\mathbf{X}_1, \mathbf{X}_2) & \mathbf{C}_{12}(\mathbf{X}_1, x^*) \\ \mathbf{C}_{12}^T(\mathbf{X}_1, \mathbf{X}_2) & \mathbf{C}_{22}(\mathbf{X}_2, \mathbf{X}_2) & \mathbf{C}_{22}(\mathbf{X}_2, x^*) \\ \mathbf{C}_{12}^T(\mathbf{X}_1, x^*) & \mathbf{C}_{22}^T(\mathbf{X}_2, x^*) & C_{22}(x^*, x^*) \end{pmatrix} \tag{11.25}$$

and

$$\boldsymbol{\Sigma}_\epsilon = \begin{pmatrix} \sigma_1^2 \mathbf{I}_{n_1} & \mathbf{0} & \mathbf{0} \\ \mathbf{0} & \sigma_2^2 \mathbf{I}_{n_2} & \mathbf{0} \\ \mathbf{0} & \mathbf{0} & \sigma_2^2 \end{pmatrix}.$$

With the joint distribution, we can use the same conditional mean and variance approach as that in (11.6) and (11.7) to provide a prediction for y_{*2}. Please note in this case, the term $\mathbf{C}(\mathbf{X},\mathbf{X})$ in (11.7) becomes the upper left 2×2 block of the $\boldsymbol{\Sigma}_{MOGP}+\boldsymbol{\Sigma}_\epsilon$ with dimension of $(n_1+n_2)\times(n_1+n_2)$, while $\mathbf{C}(\mathbf{X}^*,\mathbf{X}^*)$ becomes a scalar $C_{22}(x^*,x^*)+\sigma_2^2$. It is also interesting to note that if we only use the bottom-right 2×2 block of $\boldsymbol{\Sigma}_{MOGP}+\boldsymbol{\Sigma}_\epsilon$ to make the prediction, then it is identical to the prediction based on a single output GP model only considering $\mathbf{y}_2$. In the pairwise GP based prediction using the full covariance matrix, it is clear that the other output will also contribute to the prediction of the current output.

From the above discussion, it can be seen that once the covariance function between different outputs/signals are specified, the MOGP model estimation and prediction is very similar to that of the single output GP model. In the following, we will present an example to illustrate the pairwise GP approach for prediction.

Example 11.4 In this example, we illustrate the advantage of predicting with MOGP. Figure 11.7 shows the prediction of a simple quadratic curve using a GP. The underlying function is $x^2/2$ as shown by the solid line in the figure. The observation is this function with an addition of a zero mean error term with variance of 0.01. The circles are the observations to fit the GP with constant mean. The prediction is shown as the dashed line and the dotted lines shows the confidence interval of the prediction. The prediction results in Figure 11.7 clearly show that the prediction is erroneous when the input sites on which we want to make prediction is far from the observed input sites. In fact, it is known that in GP, the prediction will converge to the estimated mean of the process if the distance from the observed input sites get larger. This can be clearly observed from the late part of the prediction in Figure 11.7.

Now let us consider a prediction based on a two-output GP. As shown in Figure 11.8, we have the observation from two signals, represented by $+$ and $\circ$ symbols, respectively. The first signal, represented by $+$, is $y_1 = x^2 + \epsilon$, while the second signal, represented by $\circ$, is the same as the one used in Figure 11.7, $y_2 = x^2/2 + \epsilon$. We use a two-output GP to model these two signals and make

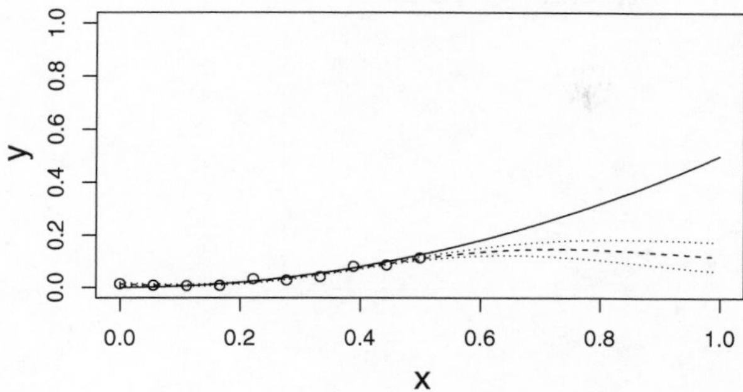

Figure 11.7 Predicting with a single output Gaussian process: "∘", observed data; ———, true signal; - - - -, predicted signal; . . . . . ., confidence interval

predictions for y_2. We assumed a constant mean for both signals and adopted a separable covariance function as shown in (11.16) for the two-output GP. Please note that in this model, we have four free parameters: one parameter associated with $\mathbf{D}$ (noting that the input space is one-dimensional so $\mathbf{D}$ is just a scalar), one cross correlation parameter $\tau_{1,2}$, one scaling parameter from correlation to covariance (not shown in (11.16)), and one parameter of error term variance (not shown in (11.16)). We can use the MLE method to estimate the parameters based on the observed data ($+$ and $\circ$ in Figure 11.8) and then make a prediction based on the estimated parameters. The predicted values for y_2 are shown as a dashed line and the confidence intervals of the prediction are shown as a dotted line. It is clear that the existence of an additional signal significantly changed the prediction. Indeed, when the prediction sites are far from any observed sites in $\mathbf{y}_2$, the prediction of y_2 no longer goes to the estimated mean. Instead, it gradually mimics y_1. Clearly, if the characteristics of y_1 are very similar to y_2, the two-output GP prediction will be much more accurate than that of a single output GP prediction.

Example 11.5 In this example, we illustrate the two-output GP modeling and prediction with non-separable covariance function. Figure 11.9 shows the prediction results. The two signals and the error term settings are the same as that shown in Figure 11.8. We use the convolution processes as shown in Figure 11.6 to construct the non-separable covariance functions. Please note that in this two-output GP model, we will have nine free parameters, two for each of the four kernel functions in Figure 11.6 and one for the variance of the error term. We can use the MLE method to estimate the parameters and then use the estimated parameters to make predictions. Clearly, the prediction results shown in Figure 11.9 are much better than that in Figure 11.8. This is due to the flexible non-separable structure of the covariance function adopted.

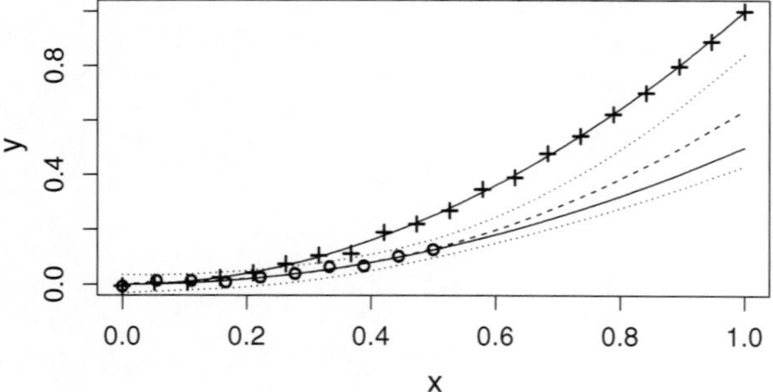

Figure 11.8 Predicting with a two-output GP with separable covariance function: "+" observed data from signal 1; "∘", observed data from signal 2; ———, true signal; - - - -, predicted signal; . . . . . ., confidence interval

Figure 11.10 shows the prediction comparison between separable and non-separable covariance functions under more complex data characteristics. In this comparison, the first signal is a sinusoidal function while the second signal is a sinusoidal function with 90 degree shift. In this case, the performances with non-separable and separable covariance functions are similar. Both have a fairly large prediction error. This indicates that the similarity between signals plays an important role in MOGP based modeling and prediction. If the two signals are quite different, then modeling them together will bring no benefit.

Assuming we have two sinusoidal functions only with magnitude difference, we can get much better prediction as shown in Figure 11.11. However, even in this case, the two-output GP with separable covariance function still has a significant prediction error.

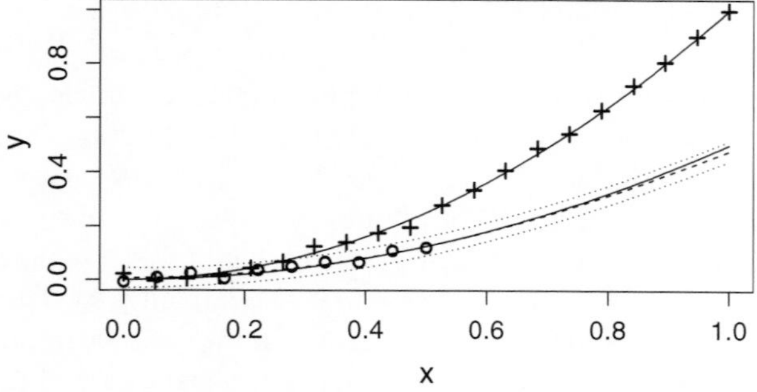

Figure 11.9 Predicting with a two-output GP with non-separable covariance function: "+", observed data from signal 1; "∘", observed data from signal 2; ———, true signal; - - - -, predicted signal; . . . . . ., confidence interval

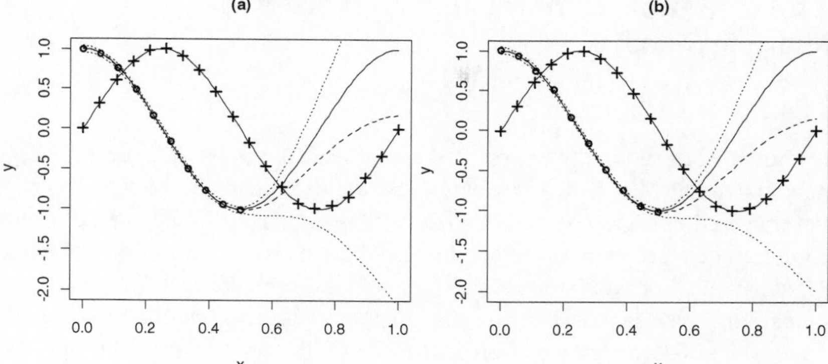

Figure 11.10 Predicting with a two-output GP for sinusoidal function with (a) separable and (b) non-separable covariance function: "+", observed data from signal 1; "∘", observed data from signal 2; ———, true signal; - - - -, predicted signal; . . . . . ., confidence interval

There is no R package currently to achieve MOGP modeling and prediction. In this example, the two-output GP model and prediction is achieved through customized code. Please refer to the book companion website for more details.

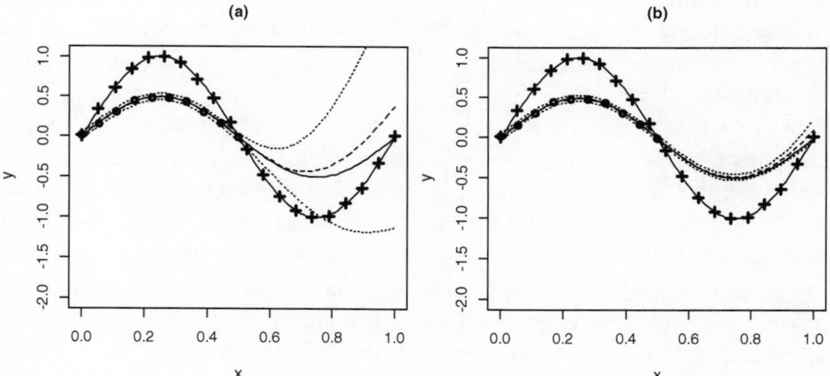

Figure 11.11 Predicting with a two-output GP for synchronous sinusoidal function with (a) separable and (b) non-separable covariance function: "+", observed data from signal 1; "∘", observed data from signal 2; ———, true signal; - - - -, predicted signal; . . . . . ., confidence interval

11.4 Multiple Output Gaussian Process for Multiple Signals

11.4.1 Model Structure

In the previous section, we presented a pairwise GP model including two signals. We mainly use such a model to illustrate the concept of MOGP. In practice, prediction based on two signals will be a fairly limited scenario and commonly we are required to conduct predictions based on multiple signals. The basic idea of GP for multiple signals is the same as that for a pairwise model. However, to reduce the number of parameters to be estimated, in a high dimensional GP model, we often only focus on the prediction of one specific signal instead of any signals within the group. We can use the non-separable convolution construction as in Figure 11.12 and the $N+1$ signals/outputs are modeled as in (11.15). More explicitly

$$\begin{aligned} y_{1s} &= \mu_1 + g_1(x_{1_s}) + \epsilon_{1_s} \\ y_{2s} &= \mu_2 + g_2(x_{2_s}) + \epsilon_{2s} \\ \vdots & \quad \vdots \quad \vdots \\ y_{Ns} &= \mu_N + g_N(x_{N_s}) + \epsilon_{Ns} \\ y_{rs} &= \mu_r + g_r(x_{rs}) + \epsilon_{rs} \end{aligned},$$

where the subscript $j \in \{1, \dots, N, r\}$, is the output index and s is the sample index from 0 to n_j. Please note we do not require that the signals are observed at the same input sites and the number of observations from different signals may be different. Similar to that in the pairwise modeling, we can collect all the observations and denote them as $\mathbf{y}_1$, $\mathbf{y}_2$, ..., $\mathbf{y}_N$, and $\mathbf{y}_r$ for each signal, respectively.

In this setting, we assume $\mathbf{y}_1,\dots,\mathbf{y}_N$ are the "historical" data, while $\mathbf{y}_r$ is the currently observed signal and we want to predict its values at other input sites. As shown in Figure 11.12, because we are not interested in predicting values for the historical data, the underlying GP for these signals is only generated through a shared white noise process. However, because we are interested in the prediction for the rth signal, the underlying GP g_r is generated from $N+1$ white noise processes, where N white noise processes $W_1,\dots,W_N$ are shared with the historical signals and one W_r is only associated with g_r. The inclusion of W_r allows unique characteristics for g_r that are not shared with the historical signals.

Considering the convolution structure in Figure 11.12 and noting W_i is independent of W_j for $i \neq j$, we can get

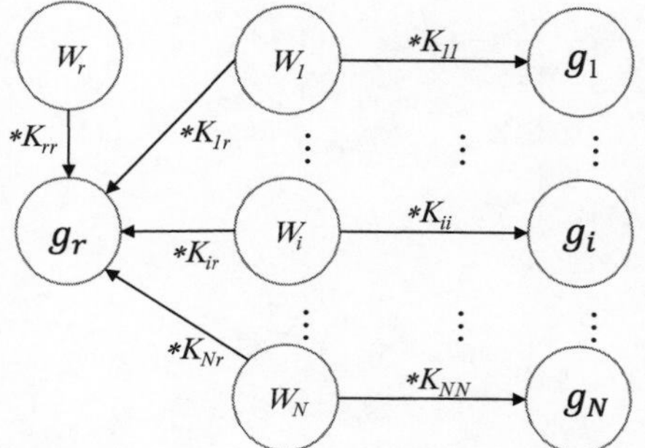

Figure 11.12 Construct MOGP model with multiple signals

$$\begin{cases} C_{ii}(\mathbf{x}, \mathbf{x}') = \int_S K_{ii}(\mathbf{u}+\mathbf{d})K_{ii}(\mathbf{u})\mathrm{d}\mathbf{u} \\ C_{ij}(\mathbf{x}, \mathbf{x}') = 0 \quad \text{for} \quad i \neq j \end{cases}, \tag{11.26}$$

where $\mathbf{d} = \mathbf{x} - \mathbf{x}'$ and $i, j \in \{1, \dots, N\}$. The covariance structure related to g_r is more complicated

$$\begin{cases} C_{ir}(\mathbf{x}, \mathbf{x}') = \int_S K_{ii}(\mathbf{u}+\mathbf{d})K_{ir}(\mathbf{u})\mathrm{d}\mathbf{u};\ i \in \{1, \dots, N\} \\ C_{rr}(\mathbf{x}, \mathbf{x}') = \int_S K_{rr}(\mathbf{u}+\mathbf{d})K_{rr}(\mathbf{u})\mathrm{d}\mathbf{u} + \sum_{i=1}^{N} \int_S K_{ir}(\mathbf{u}+\mathbf{d})K_{ir}(\mathbf{u})\mathrm{d}\mathbf{u} \end{cases} \tag{11.27}$$

Equations (11.26) and (11.27) are general expressions for the covariance functions constructed through the convolution processes shown in Figure 11.12. If the input space is one-dimensional and we select a Gaussian kernel function as shown in (11.20), we can get the explicit covariance function as

$$\begin{cases} C_{ii}(x, x') = \alpha_{ii}^2 \exp\left(-\dfrac{d^2}{4\lambda_{ii}^2}\right) \\ C_{ij}(x, x') = 0 \\ C_{ir}(x, x') = \alpha_{rr}\alpha_{ir}\sqrt{\dfrac{2|\lambda_{rr}\lambda_{ir}|}{\lambda_{rr}^2+\lambda_{ir}^2}} \exp\left(-\dfrac{d^2}{2(\lambda_{rr}^2+\lambda_{ir}^2)}\right) \\ C_{rr}(x, x') = \alpha_{rr}^2 \exp\left(-\dfrac{d^2}{4\lambda_{rr}^2}\right) + \sum_{i=1}^{N} \alpha_{ir}^2 \exp\left(-\dfrac{d^2}{4\lambda_{ir}^2}\right) \end{cases}$$

where $i, j \in \{1, \dots, N\}$ and $i \neq j$.

With the expression of the covariance function of the multiple Gaussian processes, we can get the expression of the joint distribution of $\mathbf{y} = (\mathbf{y}_1^T \dots \mathbf{y}_N^T\ \mathbf{y}_r^T)^T$ as

$$\mathbf{y} = \begin{pmatrix} \mathbf{y}_1 \\ \vdots \\ \mathbf{y}_N \\ \mathbf{y}_r \end{pmatrix} \sim \mathcal{N}\left(\begin{pmatrix} \mu_1 \cdot \mathbf{1}_{n_1} \\ \vdots \\ \mu_N \cdot \mathbf{1}_{n_N} \\ \mu_r \cdot \mathbf{1}_{n_r} \end{pmatrix}, \boldsymbol{\Sigma}_{gps} + \boldsymbol{\Sigma}_{noise} \right), \tag{11.28}$$

where

$$\boldsymbol{\Sigma}_{gps} = \begin{pmatrix} \boldsymbol{\Sigma}_{1,1} & \mathbf{0} & \cdots & \mathbf{0} & \boldsymbol{\Sigma}_{1,r} \\ \mathbf{0} & \boldsymbol{\Sigma}_{2,2} & \cdots & \mathbf{0} & \boldsymbol{\Sigma}_{2,r} \\ \vdots & \vdots & \ddots & \vdots & \vdots \\ \mathbf{0} & \mathbf{0} & \cdots & \boldsymbol{\Sigma}_{N,N} & \boldsymbol{\Sigma}_{N,r} \\ \boldsymbol{\Sigma}_{r,1} & \boldsymbol{\Sigma}_{r,2} & \cdots & \boldsymbol{\Sigma}_{r,N} & \boldsymbol{\Sigma}_{r,r} \end{pmatrix}$$

and

$$\boldsymbol{\Sigma}_{noise} = \begin{pmatrix} \sigma_1^2 \mathbf{I}_{n_1} & 0 & \mathbf{0} & \mathbf{0} \\ \vdots & \ddots & \vdots & \vdots \\ \mathbf{0} & \mathbf{0} & \sigma_N^2 \mathbf{I}_{n_N} & \mathbf{0} \\ \mathbf{0} & \mathbf{0} & \mathbf{0} & \sigma_r^2 \mathbf{I}_{n_r} \end{pmatrix}.$$

Please note in the expression for $\boldsymbol{\Sigma}_{gps}$, we used an abbreviated notation, where $\boldsymbol{\Sigma}_{i,j} = \mathbf{C}_{ij}(\mathbf{X}_i, \mathbf{X}_j)$.

11.4.2 Model Parameter Estimation and Prediction

With the expression of the joint distribution, we can write the multi-output Gaussian likelihood function and then use the MLE method to estimate the kernel function parameters. Please note that this will be a fairly high dimensional model: if we have N historical signals and one new signal we want to make prediction for, and for each signal we have n_i observations, then the dimension of the multivariate Gaussian distribution will be $n_r + \sum_{i=1}^{N} n_i$ and the number of free parameters will be $2Np + p$, where p is the number of free parameters per kernel function, e.g., $p = 2$ for a one dimensional Gaussian kernel function.

After the kernel function parameters are estimated, we can use the estimated values to evaluate the covariance function for any given input sites. Similar to the pairwise modeling case, we can write out the expanded covariance matrix similar to that in (11.24) to include the input sites we want to make predictions for. Then using the formula in (11.7), we can make predictions at those input sites. Below we will use an example to demonstrate the MOGP based prediction.

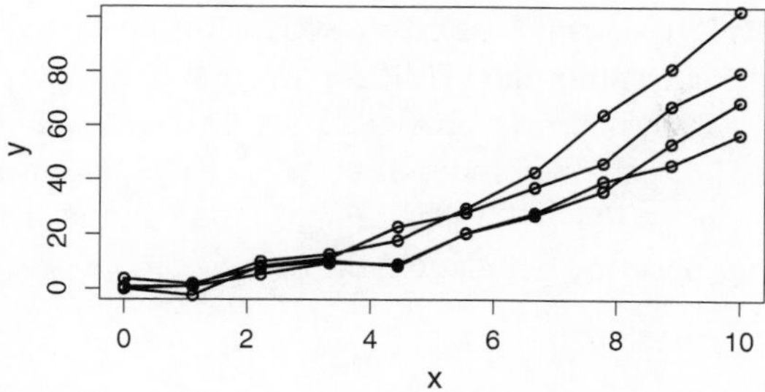

Figure 11.13 Simulation of four historical signals

Example 11.6 Consider a family of signals with the simple quadratic form $y = \beta x^2 + \epsilon$, where β is a random effect and follows normal distribution. That is, for each individual signal, the value of β is unique. Assume we have observed four signals, each with ten observations as shown in Figure 11.13. In generating these signals, the error term standard deviation is selected as three.

We can use the MOGP model introduced in this section to make a prediction for a new signal with unknown coefficient value β. The prediction results are shown in Figure 11.14 if we have three observation points from the new signal. Because we only have three observed points, the prediction performance is not satisfactory. However, the prediction using the MOGP model discussed in this section shows some interesting features. For example, it is known that the GP model is not good at extrapolation. If an input site is far from any input sites for which we have observations, then the prediction on that site will just be the

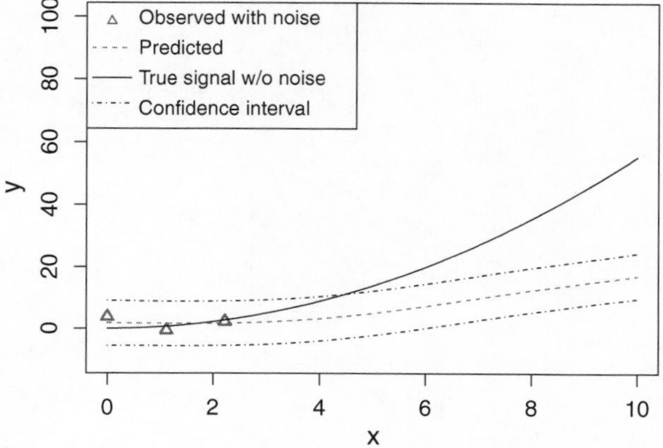

Figure 11.14 Prediction with three observed points using MOGP

mean of the process (typically 0 as in this case). However, with the MOGP model, we can see the increasing trend in the prediction as shown in Figure 11.14 even when the site is far from the three observed sites. This increasing trend is certainly the result of influence from the historical signals that have been taken into consideration by the MOGP model. Figure 11.15 shows the prediction performance with six observed points from the new signal. The prediction is clearly improved.

Example 11.7 In this example, we will consider a more complicated situation where the signals do not belong to the same functional family. As shown in Figure 11.16, we consider four historical signals and two of them are in the family of $y = \beta_0 + 0.3x^2 - 2\sin(\beta_1 \pi x) + \epsilon$ and the other two are in the family of $y = \beta_0 + 1.5x + \beta_1 \sin(x) + \epsilon$, where β_0 and β_1 are random effects and ϵ is the error term. Clearly, the behavior of these two types of signals are quite different. We can use these four signals as historical data for our MOGP model. Then we generate six observation points from a signal and use the MOGP approach to predict its future values. Please note the MOGP model for the prediction is an MOGP with five outputs, four are for the historical data and one for the new signal we want to predict. Figure 11.17 shows the prediction results for two signals from the two families, respectively. We would like to point out that the two predictions are obtained from two MOGP models, one is based on the four historical signals and six observations from a signal in the first family (circles in Figure 11.17) and the other one is based on the same four historical signals and six observations from a signal in the second family (triangles in

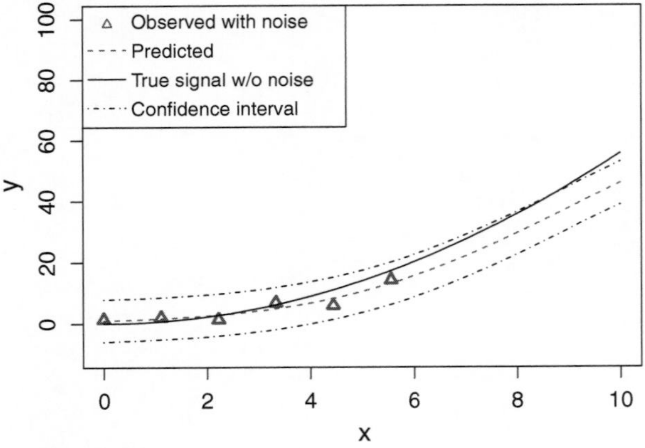

Figure 11.15 Prediction with six observed points using MOGP

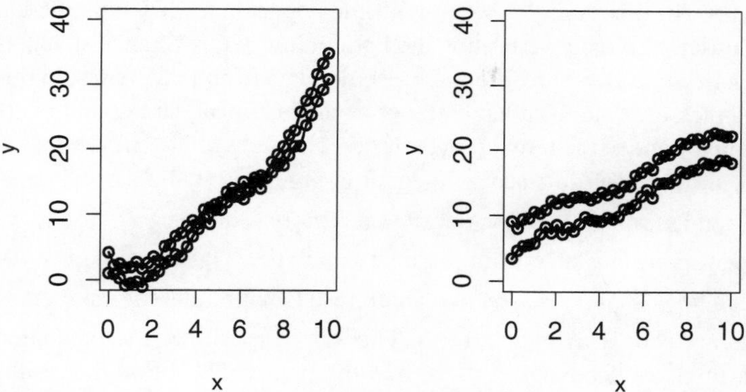

Figure 11.16 Historical signals in two different functional families

Figure 11.17). Although we only have six observations from each signal and we are using the same historical dataset, we can clearly see the predictions for these two signals are quite different: the model automatically identifies the similarities of the signals and then make predictions accordingly. This demonstrates the flexibility of MOGP based prediction and its nonparametric nature. Even if the historical dataset contains longitudinal signals with diverse characteristics, the MOGP based model can adaptively identify the characteristics consistent with the observed data from the new signals and make predictions following those characteristics.

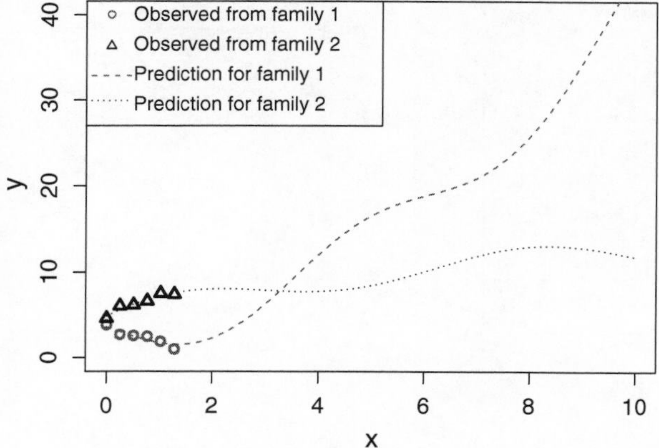

Figure 11.17 Predictions based on the observations from two families of signals

Example 11.8 In this example, we present the application MOGP prediction for a real dataset. The data we used is the battery internal resistance signal as shown in Figure 10.1. There are internal resistance data from 14 batteries. In this example, we pick the data from ten batteries as the historical dataset and make the prediction for one of the remaining batteries. Thus, the MOGP model has 11 outputs. We do not require that each signal is observed at the same inputs. Assume the total number of observations in the historical data is $\sum_{i=1}^{10} n_i$ and we have n_t observations available from the new signal, the total dimension of the fitting GP will be $n_t + \sum_{i=1}^{10} n_i$. The prediction results with different numbers of observations are shown in Figure 11.18. The left panel shows the prediction results with three available observations and the right panel shows the results with six available observations. Because the true underlying function is unknown, we plot the observed values to compare with the prediction. The result shows that MOGP prediction model is indeed very flexible and can make fairly good predictions for arbitrary signals.

In the above examples, we only showed the prediction results using MOGP models for a single signal. Systematic performance evaluation and comparison with other competing predictive analytics methods are not shown. Interested readers can refer to Kontar et al. [2018], where comprehensive performance evaluations and comparisons are presented. In general, MOGP based prediction is a very flexible and effective approach.

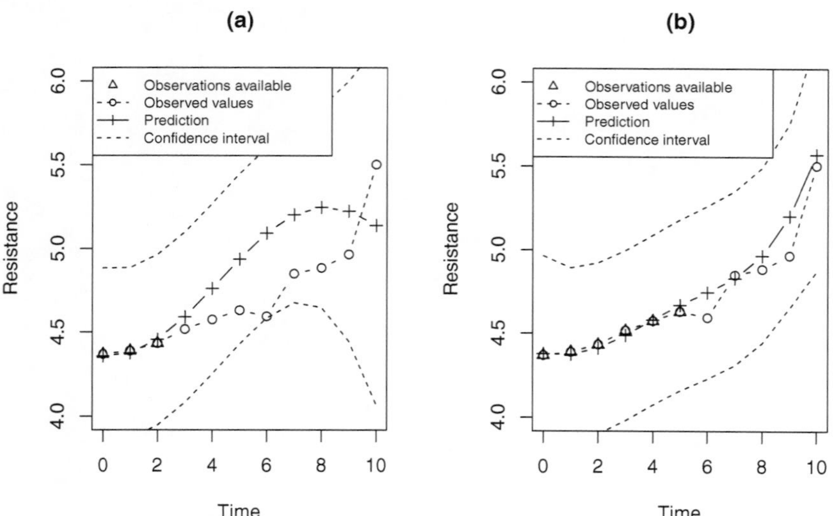

Figure 11.18 Predictions based on the observations for battery internal resistance: (a) with three observations (b) with six observations

One potential challenge in using the MOGP model is the computational overhead. As shown in this section, the final dimension of the resulting multivariate Gaussian distribution is the total number of observations across all the historical signals and the signal under study. If we want to take a large number of historical signals into consideration, the dimension can easily reach thousands or even millions. This will pose significant challenges in the maximum likelihood estimation of the model parameters. Thus, without modification, the MOGP based prediction can only be used in relatively low dimension cases, e.g., around 10 historical signals. To overcome this limitation, some techniques are proposed. For example, one clever way is to use a pairwise GP model to model each historical signal and the signal under study. We can use the pairwise model to make predictions. Then we can go through all the historical signals one by one and combine the prediction results. This is a heuristic approach that can bypass the dimensionality issue in the application of MOGP. The performance of this approach is shown to be very promising. Interested readers can refer to Kontar et al. [2020] for more details on this approach.

11.4.3 Time-to-Failure Distribution Based on GP Predictions

Similar to the settings discussed in Chapter 10, if the failure event is defined as the first time that the degradation signal surpasses a threshold, we can obtain the time to failure distribution based on the prediction of the degradation signal under the MOGP model. According to the GP prediction formula (11.7), the predicted future value of the degradation signal follows normal distribution. In other words, if for a specific unit p, we have the observation of its degradation signal y_p up to t^*, then the prediction of its future value $\hat{y}(t^* + t)$ follows a normal distribution, whose mean and variance is given by (11.7). Please note that (11.7) can be viewed as a general GP based prediction formula that can be applied to both single output GP prediction and MOGP prediction. For MOGP, the mean and covariance functions in (11.7) include the mean and covariance functions for all the GPs involved as shown in (11.28).

Because the predicted degradation signal value follows GP, the results of time to failure distribution obtained in Section 10.3 still hold. In fact, the formula (10.13) can be used here. We only need to plug in the mean and variance of $\hat{y}(t^* + t)$ obtained in the MOGP prediction procedure to obtain the time to failure distribution.

Bibliographical Notes

The Gaussian process, also known as Kriging method, is a very flexible yet effective nonparametric model to describe smooth functional data. Popular books covering the general framework of GP include Rasmussen and Williams

[2006], Shi and Choi [2011], Gramacy [2020], Santner et al. [2003]. The GP modeling approach has found broad applications in industrial data analytics. Various statistical monitoring techniques for functional data, also known as profile data in statistical process control literature, have been reported [Zhang et al., 2014; Colosimo et al., 2014; Kontar et al., 2017b; Jahani et al., 2018; Li et al., 2018; Cho et al., 2019; Zhang et al., 2016; Tapia et al., 2016; Wang et al., 2018]. The GP model has also been used for modeling system degradation signals and for failure prognosis [Baraldi et al., 2015; Padgett and Tomlinson, 2004; Richardson et al., 2017; Li and Xu, 2015; Tagade et al., 2020; Li et al., 2020; Liu et al., 2019]. MOGP is a significant extension of single output GP where the correlations among different outputs are modeled and captured. Good references on MOGP include Álvarez and Lawrence [2011], Boyle and Frean [2005], Wilson et al. [2011], Bilionis et al. [2013], Zhao and Sun [2016]. MOGP model has also been used for degradation signal modeling and prognosis in engineering field [Liu et al., 2017; Duong et al., 2018; Tanwar and Raghavan, 2020]. Parts of this chapter are based on Kontar et al. [2018].

Exercises

1. (a) Simulate multiple single output GP with 10 data points and the squared exponential covariance function under three settings. Let $x_i \in (0, 1)$ and assume a noiseless process with zero mean.
   (a) $l = 0.1$
   (b) $l = 0.3$
   (c) $l = 1$

   (b) Comment on the effect of l.

   (c) Using (11.19) and (11.22), simulate multiple two-output GP $y_1(\mathbf{x})$ and $y_2(\mathbf{x})$. Assume $\alpha_{0i} = \alpha_{0j} = \alpha_{ii} = \alpha_{jj} = 1$ and $\lambda_{0i} = 1$, $\lambda_{0j} = 2$, $\lambda_{ii} = \lambda_{jj} = 0.5$, $\sigma_1 = \sigma_2 = 0.01$. Also let $x_i \in (0, 1)$ and assume the mean is zero.
2. Given a data set with $\mathbf{X} = (0\ 0.2\ 0.4\ 0.6\ 0.8\ 1)^T$ and $\mathbf{y} = (-1.6\ -1\ -0.5\ 0\ 0.5\ 0.7)^T$.
   (a) Assuming a squared exponential covariance function, write down $\mathbf{C}(\mathbf{X}; l)$. Assuming $\mu = 0$. Hint: each entry of the 6×6 matrix is a function of l.
   (b) Fit the GP and plot the mean and 95% confidence interval for 100 input points uniformly distributed between 0 and 1, `x=seq(0,1, length.out=100)`. Assume $\mu = 0$.
   (c) Refit and plot the GP with $\mu = 5$. Comment on the results.

3. Consider two outputs $\mathbf{y}_1 = (0\ 25\ 100)^T$, $\mathbf{y}_2 = (0\ 12.5\ 50)^T$ with the same input points $\mathbf{X} = (0\ 5\ 10)^T$. Let $\mathbf{x}^* = (2.5\ 7.5)^T$ and assume the covariance is

$$\mathbf{C}(\mathbf{X}) = \begin{pmatrix} 1767.59 & 676.35 & 37.89 & 883.79 & 338.17 & 18.95 \\ 676.35 & 1767.59 & 676.35 & 338.18 & 883.79 & 338.17 \\ 37.89 & 676.35 & 1767.59 & 18.95 & 338.18 & 883.79 \\ 883.79 & 338.18 & 18.95 & 441.90 & 169.09 & 9.47 \\ 338.17 & 883.79 & 338.18 & 169.09 & 441.90 & 169.09 \\ 18.95 & 338.17 & 883.79 & 9.47 & 169.09 & 441.90 \end{pmatrix}$$

$$\mathbf{C}(\mathbf{X}, \mathbf{x}^*) = \begin{pmatrix} 695.10 & 695.1 & 101.77 & 347.55 & 347.55 & 50.89 \\ 101.77 & 695.1 & 695.10 & 50.89 & 347.55 & 347.55 \end{pmatrix}$$

Predict $E(y_2(x = 2.5) \mid \mathbf{y}_1, \mathbf{y}_2, \mathbf{X})$ and $E(y_2(x = 7.5) \mid \mathbf{y}_1, \mathbf{y}_2, \mathbf{X})$

4. Consider two outputs $\mathbf{y}_1 = (0\ 25\ 100)^T$, $\mathbf{y}_2 = (0\ 12.5\ 50)^T$ with the same input points $\mathbf{X} = (0\ 5\ 10)^T$.
Fit the convolution based MOGP and plot the mean and 95% confidence interval for 100 input points uniformly distributed between 0 and 100.
5. Consider an MOGP with three outputs and covariance structure given in (11.27). Let $\mathbf{y}_1 = (0.0\ 6.25\ 25.00\ 56.25\ 100.00)^T$, $\mathbf{y}_2 = (0.0\ 4.17\ 16.67\ 37.5\ 66.67)^T$ $\mathbf{y}_3 = (0.0\ 3.125\ 12.5\ 28.125\ 50.0)^T$ with the same input points $\mathbf{X} = (0.0\ 2.5\ 5.0\ 7.5\ 10.0)^T$. Fit the convolution based MOGP and plot the mean and 3-sigma (99.7%) confidence interval for 50 input points uniformly distributed between 0 and 100.

12

Prognosis Through Mixed Effects Models for Time-to-Event Data

In Chapter 10, we use the value of a degradation signal to indicate the system health status. The failure event is defined as the degradation signal hitting some pre-specified threshold. The failure prediction is achieved through comparing the forecast of the degradation signal and the threshold. This method is conceptually easy to follow and effective if a good indicator signal for the system degradation is available. However, in many practical situations, such a clear indicator signal may not be available. In such cases, the degradation signal based method cannot be applied. Another very important type of data regarding system failure is time-to-event data, which records the occurrence time of failure event from multiple subjects and/or the repeated failure events from a single subject. Time-to-event data contains rich information about life time distribution and has been extensively used in prediction of system failure.

Table 12.1 shows the time-to-failure data of 14 batteries collected from a battery life test experiment. The same data is also plotted in Figure 12.1.

The batteries belong to two different product models, which are indicated by the "Battery model" column. The "Status" column of the table indicates the failure condition of the battery. If the battery is in a failure state, the corresponding "Status" value is 1; otherwise, it is 0. Because the experiment is run to failure, the failure event of all the batteries is observed. Thus, the value of the "Status" column is always 1 in the table. However, in many practical situations, there is often "censoring" in the failure event observations. Censoring is a condition in which the value of a measurement or observation is only partially known. One of the common censoring schemes is "right" censoring, in which the exact value of the observation is unknown but we know it is larger than a known value. For example, for the data in Table 12.1, if we stop the experiment at time instance 14, then we will not be able to observe the failure

Industrial Data Analytics for Diagnosis and Prognosis: A Random Effects Modelling Approach, First Edition. Shiyu Zhou and Yong Chen.

Table 12.1 Battery failure time data

ID	Battery model	Failure time	Status
01	A	11	1
02	A	9	1
03	A	9	1
04	A	12	1
05	A	12	1
06	A	14	1
07	A	7	1
08	B	13	1
09	B	11	1
10	B	14	1
11	B	10	1
12	B	15	1
13	B	14	1
14	B	13	1

event for battery `ID12` but we know it occurs after time 14. Thus its failure time will be right censored and the corresponding value in the "Status" column will be 0. Stopping the observation or a subject "dropping-out" from the study are the common mechanisms for right censoring. The other two common censoring mechanisms are "left" censoring and "interval" censoring. For left censored observations, we know the failure occurs before certain time but we do not know the exact time. Similarly, for interval censored observations, we know the failure occurs within a known interval.

In this chapter, we focus on the statistical models and analytics methods for time-to-event data. In Section 12.1, both nonparametric and parametric statistical models for time-to-event data are introduced. The typical models with covariates for time-to-event data, also called survival regression model, are introduced in Section 12.2. The basic event prediction techniques based on these models are also introduced respectively. In Section 12.3, we introduce a joint modeling approach to model the time-to-event data and the longitudinal degradation data together. In a broad sense, the joint model is a mixed effects model, where the time-to-event data is described using a survival regression model and the longitudinal degradation data is treated as the covariates of the regression model and described as a linear mixed effects model as presented in Chapter 10. An important mixed effects model for time-to-event data, called the frailty model, will be introduced in Section 12.4. The frailty model allows

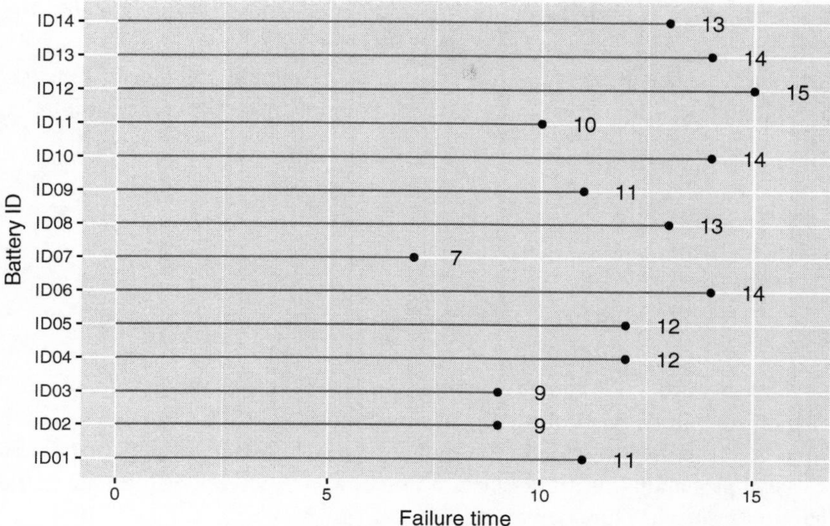

Figure 12.1 Failure time of 14 batteries

unit-to-unit variation in time-to-event data. The prediction of event occurrence of an individual unit using the frailty model is also introduced.

12.1 Models for Time-to-Event Data Without Covariates

In time-to-event data modeling and analysis, we often use T as the random variable representing the time-to-event. It is also called the survival time. Suppose there are n independent samples of T, denoted as $t_1, t_2, \ldots, t_n$. Please note that these n samples could be collected from n different units – each unit has a failure event; or they could be collected from one unit that failed n times. For the latter case, we should ensure that the assumption of the independence of these events is reasonable. In practice, if a specific failure occurs, a diagnosis will be conducted and the responsible component will be replaced by a new one. In such a situation, the independent event occurrence is often a reasonable assumption. In the analysis, we often arrange the distinct time-to-event in ascending order: $t_{(1)} < t_{(2)} < \cdots < t_{(r)}$, where $r \leq n$ because there may be ties. As mentioned above, the observation may be censored. In this book, we only consider the right censoring. We use a binary variable δ_i to indicate if the observation is censored or not. Specifically, $\delta_i = 1$ if t_i is the observed event time and

$\delta_i = 0$ if t_i is a censored time (i.e., the time we stop observing). Then, the complete observed data can be written as $\{(t_i, \delta_i), i = 1, \ldots, n\}$.

In studying the distribution of T, the survival function and hazard function are often used. The survival function $S(t)$ is simply one minus the cumulative distribution function of T, $F(t)$, i.e.,

$$S(t) = \Pr(T > t) = 1 - F(t), t > 0.$$

The hazard function is defined as

$$h(t) = \lim_{\Delta t \to 0} \frac{\Pr(t \le T \le t + \Delta t \mid T \ge t)}{\Delta t}, \quad t > 0.$$

The hazard function can be viewed as the risk/hazard or the "instantaneous" risk of the failure event occurring at time t given that the system survives to time t. Obviously, higher $h(t)$ implies higher failure probability and less survival probability. With the hazard function $h(t)$, we can also calculate the cumulative hazard function as

$$H(t) = \int_0^t h(u)\mathrm{d}u, \quad t > 0.$$

The cumulative hazard function has a direct relationship with survival function. It also has an interesting interpretation as the expected death number up to time t if we imagine that after each death, the unit comes back to life and continues to follow the same hazard trajectory [Cleves, 2008].

The survival function, hazard function, and the cumulative hazard function are mathematically related. Using the conditional probability formula, we have

$$\begin{aligned} h(t) &= \lim_{\Delta t \to 0} \frac{\Pr(t \le T \le t + \Delta t \mid T \ge t)}{\Delta t} \\ &= \lim_{\Delta t \to 0} \frac{\Pr(t \le T \le t + \Delta t, T \ge t)}{\Pr(T \ge t)\Delta t} = \frac{\lim_{\Delta t \to 0} \dfrac{\Pr(t \le T \le t + \Delta t)}{\Delta t}}{1 - F(t)} \\ &= \frac{f(t)}{S(t)}, \end{aligned}$$

where $f(t)$ is the probability density function of T. From the definition of $S(t)$, we have

$$\frac{\mathrm{d}S(t)}{\mathrm{d}t} = \frac{\mathrm{d}(1 - F(t))}{\mathrm{d}t} = -\frac{\mathrm{d}F(t)}{\mathrm{d}t} = -f(t).$$

Then using the chain rule of derivatives, we have

$$-\frac{\mathrm{d}(\log S(t))}{\mathrm{d}t} = -\frac{\frac{\mathrm{d}S(t)}{\mathrm{d}t}}{S(t)} = \frac{f(t)}{S(t)} = h(t).$$

Taking the integral on both sides from 0 to t, we have

$$-\log S(t) = \int_0^t h(u)\mathrm{d}u = H(t).$$

Summarizing these results, we have

$$f(t) = h(t)S(t), \quad h(t) = -\mathrm{d}(\log S(t))/\mathrm{d}t, \quad H(t) = -\log S(t).$$

Clearly, by knowing one of hazard function or survival function or the density function of T, we can get the other two.

With these functions, we can compute the mean value of the time-to-event as

$$\mu_T = \int_0^\infty S(t)\mathrm{d}t \tag{12.1}$$

and the medium point of the time-to-event as $t_{0.5}$ such that $S(t_{0.5}) = 0.5$.

Now we turn our attention to the estimation of these functions based on the observed dataset $\{(t_i, \delta_i), i = 1, \ldots, n\}$. There are three types of model for the distribution of time-to-event data, namely, the parametric model, the non-parametric model, and the semi-parametric model. We will introduce the parametric model and the nonparametric model below, and the semi-parametric model will be introduced in Section 12.2.

12.1.1 Parametric Models for Time-to-Event Data

The parametric model uses an explicit mathematical function as the hazard (or survival or density) function of the time-to-event data. Three popular functions are listed in Table 12.2.

For Weibull and log-logistic distribution, the parameters λ and k are called the scale and the shape parameter, respectively. The probability density function for exponential, Weibull, and log-logistic distribution under different parameter settings are shown in Figures 12.2, 12.3, and 12.5, respectively.

The hazard function for exponential distribution is constant and does not change with time. Thus, exponential distribution is often used to model relatively simple time-to-event data. One interesting fact of exponential distribution is that if the time-to-event follows exponential distribution, then the count of events will follow a Poisson distribution. Also, the density function for exponential distribution is always a convex function.

Table 12.2 Important parametric models for time-to-event data ($t \geq 0$)

Model	Hazard function $h(t)$	Survival function $S(t)$	Density function $f(t)$
Exponential ($\lambda > 0$)	λ	$\exp(-\lambda t)$	$\lambda\exp(-\lambda t)$
Weibull ($\lambda > 0, k > 0$)	$\frac{k}{\lambda}\left(\frac{t}{\lambda}\right)^{k-1}$	$\exp\left(\left(-\frac{t}{\lambda}\right)^{k}\right)$	$\frac{k}{\lambda}\left(\frac{t}{\lambda}\right)^{k-1}\exp\left(-\left(\frac{t}{\lambda}\right)^{k}\right)$
Log-logistic ($\lambda > 0, k > 0$)	$\frac{(k/\lambda)(t/\lambda)^{k-1}}{1+(t/\lambda)^{k}}$	$\frac{1}{1+(t/\lambda)^{k}}$	$\frac{(k/\lambda)(t/\lambda)^{k-1}}{(1+(t/\lambda)^{k})^{2}}$

The Weibull distribution is much more flexible than the exponential distribution. With different parameters, we can have both convex and concave shaped probability density functions as shown in Figure 12.3. Figure 12.4 shows different types of hazard function for Weibull distribution. Clearly, for a Weibull distribution, the hazard function could be constant, monotonic increasing, or monotonic decreasing. The log-logistic distribution is quite

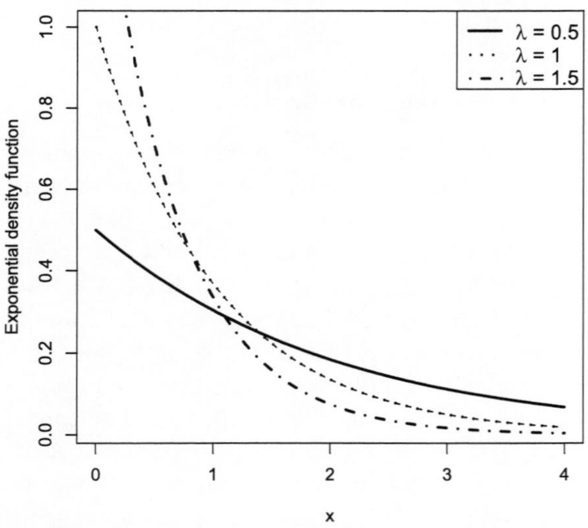

Figure 12.2 Probability density function of exponential distribution

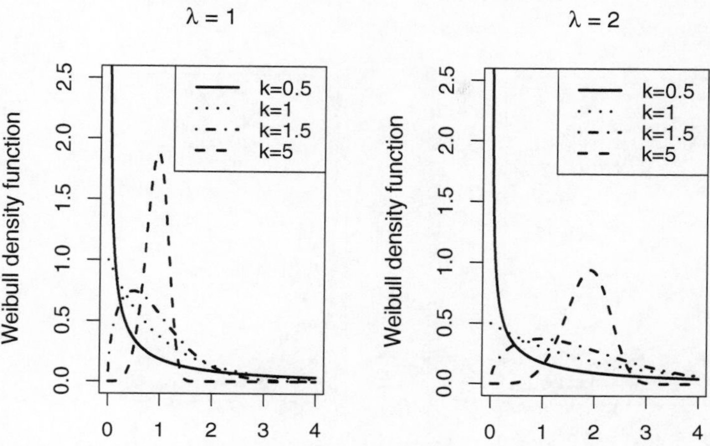

Figure 12.3 Probability density function of Weibull distribution

flexible as well. In fact, one feature of log-logistic distribution is that its hazard function could be increasing at early stages and then decreasing at later stages as shown in Figure 12.6. When the physical process exhibits this feature, log-logistic distribution is often selected.

Given a time-to-event dataset $\{(t_i, \delta_i), i = 1, \ldots, n\}$, we can use these parametric models to fit the data by estimating the model parameters using the observed

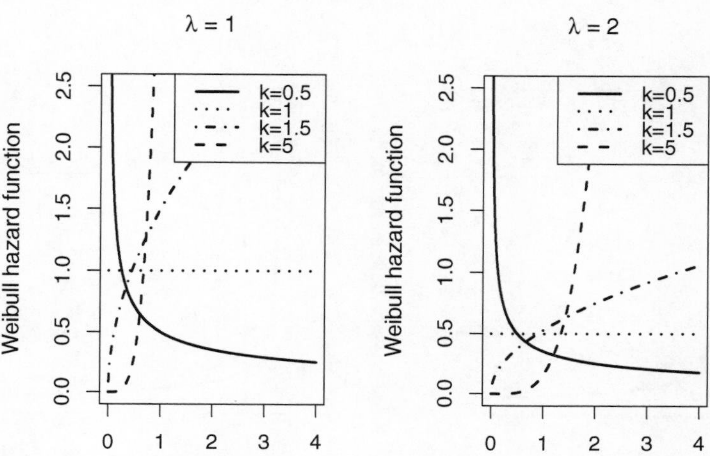

Figure 12.4 Hazard function of Weibull distribution

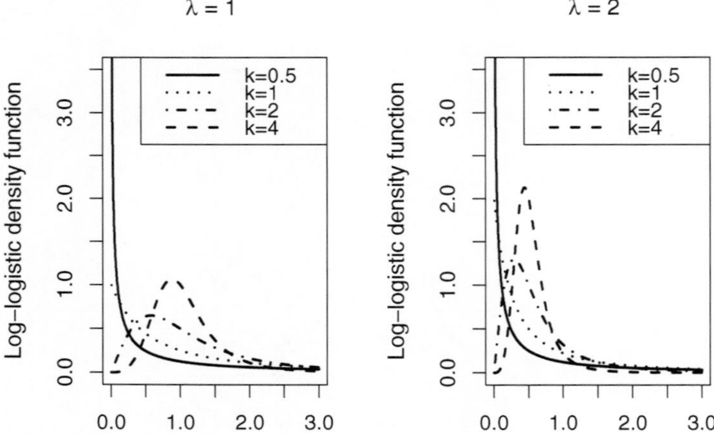

Figure 12.5 Probability density function of log-logistic distribution

data. MLE is the most commonly used parameter estimation method. If the dataset contains right censored observations, the likelihood function can be written as

$$L(\boldsymbol{\theta}) = \prod_{i=1}^{n} f(t_i; \boldsymbol{\theta})^{\delta_i} (1 - F(t_i; \boldsymbol{\theta}))^{1-\delta_i}, \tag{12.2}$$

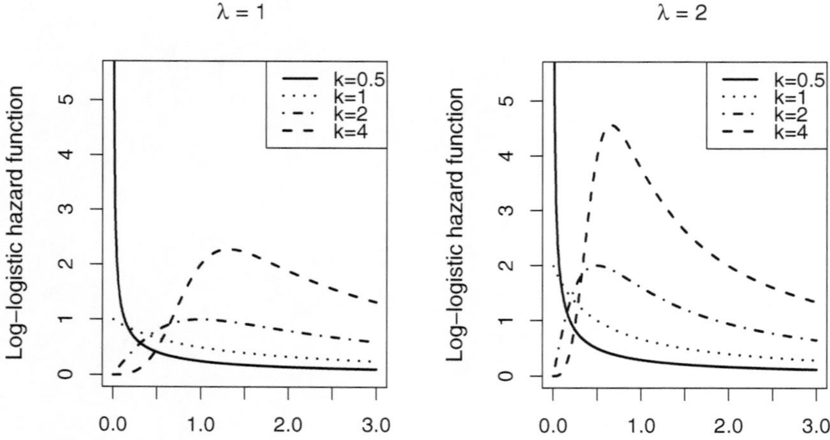

Figure 12.6 Hazard function of log-logistic distribution

where $\boldsymbol{\theta}$ is a vector containing all the unknown parameters of the distribution. Equation (12.2) essentially says if an observation t_i is not censored, then its contribution to the likelihood function is $f(t_i; \boldsymbol{\theta})$, while if it is right censored, its contribution to the likelihood function is $(1 - F(t_i; \boldsymbol{\theta}))$. With the likelihood function, the parameter estimation is obtained through maximizing the logarithm of the likelihood function (or minimizing the negative logarithm of the likelihood function),

$$\hat{\boldsymbol{\theta}} = \underset{\boldsymbol{\theta}}{\operatorname{argmax}} \log L(\boldsymbol{\theta}). \tag{12.3}$$

The logarithm function is used to convert the multiplicative operations in $L(\boldsymbol{\theta})$ into additive operations, which is more numerically stable. Please note that because the logarithm function is a monotonic function, the solution in (12.3) is identical to the solution of maximizing $L(\boldsymbol{\theta})$ directly. For many distributions, there is no closed form solution to (12.3) and numerical optimization techniques are needed to get $\hat{\boldsymbol{\theta}}$.

Example 12.1 Consider the data in Table 12.1. We can fit the data to exponential, Weibull, and log-logistic distribution, respectively.

The estimation of the parametric distributions from time-to-event data has been implemented in multiple packages in `R` and can be easily used. Here we provide an example using the `flexsurvreg` function from the `flexsurv` package. The following code implements the estimation of Weibull distribution using the data stored in `bfdata`:

```
su.obj <- Surv(bfdata$Failure.time , bfdata$Status)
fit.w <- flexsurvreg(su.obj ~ 1, data = bfdata ,
                     dist = "weibull")
fit.ll <- flexsurvreg(su.obj ~ 1, data = bfdata ,
                      dist = "llogis")
```

In this code, `Surv` function creates the response object using the failure time data in the `bfdata` dataset, `bfdata$Failure.time` and the censoring indicator `bfdata$Status`, where 1 indicates an observed failure and 0 indicates a censored observation. The function `flexsurvreg` fits the parametric survival model using the `data` and the specified distribution. The first input `su.obj ~ 1` indicates that there is no covariate in the model and we want to fit the distribution only. The fitted result is stored in the model object as the output of the function, i.e., `fit.w` and `fit.ll`. The fitted distribution, hazard, and survival functions can be easily obtained from the output objects using the `summary` method. For example, by default, `summary(fit.w)`

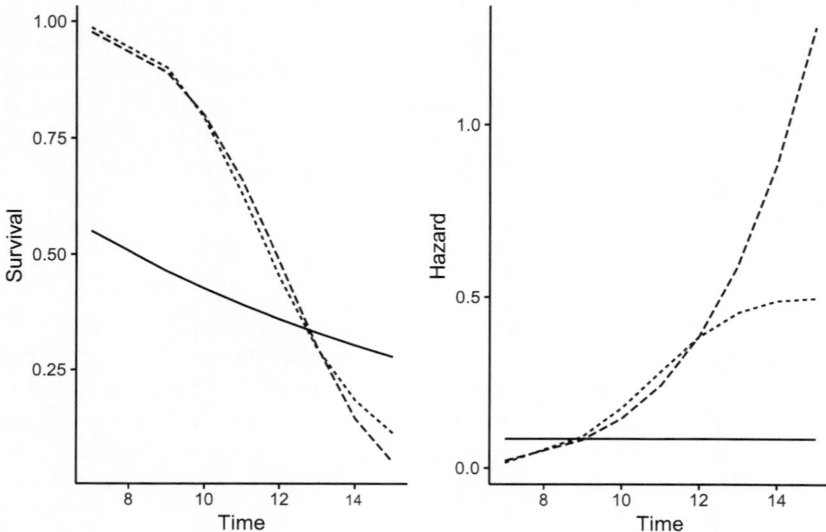

Figure 12.7 Fitted survival and hazard function: ——, exponential; ------, Weibull; ······, log-logistic

gives the survival function and `summary(fit.w,type="hazard")` gives the hazard function.

We can see that the results in Figure 12.7 for the Weibull and log-logistic distribution are similar while the result for exponential distribution is quite different. We can use the mean value of the fitted time-to-event distribution to predict the failure time for a unit in the population. The mean values for exponential, Weibull, and log-logistic distributions are 11.71, 11.75, 12.00, respectively. It is interesting to note that although the shape of the survival function is different, the mean of the fitted exponential and the Weibull distribution are close.

In the dataset, we do not have any censored data points. If we add some censored data point at time 16, we will see these points will push the survival curve upward because we have some units which survive beyond time 16. The result is shown in Figure 12.8. The mean of the fitted distributions are 17.42, 13.33, and 13.80, respectively. Obviously the estimated mean is larger than the estimates when no censored data is included.

The parametric models have explicit mathematical forms and are typically tractable and easy to be interpreted. However, selection of a parametric form often means limiting the distribution form to a specific family of functions.

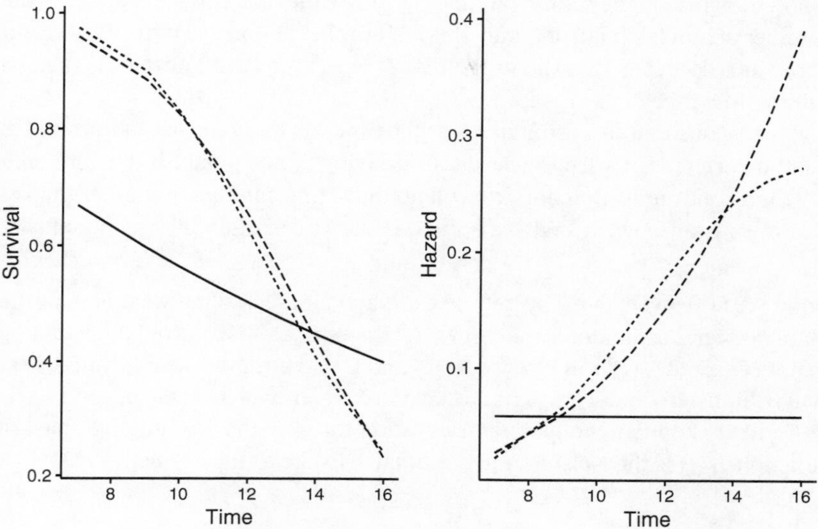

Figure 12.8 Fitted survival and hazard function with five additional right censored observations at time 16: ———— Exponential; - - - - - - - - - -, Weibull; ······, log-logistic

This will lead to limited flexibility and expressiveness of the functions. Non-parametric models can be used to address this issue.

12.1.2 Non-parametric Models for Time-to-Event Data

The most commonly used nonparametric estimator for the survival function using time-to-event data with right censoring is the Kaplan–Meier estimator [Kaplan and Meier, 1958]. It is also called Product–Limit estimator. For a time-to-event dataset with n observations (t_i, δ_i), $i = 1, \ldots, n$, the Kaplan–Meier estimator is

$$\hat{S}(t) = \begin{cases} 1 & \text{if } t < t_1 \\ \prod_{t_i \leq t}[1 - d_i / Y_i] & \text{if } t_1 \leq t \end{cases}, \tag{12.4}$$

with its point-wise variance estimation as

$$\hat{\sigma}_{\hat{S}}^2(t) = \hat{S}(t)^2 \sum_{t_i \leq t} \frac{d_i}{Y_i(Y_i - d_i)},$$

where t_1 is the smallest survival time (i.e., the time of the first event), d_i is the number of units that fail at t_i and Y_i is the number of units is *at risk* at t_i. A unit being at risk means that the unit either fails at t_i or is still alive at t_i (i.e., its failure time is beyond t_i).

It is obvious that this estimator is a step function that changes its value once a failure event occurs. The basic idea underlying this estimator is that although $S(t)$ is a continuous function, we will not have information on it other than at the time instance when a failure occurs. Thus, we can discretize the function at t_is. Consider $\frac{S(t_i)}{S(t_{i-1})} = \hat{\Pr}(T > t_i | T \geq t_i)$ and note $S(t_i) = \frac{S(t_i)}{S(t_{i-1})}\frac{S(t_{i-1})}{S(t_{i-2})}\cdots\frac{S(t_1)}{S(0)}S(0)$ with $S(0) = 1$, we can easily get the estimator in (12.4). One weakness of the Kaplan–Meier estimator is that for $t > t_n$, the estimator is not well defined. If t_n is not censored, $\hat{S}(t)$ will be 0 for $t > t_n$ and if t_n is a right censored value, $\hat{S}(t)$ will be non-zero and we will not be able to differentiate its value beyond t_n.

Another popular nonparametric estimator for the cumulative hazard function $H(t)$ is the Nelson–Aalen estimator [Nelson, 1972, Aalen, 1978]

$$\widetilde{H}(t) = \begin{cases} 0 & \text{if } t < t_1 \\ \sum_{t_i \leq t} \frac{d_i}{Y_i} & \text{if } t_1 \leq t \end{cases}, \tag{12.5}$$

with its point-wise variance estimation as

$$\hat{\sigma}^2_{\widetilde{H}}(t) = \sum_{t_i \leq t} \frac{d_i}{Y_i^2}.$$

We use tilde rather than a hat on $H(t)$ to differentiate with the Kaplan–Meier estimator because using $\hat{S}(t)$ in (12.4), we can also get an estimation of the cumulative function as $\widehat{H}(t) = -\ln\hat{S}(t)$. The Nelson–Aalen estimator is often viewed as having better performance when the sample size is small.

Example 12.2 Consider the data in Table 12.1. We can fit the data using a nonparametric model based on the Kaplan–Meier and the Nelson–Aalen method, respectively. The fitted results without and with the censored data are shown in Figures 12.9 to 12.12, respectively. The fitted mean failure time for these two methods without censoring data is 11.71 and 11.94, respectively. If extra censoring data are considered, the fitted mean failure time will be 12.84 and 13.00, respectively. The shaded area in the figures are the confidence intervals of the estimation, which is obtained based on the estimated variance. It is observed that for both methods, the estimator is not defined after the largest survival time. The implementation of these estimators are realized by `survfit` function in the `survival` package in `R`.

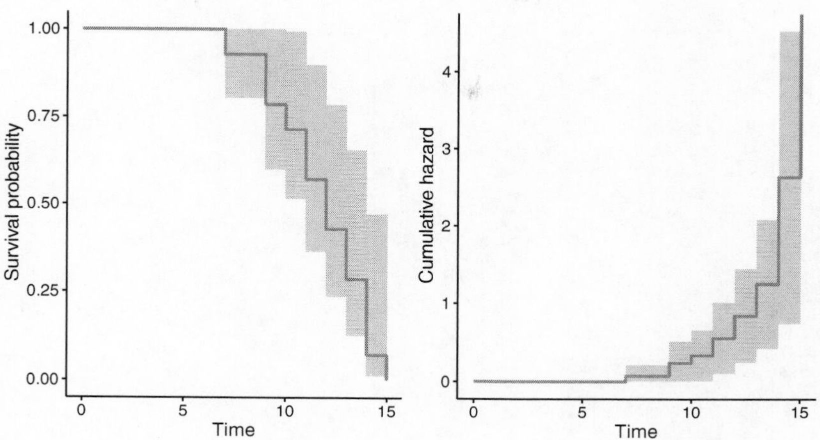

Figure 12.9 Fitted survival and cumulative hazard function using Kaplan–Meier method. The shaded area is the confidence interval.

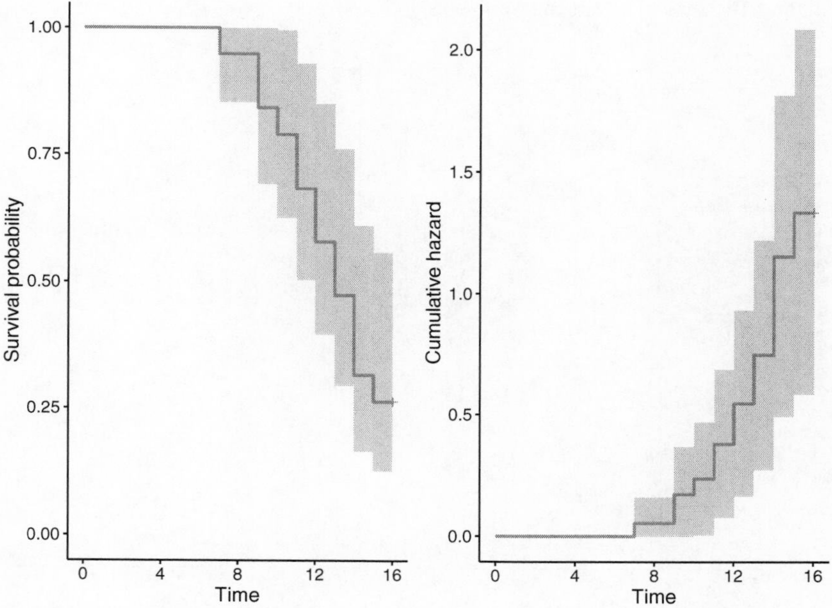

Figure 12.10 Fitted survival and cumulative hazard function for data with censoring using Kaplan–Meier method. The shaded area is the confidence interval.

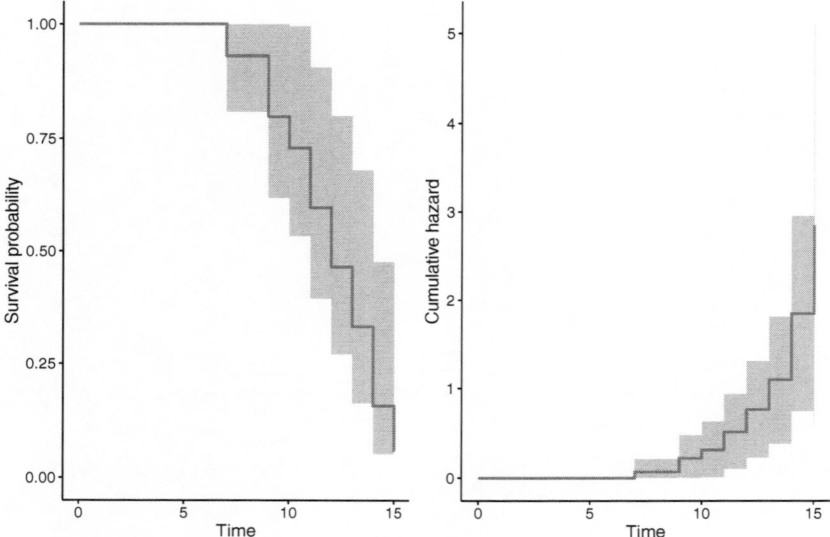

Figure 12.11 Fitted survival and cumulative hazard function using Nelson–Aalen method. The shaded area is the confidence interval.

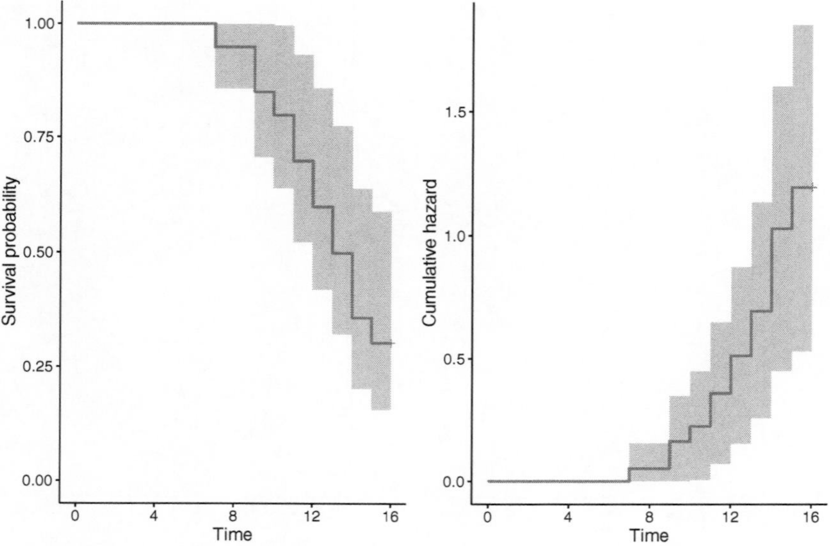

Figure 12.12 Fitted survival and cumulative hazard function for data with censoring using the Nelson–Aalen method. The shaded area is the confidence interval.

12.2 Survival Regression Models

The models introduced in Section 12.1 do not have any covariates, which means that they can only be used to describe a homogeneous population and provide a population level description. They cannot be used to differentiate the characteristics of units from different groups. However, in practice, we often know some specific extra information regarding the unit (or group of units) under study. We need to adjust the survival function to take this extra information into consideration. For example, for the data in Table 12.1, we may want to compare the reliability of different battery models. In the examples given in the previous section, the "Battery model" information given in the table is not considered. For a more sophisticated analysis, we may also want to predict the reliability of a battery based on its time varying usage pattern and the working environment. Survival regression models are developed to achieve this goal. Similar to classical regression, in survival regression, the extra information is often called covariates, independent variable, predictors, or explanatory variables etc. Depending on the response variable used, we can put the survival regression models into two major categories.

The first type of model uses the logarithm of the survival time as the response variable and is called the accelerated failure time model. The logarithm operation is the typical transformation used in classical regression models to convert positive variables to variables on the entire real domain. Let $Y_i = \ln(T_i)$ and denote $\mathbf{x}_i$ as the covariates observed for unit i. The accelerated failure time model is in the form

$$Y_i = \gamma_0 + \mathbf{x}_i^T \gamma + \epsilon_i.$$

The term ϵ_i represents the model error and γ_0 is the model intercept. The common choices of the error distribution include the standard normal distribution, extreme value distribution, and logistic distribution. This model is called accelerated failure time model because the survival function of this model is a time scaled version of the survival function when the covariate $\mathbf{x}_i$ is zero. To see this, let $S_0(t)$ be the survival function given $\mathbf{x} = \mathbf{0}$. Then we have

$$\begin{aligned} S_0(t) = \Pr(T > t|\mathbf{x} = \mathbf{0}) &= \Pr(e^Y > t|\mathbf{x} = \mathbf{0}) \\ &= \Pr(e^{\gamma_0 + \epsilon} > t). \end{aligned}$$

However, for a general case with given covariates $\mathbf{x}$, we have

$$\begin{aligned} S(t) = \Pr(T > t|\mathbf{x}) &= \Pr(e^Y > t|\mathrm{x}) \\ &= \Pr(e^{\gamma_0 + \mathbf{x}^T\gamma + \epsilon} > t) \\ &= \Pr(e^{\gamma_0 + \epsilon} > te^{-\mathbf{x}^T\gamma}). \end{aligned}$$

Thus, we have

$$S(t) = S_0(te^{-\mathbf{x}^T\gamma}). \tag{12.6}$$

Clearly, (12.6) indicates that the survival function of the accelerated failure time model is time scaled version of the base line survival function when all the covariates are zeros. The accelerated failure time model is an extension of the classical regression model and the parameters of the model can be estimated using the maximum likelihood estimation method.

The second type of model is called a hazard rate model, in which we use the hazard function as the response. There are two sub-types in this category: the multiplicative hazard model and the additive hazard model. As the name implies, the multiplicative hazard model is in a form of product of baseline hazard function and a non-negative function of covariates

$$h(t|\mathbf{x}) = h_0(t)c(\mathbf{x}), \tag{12.7}$$

while the additive model is in a form of addition of a baseline hazard function and a regression function as

$$h(t|\mathbf{x}) = h_0(t) + \mathbf{x}^T\gamma(t).$$

Please note that typically in an additive model, the coefficient γ is allowed to be time varying.

From (12.7), it is easy to see that if the covariates are fixed for two units, then their hazard functions are proportional to each other and do not change with time. Thus, the multiplicative hazard rate model is also called a proportional hazard model. A special case of the multiplicative model is the Cox proportional hazard (PH) model, where the function $c(\cdot)$ is selected as an exponential function $\exp(\mathbf{x}^T\gamma)$.

Cox PH model is quite flexible and has been widely used in biostatistics and reliability engineering. In this section, we will focus on introducing Cox PH model.

12.2.1 Cox PH Model with Fixed Covariates

The Cox PH model is first introduced in Cox [1972]. With fixed covariates, the Cox PH model can be written as

$$h_i(t) = h_0(t)\exp(\mathbf{x}_i^T\gamma). \tag{12.8}$$

Noting the relationship $S_i(t)=\exp(-H_i(t))$, we can easily get $S_i(t)=\exp(-H_0(t)\cdot\exp(\mathbf{x}_i\gamma))$ under Cox PH model with time fixed covariates $\mathbf{x}_i$, and $S_0(t)=\exp(-H_0(t))$, hence

$$S_i(t)=S_0(t)^{\exp(\mathbf{x}_i^T\gamma)}. \tag{12.9}$$

The baseline hazard function $h_0(t)$ should be non-negative and can be in either parametric form or nonparametric form. If a nonparametric form is used, the model in (12.8) is also called semi-parametric proportional hazard model because the regression part is still in parametric form. If we want to use a parametric form for $h_0(t)$, Weibull distribution is a popular choice, i.e., we can select $h_0(t)=\frac{k}{\lambda}(\frac{t}{\lambda})^{k-1}$. It is interesting to note that if Weibull distribution is selected for $h_0(t)$, then the Cox PH model can be put in an accelerated failure time model format (refer to p. 48 of Klein and Moeschberger [2005]). The parametric model has fewer number of unknown parameters to estimate and thus we can have more precise estimates of the parameters. However, one limitation of the parametric model is that it is more restrictive. If the data does not follow the selected parametric form, the estimation result will be misleading.

The estimation of the model parameters including the regression coefficients γ and the baseline hazard function $h_0(t)$ can be obtained through the maximum likelihood estimation method. Let the observed dataset with n observations be $\{t_i, \delta_i, \mathbf{x}_i\}$, $i=1,\ldots,n$. Then the full log-likelihood for observation i is

$$l_i(h_0(t_i),\gamma)=\delta_i\ln(f(t_i|\mathbf{x}_i,\gamma))+(1-\delta_i)\ln(S(t_i|\mathbf{x}_i,\gamma)). \tag{12.10}$$

If the observations are independent, the full log-likelihood for n observations is the summation of l_i for $i=1,\ldots,n$. If a parametric form is selected for the baseline hazard function $h_0(t)$ such as Weibull, we can estimate the parameters including the parameters of $h_0(t)$ and the regression coefficients γ by maximizing the log-likelihood function. However, if a nonparametric form of $h_0(t)$ is used, then the conventional MLE of maximizing (12.10) cannot be implemented. Instead, we can estimate the regression parameters γ without estimating the baseline function. In other words, if we are only interested in comparing the hazard of different units using the Cox PH model, we can simply estimate the regression coefficients and need not to estimate the baseline function. This is achieved through maximizing the partial likelihood.

First, if we assume that all the failure times in $\{t_i, \delta_i, \mathbf{x}_i\}$, $i=1,\ldots,n$ are distinct, then the partial likelihood is

$$L(\gamma)=\prod_{i=1}^{n}\left(\frac{\exp(\mathbf{x}_i^T\gamma)}{\sum_{j\in\mathcal{G}(t_i)}\exp(\mathbf{x}_j^T\gamma)}\right)^{\delta_i}, \tag{12.11}$$

where the set $\mathcal{G}(t_i)$ includes all the units whose failure event (or censoring) time is at or larger than t_i. The intuition of the likelihood is that for $\delta_i = 1$, the term $\frac{\exp(\mathbf{x}_i^T\gamma)}{\sum_{j\in\mathcal{G}(t_i)}\exp(\mathbf{x}_j^T\gamma)}$ actually is the probability that unit i fails at t_i given that only one unit fails at t_i.

By maximizing the partial likelihood (or its logarithm) in (12.11) using popular optimization techniques such as the Newton–Raphson method, we can obtain the estimate of γ. The standard results regarding the properties of MLE still hold for partial MLE. For example, the estimation is consistent and the estimates asymptotically follow normal distribution.

The likelihood function in (12.11) needs the assumption that the failure times are distinct. If there are ties in the data, different methods have been proposed. One method used in Breslow [1974] is to use (12.11) unmodified even ties exist. Another method is proposed by Efron [1977], which is close to Breslow's method but is considered to give better results. Let d_i be the number of tied observations at t_i and the set $\mathcal{F}(t_i)$ includes that all units fail at t_i. The partial likelihood proposed by Efron is

$$L(\gamma) = \prod_{i=1}^{n}\left(\frac{\prod_{k=1}^{d_i}\exp(\mathbf{x}_k^T\gamma)}{\prod_{k=1}^{d_i}\left(\sum_{j\in\mathcal{G}(t_i)}\exp(\mathbf{x}_j^T\gamma) - \frac{k-1}{d_i}\sum_{l\in\mathcal{F}(t_i)}\exp(\mathbf{x}_l^T\gamma)\right)}\right)^{\delta_i} \quad (12.12)$$

Clearly, if the number of tied observations is small, (12.12) will be very close to (12.11).

If the purpose of the modeling is to compare the hazard rate of units in different groups, then the estimate of γ will be sufficient. However, in many cases, we need to obtain the hazard and/or survival function for a specific unit. In such cases, the baseline hazard function $h_0(t)$ needs to be estimated.

Breslow's estimator is a popular method to estimate the baseline hazard function. The basic idea is to plug the estimated $\hat{\gamma}$ from the partial likelihood into the full likelihood (i.e., summation of (12.10)) and then find the values of $h_0(t_i)$. Through some algebraic manipulation, it is not difficult to see that the optimal $h_0(t)$ will be zero everywhere except at the time instances when a failure occurs. In other words, $h_0(t)$ will be non-zero at t_is when $\delta_i = 1$. Specifically, we have

$$\hat{h}_0(t_i) = \frac{1}{\sum_{j\in\mathcal{G}(t_i)}\exp(\mathbf{x}_j^T\hat{\gamma})}. \quad (12.13)$$

and the corresponding cumulative baseline hazard function is

$$\widehat{H}_0(t) = \sum_{t_i \le t} \frac{1}{\sum_{j \in \mathcal{G}(t_i)} \exp(\mathbf{x}_j^T \hat{\gamma})}. \tag{12.14}$$

Please note that although the hazard function only takes non-zero values at time instance when an event occurs, the censored observations are considered because they will influence the set $\mathcal{G}(t_i)$. The above result is for data without tied event observations. When there are tied events, (12.13) and (12.14) can still be used and thus we have

$$\widehat{H}_0(t) = \sum_{t_i \le t} \frac{d_i}{\sum_{j \in \mathcal{G}(t_i)} \exp(\mathbf{x}_j^T \hat{\gamma})}, \tag{12.15}$$

where d_i is the number of events observed at t_i. With $\widehat{H}_0(t)$, we can obtain $\hat{S}_0(t) = \exp(-\widehat{H}_0(t))$. Then for a unit with covariates $\mathbf{x}_0$ and considering the relationship (12.9), we can get its survival function as

$$\hat{S}(t \mid \mathbf{x}_0) = \hat{S}_0(t)^{\exp(\mathbf{x}_0^T \hat{\gamma})}. \tag{12.16}$$

12.2.2 Cox PH Model with Time Varying Covariates

In addition to fixed covariates, some covariates change before the event occurs. For example, consider the failure event of a battery. The vendor and model of the battery are the fixed covariates that may influence the battery survival. However, there are important time varying covariates such as the capacity and internal resistance, which are good indicators of battery survival probability.

The time varying covariates can be included in Cox PH model. If we treat the fixed covariates as a special case of time varying covariates, we can represent the observed data by $\{t_i, \delta_i\, [\mathbf{x}_i(t), 0 \le t \le t_i]\}$, $i = 1, \ldots, n$. Now the Cox PH model can be written as

$$h_i(t) = h_0(t)\exp(\mathbf{x}_i^T(t)\gamma)$$

and the partial likelihood is in the same form as (12.11)

$$L(\gamma) = \prod_{i=1}^{n} \left(\frac{\exp(\mathbf{x}_i^T(t_i)\gamma)}{\sum_{j \in \mathcal{G}(t_i)} \exp(\mathbf{x}_j^T(t_i)\gamma)} \right)^{\delta_i}.$$

The estimation and testing can proceed as the fixed covariate case. If ties in event times exist, similar extension as that in (12.12) can be used.

With the estimated regression coefficients, we can also simply extend (12.13) and (12.15) by replacing fixed $\mathbf{x}_j$ to time varying $\mathbf{x}_j(t_i)$ to estimate the hazard and cumulative hazard function. However, the survival function estimation becomes more complicated. We cannot estimate the survival function by extending (12.16) by replacing $\mathbf{x}_0$ with $\mathbf{x}_0(t)$. The reason is that with time varying covariates, we can no longer separate the baseline hazard function and the regression part in computing the cumulative hazard function. Specifically, for a unit with covariates $\mathbf{x}_0(t)$, we have $\hat{h}(t \mid \mathbf{x}_0(t)) = \hat{h}_0(t)\exp(\mathbf{x}_0^T(t)\hat{\gamma})$ and thus

$$\widehat{H}(t \mid \mathbf{x}_0(t)) = \int_0^t \hat{h}_0(u)\exp(\mathbf{x}_0^T(u)\hat{\gamma})du \neq \widehat{H}_0(t)\exp(\mathbf{x}_0^T(t)\hat{\gamma})$$

and

$$\hat{S}(t \mid \mathbf{x}_0(t)) = \exp\left(-\widehat{H}(t \mid \mathbf{x}_0(t))\right). \tag{12.17}$$

A numerical integration is needed to compute $\widehat{H}(t \mid \mathbf{x}_0(t))$ and then $\hat{S}(t \mid \mathbf{x}_0(t))$. The computational load is obviously higher than that in the fixed covariates case. Also, from (12.17), it is clear that when we have time varying covariates, we cannot predict the future survival function and predict the future event occurrence unless we know the future values of the covariates.

12.2.3 Assessing Goodness of Fit

After a Cox PH model has been fitted to the observed dataset, we need to check if the fitted model is adequate to describe the data at hand. Many model checking procedures in statistics theory are based on the residuals. Model residual is typically the difference between the model prediction and the observed data. If the fitted model is adequate, the residual will follow a certain distribution. Then by comparing with the observed residuals, we can tell the adequacy of the fitted model. The Cox PH model is for time-to-event data with possible censoring. The model checking procedure is more complicated than that for regular linear regression models. In this section, we will briefly introduce several residual based model checking techniques, including Cox–Snell residuals, Martingale residuals, and Schoenfeld residuals.

1. Cox–Snell residual
Given the observed dataset with n units $\{t_i, \delta_i, \mathbf{x}_i\}$, $i = 1, \ldots, n$ and the fitted model for the ith unit as

$$\hat{h}_i(t) = \hat{h}_0(t)\exp(\mathbf{x}_i^T\hat{\gamma}).$$

Then the Cox–Snell residual for the ith unit is given as

$$r_{ci} = \widehat{H}_0(t_i)\exp(\mathbf{x}_i^T\hat{\gamma}),$$

where $\widehat{H}_0(t_i)$ is the estimated baseline cumulative hazard function evaluated at t_i. Noting the relationship in (12.16), we can infer that the Cox–Snell residual for the ith unit is actually the cumulative hazard function evaluated at t_i,

$$r_{ci} = \widehat{H}(t_i) = -\ln[\hat{S}_i(t_i)].$$

Through analyzing the distribution of a function of a random variable, we can obtain that if T is a random variable representing time-to-event of a unit, and $S(T)$ is the corresponding survival function, then the random variable $Y = -\ln[S(T)]$ follows an exponential distribution with unit mean. With this result, we can argue that if the fitted model is adequate and satisfactory, then the estimated survival function $\hat{S}_i(t)$ will be close to the true underlying survival function and thus the distribution of the observed Cox–Snell residuals should be close to the exponential distribution with unit mean. Please note that if the observed survival time is censored, then the corresponding value of the Cox–Snell residual is also censored. Thus, the resulting distribution of the Cox–Snell residual will be a censored exponential distribution with unit mean. To check if $\{r_{ci}, \delta_i\}$, $i = 1, \ldots, n$ follows the censored exponential distribution with unit mean, we can compute the cumulative hazard function for $\{r_{ci}, \delta_i\}$ (for example, using the Nelson–Aalen estimator in (12.5)) as $\widehat{H}_r(r)$ and then plot r_{ci} versus $\widehat{H}_r(r_{ci})$. The plot should be a straight line with slope 1. If the plot deviates, then we can say the overall fitting of the model is not satisfactory.

The Cox–Snell residual is used widely in testing the overall fit of the model. One weakness of this method is that if the Cox–Snell residual indicates a bad fit, it is hard to get insights into the cause of the lack of fit.

2. Martingale residual

The Martingale residual is another popular model residual used for model checking, which is defined as

$$r_{mi} = \delta_i - r_{ci},$$

where δ_i is the indicator of event occurrence, i.e., it is one when there is no censoring and zero when the observation is a censored observation. The rigorous derivation of the Martingale residual needs the theory of counting stochastic processes and is omitted here. If we note r_{ci} is actually the cumulative hazard function obtained from the fitted model and the cumulative hazard function can be viewed as the expected death number up to time t_i for unit i, then we can get an intuitive interpretation of the Martingale residual: it can be viewed as the difference between the observed number of death for unit i up to t_i, which is actually δ_i, and the expected number of death for unit i predicted by the model, which is r_{ci}. The properties of the Martingale residual are similar to that of the residual we used in conventional linear regression. The Martingale residuals sum to zero, are uncorrelated, and have zero mean.

The Martingale residual is often used to find the function form for the covariates in the Cox PH model. The basic procedure is to plot the covariate under study versus the martingale residuals for the Cox PH model without this specific covariate. Then the smoothed curve of the plot suggest the preferred function form for the covariate. The detailed procedures can be found in Klein and Moeschberger [2005].

3. Schoenfeld residual

The Schoenfeld residual is often used to check the proportional hazard assumption for the Cox PH model. It does not require the calculation of the cumulative hazard function. Another important feature of the Schoenfeld residual is that for each covariate, we will have a Schoenfeld residual. Specifically, let x_{ji} be the jth covariate for unit i. The Schoenfeld residual is defined as

$$r_{sji} = \delta_i \left(x_{ji} - \frac{\sum_{l \in \mathcal{G}(t_i)} x_{jl} \exp(\mathbf{x}_l^T \hat{\gamma})}{\sum_{l \in \mathcal{G}(t_i)} \exp(\mathbf{x}_l^T \hat{\gamma})} \right).$$

The summation of r_{sji} over all the units $i = 1, \ldots, n$ is actually an estimate of the first derivative of the logarithm of the likelihood function with respect to the jth covariate. The Schoenfeld residuals sum to zero. In large samples, these residuals are uncorrelated and have zero mean. In practice, the scaled Schoenfeld residual is often used, which is defined as follows:

$$\mathbf{r}_{si}^* = r \cdot \text{cov}(\hat{\gamma}) \mathbf{r}_{si},$$

where r is the number of death among the n units, $\mathbf{r}_{si}$ is a vector consisting of all the Schoenfeld residuals for all the covariates, and $\text{cov}(\hat{\gamma})$ is the covariance matrix of the estimates of γ.

The scaled Schoenfeld residuals can be used to check the proportional hazard assumption for the Cox PH model. If we plot the event time versus the residuals, the points should be scattered randomly around zero and should not exhibit any patterns. Otherwise, the proportional hazard assumption does not hold for the corresponding covariate.

All the above residuals are derived under the fixed covariates case for the Cox PH model. These residuals can be extended to the Cox PH model with time-varying covariates. However, the expression are more complicated under such cases. Limited by the scope of this book, we will not provide more details here. Interested readers can refer to literature that are specialized on survival regression. In the following, we will present a couple of examples to illustrate the fitting, model checking, and event prediction using the Cox PH model.

Example 12.3 Consider the data in Table 12.1 again. We can build a Cox PH model using the "Battery model" as a fixed covariate. Because we only have two battery models, it is natural to encode it as a covariate x as

$$x_i = \begin{cases} 0, & \text{battery } i \text{ is of model A} \\ 1, & \text{battery } i \text{ is of model B} \end{cases}. \tag{12.18}$$

One point that needs to be mentioned is that if we have more than two categories, it is not appropriate to use one single covariate to encode all the categories by assigning each category a different numerical value. For example, if we have an additional category C of battery model, we cannot assign x_i as 2 to represent model C. Instead, we should create an additional covariate to represent different models. We will have two covariates x_{i1} and x_{i2} encoded as

$$x_{i1} = 1 \text{ if battery } i \text{ is of model B; 0 otherwise}$$
$$x_{i2} = 1 \text{ if battery } i \text{ is of model C; 0 otherwise.}$$

Using the covariate encoding in (12.18), we can fit a Cox PH model to the data in Table 12.1. In R, the function `coxph` in the `survival` package is widely used for Cox PH model fitting. Assume that the data in Table 12.1 is saved in `ft.csv` file. Then we can use the following code to fit and plot the fitted survival function for different covariate values.

```
bfdata <- read.table("ft.csv",sep=",",header=TRUE)
su.obj <- Surv(bfdata$Failure.time, bfdata$Status)
fitph <- coxph(su.obj~Model ,data=bfdata)
survfit1 <-survfit(fitph,
                  newdata=data.frame(Model=c(0,1)));
plot(survfit1 ,lty=c(1,2),xlab='Time',
      ylab='Estimated Survival Function S(t)')
legend('topright ',c('Type A','Type B'),lty=c(1,2))
```

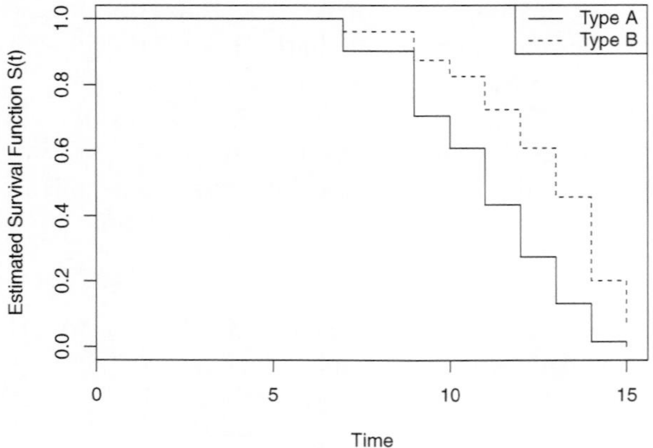

Figure 12.13 Fitted survival function for Cox PH model: Type A: $x_i = 0$; Type B: $x_i = 1$

The fitted survival function is shown in Figure 12.13.

Please note that when $x_i = 0$, the survival function is actually the baseline survival function in this example because we only have one covariate in the model. The summary of the fitted Cox PH model, `fitph` is

```
        coef exp(coef) se(coef)     x     p
Model -0.954     0.385    0.575 -1.66 0.097
Likelihood ratio test=2.72  on 1 df, p=0.0988
n= 14, number of events= 14
```

The negative value of the fitted coefficient indicates that Model B is more reliable than Model A. The p value of 0.097 indicates that the covariate of battery model has a certain level of significance. Using the `print(survfit1,print.rmean=TRUE)` function, we can get the mean and median of time-to-event for $x_i = 0$ and $x_i = 1$, respectively. For $x_i = 0$, they are both 11, while for $x_i = 1$, they are 12.6 and 13, respectively. Not surprisingly, the mean and median life for Model B is longer than that of Model A.

We can also use the Schoenfeld residual to test the proportional hazard assumption. In `R`, the function `cox.zph(fitph)` can be used to compute the residuals for a given fitted Cox PH model `fitph`. The output of `cox.zph` can be plotted by the function `ggcoxzph` as shown in Figure 12.14.

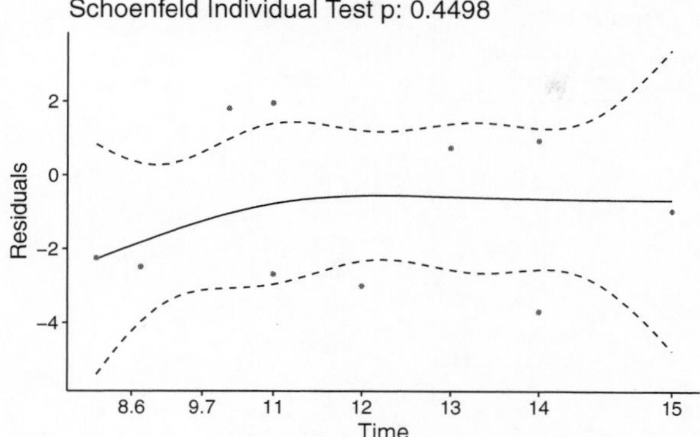

Figure 12.14 Schoenfeld residuals

From this figure, no clear pattern is shown and thus we cannot reject the proportional hazard assumption.

Example 12.4 In this example, we demonstrate the Cox PH model with time varying covariates. The dataset is an extension of that in Table 12.1. For each unit, we add a history of temporal observations of the resistance of the battery. It is known that the resistance value of a battery is an important indicator of the battery health. Thus, we would like to incorporate it in the model as a time varying covariate. The data for one battery is shown in Table 12.3. Please note that we will have the history of resistance value for all 14 batteries.

From this table, we can see that to consider the time varying covariates, we need to split the life time of the battery into different segments indicated by the "Start" and "Stop" column and the value of the time varying covariate is given for each period. The "Status" column indicates if the failure event occurs in the corresponding time interval.

Using the above dataset, we can create the survival object and then fit a Cox PH model using the following code:

```
surv_obj = Surv(bftdata$start, bftdata$stop,
                bftdata$Status)
fitsurv = coxph(surv_obj~Model+Resis,data=bftdata)
```

Table 12.3 Resistance data for one battery

ID	Model	Resis.	obstime	Start	Stop	Status
01	A	4.84	1	1	2	0
01	A	4.67	2	2	3	0
01	A	4.78	3	3	4	0
01	A	4.92	4	4	5	0
01	A	5.12	5	5	6	0
01	A	5.15	6	6	7	0
01	A	5.37	7	7	8	0
01	A	5.42	8	8	9	0
01	A	5.61	9	9	10	0
01	A	5.77	10	10	11	1

The summary of the fitted Cox PH model is as follows. It can be seen that the time varying covariate `r` has smaller p-value comparing with the fixed covariate x and it seems to be more significant.

```
        coef     exp(coef) se(coef)     x    p
x     -1.150     0.317     0.634     -1.81 0.070
r      0.512     1.669     0.266      1.92 0.055
Likelihood ratio test=6.62 on 2 df, p=0.0365
n= 149, number of events= 14
```

The estimated survival function for the population is given in Figure 12.15.

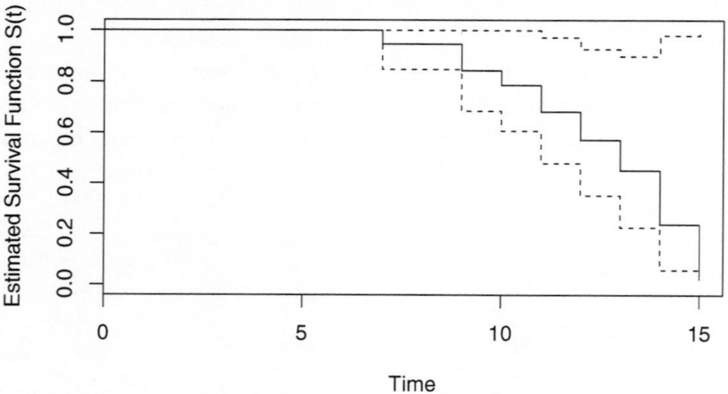

Figure 12.15 Survival function for Cox PH model with time varying covariate: ——, survival function; - - - -, confidence interval

We should be cautious about using the Cox PH model with time varying covariates to predict the future survival probability. Obviously, the future values of the time varying covariates should be given to compute the survival function using such a Cox PH model. Thus, a prediction model for the time varying covariates is also needed for the prediction. Recently, a joint modeling approach has been proposed to achieve such a goal. We will introduce the joint model in the following section.

12.3 Joint Modeling of Time-to-Event Data and Longitudinal Data

In Chapter 10, we presented the mixed effects modeling approach for longitudinal data. In this chapter, we presented various statistical models for time-to-event data. In practice, we may need to model the time-to-event data and the longitudinal data jointly for event prediction purposes. One obvious reason is that as shown in Section 12.2.2, if we want to use a Cox PH model with time varying covariate for prediction, the future values of the time varying covariates need to be predicted as well. Plugging these predicted values into the Cox PH model, we can obtain the survival function at a future time and thus obtain the prediction for the failure. Clearly a model for longitudinal data as shown in Chapter 10 that can forecast is needed.

On the other hand, the prognosis methods introduced in Chapter 10 is solely based on the value of a degradation or condition monitoring signal. The failure event is defined by the degradation signal hitting some pre-determined threshold. Such a failure event is often called *soft* failure. However, in general, soft failure will not always coincide with the real hard failure, which makes the system break down. In many applications, we may know that some signals are related to the system degradation but it is not reasonable to use such signals directly to define the failure event. Specifically, it may be difficult to determine one fixed threshold for the whole population because it depends on the unit's individual characteristics, which varies randomly from unit to unit. The difficulty in determining a fixed threshold poses challenges to those prognosis methods that depend solely on longitudinal data. If the time-to-event data are available besides longitudinal signals, a more reasonable way is to model the longitudinal data and the time-to-event data together and use the longitudinal signal as a time varying covariate in the survival regression model.

The joint modeling framework essentially integrates the mixed effects model for longitudinal data introduced in Chapter 10 and the survival regression model for time-to-event data introduced in Section 12.2. The failure prognosis based on the joint modeling framework is composed of two stages: the offline modeling stage, and the online prediction stage. During the offline stage, the parameters of the joint model are estimated based on historical data of event times, degradation signal and possibly other time-fixed covariates (such as manufacturer, device type, etc.) for a number of similar units. During the online stage, the degradation signal of the in-service unit that we want to make prediction for is obtained and used for the prediction of the unit's remaining useful life based on the model parameters estimated during the offline stage. The predictions made for an in-service unit can be updated as more observations on the degradation signal are obtained during its use. The techniques for these two stages are introduced in the following two sub-sections, respectively.

12.3.1 Structure of Joint Model and Parameter Estimation

Assume we have historical data of N similar units. For the ith unit, its associated data is denoted by $\mathcal{D}_i = \{t_i, \delta_i, \mathbf{y}_i, \mathbf{x}_i\}$, where t_i is the event or censoring time for unit i, δ_i is the indicator for event occurrence, $\mathbf{y}_i$ is the observations of longitudinal signal along time until t_i, and $\mathbf{x}_i$ are the time fixed covariates for unit i. Based on the nature of the data, the joint modeling approach involves two distinct sub-models: longitudinal model for the degradation signal and survival regression model for the time-to-event data.

Sub-model I: Longitudinal model for the degradation signal

We use the mixed effects model in (10.1) for the longitudinal data. For the sake of convenience, we copy the model here in scalar form as:

$$y_i(t) = \mathbf{z}^T(t)\mathbf{b}_i + \epsilon_i(t), \tag{12.19}$$

where $\mathbf{z}^T(t)\mathbf{b}_i$ is the true but unobservable value of the degradation signal, $\epsilon_i(t)$ is the error term due to measurement noise which is assumed to be independent and follows normal distribution $\mathcal{N}(0, \sigma_\epsilon^2)$, $\mathbf{z}(t)$ contains the intercept and time-dependent regression functions and $\mathbf{b}_i$ is a vector of random effects. As discussed in Chapter 10, with the random effects, the model in (12.19) assumes the units have distinct but similar degradation signal paths. If we assume $\mathbf{b}_i$ has non-zero mean, then the mean of $\mathbf{b}_i$ being the fixed effects and $\mathbf{z}^T(t)$ being the regressors are included in the model implicitly as well. One limitation of the model in (12.19) comparing with a general mixed effects model is that the design matrices for the fixed and the random effects are always the same. In cases where nonlinear behaviors of degradation signal are evident, we may add higher order polynomial terms or splines in $\mathbf{z}^T(t)$ as shown in Example 10.3. We may also include certain specific nonlinear functions of time (e.g., $\log(t)$ or $t^{0.5}$ in the regressors $\mathbf{z}^T(t)$, if they are believed to be the correct forms based on either physical knowledge or empirical evidence).

Sub-model II: Survival regression model for time-to-event data

If we let $y_i'(t) = \mathbf{z}^T(t)\mathbf{b}_i$ as the value of the degradation signal without measurement noise at time t, then a typical joint survival regression model is in the form

$$h_i(t) = h_0(t)\exp(y_i'(t)\gamma_1 + \mathbf{x}_i^T\gamma_2). \tag{12.20}$$

This model is in the form of the general Cox PH model in (12.8) with additional time varying covariates. In this model, the longitudinal degradation signal $y_i'(t)$ is associated with a single coefficient to reflect its impact on the hazard rate. However, in many physical systems, the signal's initial condition and the increments of the signal play different roles. Thus, for greater flexibility, we split $y_i'(t)$ into two components, its initial value $y_i'(0)$ and its increment $y_i'(t) - y_i'(0)$, and associate them with two different coefficients. This allows the two components to have different impacts on the hazard rate. Note $y_i'(0)$ is simply the intercept term of the model in (12.19) and if we split $\mathbf{z}^T(t)$ into $\mathbf{z}^T(t) = (1\ \mathbf{z}_1^T(t))$ and $\mathbf{b}_i$ into $\mathbf{b}_i^T = (b_{i0}\ \mathbf{b}_{i1}^T)$, we have

$$h_i(t) = h_0(t)\exp(b_{i0}\gamma_0 + \mathbf{z}_1^T(t)\mathbf{b}_{i1}\gamma_1 + \mathbf{x}_i^T\gamma_2). \tag{12.21}$$

If imposing the constraint $\gamma_0 = \gamma_1$, the model (12.21) will reduce to the typical formulation in (12.20) where the entire degradation signal is associated with a single parameter γ_1. The joint model essentially includes the degradation signal as a covariate for the hazard rate function.

In model (12.21), $h_0(t)$, γ_0, γ_1, γ_2 do not depend on the unit index i; they represent population characteristics while individual specific information is contained in $\mathbf{x}_i$ and $\mathbf{b}_i$. The baseline hazard rate $h_0(t)$ can be assumed to have a parametric form such as Weibull, but can also be nonparametric for better flexibility. The two sub-models, (12.19) and (12.21) constitute the joint modeling framework for prognosis.

We use $\boldsymbol{\theta} = \{\gamma_0, \gamma_1, \gamma_2, \mathbf{b}, \boldsymbol{\Sigma}_{\mathbf{b}}, \sigma_\epsilon^2\}$ to collectively denote all the unknown parameters in the joint model, where $\mathbf{b}, \boldsymbol{\Sigma}_{\mathbf{b}}$ are the mean and covariance of the random effects $\mathbf{b}_i$. In the offline stage, we need to estimate $\boldsymbol{\theta}$ and the baseline hazard function $h_0(t)$ using the historical data $\mathcal{D}_i$, for $i = 1, \ldots, N$.

In principle, the MLE method can be used to estimate these parameters. Specifically, the observed data likelihood can be written as

$$\begin{aligned} L(t_i, \delta_i, \mathbf{y}_i, \mathbf{x}_i; \boldsymbol{\theta}) &= \int f(t_i, \delta_i, \mathbf{y}_i \mid \mathbf{x}_i, \mathbf{b}_i; \boldsymbol{\theta}) f(\mathbf{b}_i; \boldsymbol{\theta}) d\mathbf{b}_i \\ &= \int f(t_i, \delta_i \mid \mathbf{x}_i, \mathbf{b}_i; \boldsymbol{\theta}) f(\mathbf{y}_i \mid \mathbf{b}_i; \boldsymbol{\theta}) f(\mathbf{b}_i; \boldsymbol{\theta}) d\mathbf{b}_i. \end{aligned} \tag{12.22}$$

Noting the general likelihood function for the time-to-event data given in (12.2) and the relationship between density function, hazard function, and the survival function for time-to-event random variable as $f(t) = h(t)S(t)$, we have

$$\begin{aligned} f(t_i, \delta_i | \mathbf{x}_i, \mathbf{b}_i; \theta) &= h_i(t_i | \mathbf{x}_i, \mathbf{b}_i; \theta)^{\delta_i} S_i(t_i | \mathbf{x}_i, \mathbf{b}_i; \theta) \\ &= h_i(t_i | \mathbf{x}_i, \mathbf{b}_i; \theta)^{\delta_i} \exp\left(-\int_0^{t_i} h_i(u \mid \mathbf{x}_i, \mathbf{b}_i; \theta) \mathrm{d}u \right), \end{aligned}$$

where the expression of $h_i(t)$ given $\mathbf{x}_i$ and $\mathbf{b}_i$ is given in (12.21).

The second term $f(\mathbf{y}_i | \mathbf{b}_i; \theta)$ is the likelihood for a linear mixed effects model that is similar to that given in (10.4) as

$$\begin{aligned} f(\mathbf{y}_i | \mathbf{b}_i; \theta) &= \prod_{j=1}^{n_i} f(y_{ij} | \mathbf{b}_i; \theta) \\ &= \prod_{j=1}^{n_i} \frac{1}{\sqrt{2\pi\sigma_\epsilon^2}} \exp\left(-\frac{(y_{ij} - \mathbf{z}^T(t_{ij})\mathbf{b}_i)^2)}{2\sigma_\epsilon^2} \right), \end{aligned}$$

where n_i is the number of longitudinal observations and y_{ij} is the observation of $y_i(t)$ at t_{ij}, $j = 1, \ldots, n_i$.

The third term $f(\mathbf{b}_i; \theta)$ is simply a multivariate normal likelihood as

$$f\left(\mathbf{b}_i; \theta\right) = \frac{1}{\sqrt{(2\pi)^k \left|\boldsymbol{\Sigma}_\mathbf{b}\right|}} \exp\left(-\frac{1}{2}(\mathbf{b}_i - \mathbf{b})^T \boldsymbol{\Sigma}_\mathbf{b}^{-1}(\mathbf{b}_i - \mathbf{b}) \right),$$

where k is the dimension of $\mathbf{b}_i$.

There are two ways for parameter estimation using the likelihood functions defined in (12.22). One is directly maximizing the joint log-likelihood function defined as $\sum_{i=1}^N \ln[L(t_i, \delta_i, \mathbf{y}_i, \mathbf{x}_i; \theta)]$. We could also employ the EM algorithm by treating the random effects as missing variable [Wulfsohn and Tsiatis, 1997]. However, it is obvious that the joint likelihood function is very complex, making the maximization step computationally demanding or even infeasible due to convergence issue. Another method is an approximated "two-step" procedure, namely we estimate the parameters in the mixed effects longitudinal model first, then estimate the parameters in the Cox PH model by treating the mixed effects model as known. In this way, the mixed effects model provides "true" values of the time-dependent covariate at any time point to facilitate the estimation of the Cox PH model. It is known that this approximation may cause bias in the estimation but it has been shown that the difference is not significant in most cases [Wulfsohn and Tsiatis, 1997]. Thus, the two-step procedure is a well-accepted approach for joint model estimation.

12.3.2 Online Event Prediction for a New Unit

Based on the estimated offline model, predictions can be made for the RUL of a new in-service unit p during the online stage. This new unit is similar to those in the historical data, and it can be considered as individuals sampled from the same population. We assume that unit p is still in use and its degradation signal is obtained continuously during its usage.

To make a RUL prediction, we need to evaluate the hazard function given in (12.21) for future time t. We will need the population level parameters in (12.21). However, we also need to provide a specific estimation of the individual specific parameter $\mathbf{b}_p$. We can update the distribution of $\mathbf{b}_p$ based on the observed degradation signal from unit p in order to make the prediction. We can use the Bayesian update scheme introduced in Chapter 10 to get an estimation for $\mathbf{b}_p$. Indeed, (10.5) is the closed form updating expression for the updated mean and variance of $\mathbf{b}_p$ with the observed degradation signal up to time t^*, denoted as $\mathbf{y}_p^*$.

With the updated distribution of $\mathbf{b}_p$, we can obtain the distribution of the survival function of unit p. Assume the unit has survived to t^* and we want to make RUL prediction at t^*. The conditional survival function is

$$\begin{aligned} S(t|t^*, \mathbf{x}_p, \mathbf{b}_p; \hat{\theta}, \hat{h}_0(t)) &= S(t \mid \mathbf{x}_p, \mathbf{b}_p; \hat{\theta}, \hat{h}_0(t))/S(t^* \mid \mathbf{x}_p, \mathbf{b}_p; \hat{\theta}, \hat{h}_0(t)) \\ &= \exp\left(-\int_{t^*}^{t} \hat{h}_0(u)\exp(b_{p0}\hat{\gamma}_0 + \mathbf{z}_1^T(t)\mathbf{b}_{p1}\hat{\gamma}_1 + \mathbf{x}_p^T\hat{\gamma}_2)\mathrm{d}u\right). \end{aligned} \tag{12.23}$$

Please note $\mathbf{b}_p^T = (b_{p0}\ \mathbf{b}_{p1}^T)^T$ follows a multivariate normal distribution with updated mean and variance given in (10.5).

The survival function in (12.23) has to be evaluated numerically. We can sample the value of $\mathbf{b}_p$ from the corresponding multivariate normal distribution and plug it into (12.23) to obtain a survival curve. With the survival curve, we can obtain the RUL prediction as the mean time-to-event given in (12.1). We could also use the median time-to-event as the prediction of RUL. We can repeat this process by sampling a different value of $\mathbf{b}_p$ to get multiple predictions of the RUL. The point prediction of RUL can be the mean of individual predictions and its corresponding point-wise Bayesian credible interval can also be obtained based on the individual predictions. Another way of obtaining a point prediction of RUL is through integration of the survival function over the distribution of $\mathbf{b}_p$ as

$$S(t|t^*, \mathbf{x}_p, \mathbf{y}_p^*; \hat{\theta}, \hat{h}_0(t)) = \int S(t|t^*, \mathbf{x}_p, \mathbf{b}_p; \hat{\theta}, \hat{h}_0(t)) f(\mathbf{b}_p|\mathbf{y}_p^*) d\mathbf{b}_p. \quad (12.24)$$

The numerical integration of (12.24) is challenging if the dimension of $\mathbf{b}_p$ is high. The Gauss–Hermite quadrature method can be used to take advantage of the multivariate normal distribution of $\mathbf{b}_p$. Some fast approximation method is also introduced in Zhou et al. [2014]. Equation (12.24) provides a point estimation for the survival function. With it, a point prediction of RUL can be obtained.

Example 12.5: Consider the data shown in Table 12.3 again. We can build a joint prognostic model using the dataset. In `R` language, the package `JM` provides implementation of a joint model through many useful functions. In the simplest case, fitting a joint model can be implemented by the codes as follows:

```
fitlme <- lme(r~obstime+I(obstime ^2),
                random=~obstime+I(obstime ^2)|id,
                data=bdata)
bdata.surv<-Surv(bdata.id$eventtime, bdata.id$Status)
fitsurv <- coxph(bdata.surv~x,data=bdata.id, x=TRUE)
fitJOINT <- jointModel(fitlme, fitsurv,
                timeVar = "obstime",
                method = "weibull -PH -GH")
```

Please note that in the above code, two data tables are needed, namely `bdata` and `bdata.id`. The `bdata` table is similar to Table 12.3 containing all the temporal observations of the longitudinal signal, which are used to build the longitudinal mixed effects model. The `bdata.id` table is a reduced data table: each row corresponds to one unit in the table and contains the failure or censoring occurrence time for the unit. Also, the name of each column of these two data tables should be the same. The function `jointModel` fits the joint model using the two sub-models, i.e., the longitudinal mixed effects model and the Cox PH model for the event occurrence. Once the model is fitted, we can provide some new data and obtain the predicted survival function as follows:

```
sfit1 <- survfitJM(fitJOINT,
        newdata = newDatabat[1:2, ],idVar = "id")
sfit2 <- survfitJM(fitJOINT ,
        newdata = newDatabat[1:7, ],idVar = "id")
```

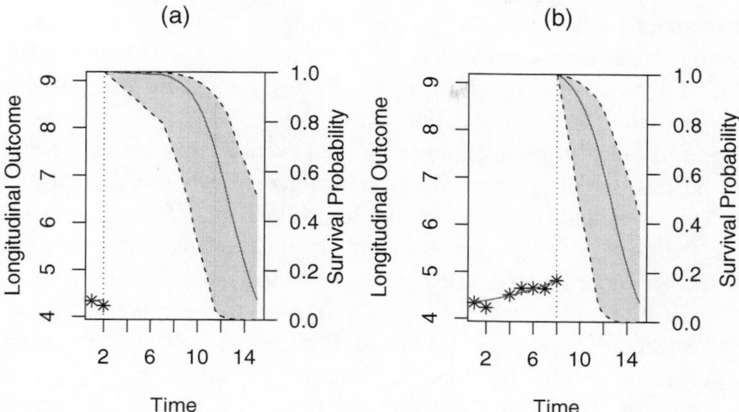

Figure 12.16 Survival function prediction with the joint model. (a) predicted survival function with 2 observations (b) predicted survival function with 7 observations. ——, predicted survival function; - - - -, confidence interval; * observed longitudinal data point

The `newDatabat` data table should have the same format as `bdata`. In the first line of the code, we make a prediction with two observed data points while in the second line of the code, we make a prediction with seven observed data points. The predicted survival functions can be plotted as shown in Figure 12.16.

With the predicted survival probability at the future times, it is easy to obtain the mean or median of the survival time, which can be used as a point estimate of the remaining useful life of the specific unit.

12.4 Cox PH Model with Frailty Term for Recurrent Events

As discussed in Section 12.2, the Cox PH model provides a simple way for event prediction: we can use the historical data to estimate the model parameters and then plug in the time-fixed covariates of a new unit to obtain the hazard and hence the survival function for the new unit. One limitation of the Cox PH model is that it is a population-level model and does not take the individual characteristic of each unit into consideration. The Cox PH model assumes that all the units behave identically and thus, if two different units have the same values for the covariates, then the prediction for these two units will be identical. However, in practice, we often have some unobserved distinct characteristics among the units and it is desirable to capture such individual characteristic using the model. In this section, we will introduce the "frailty" model which renders the heterogeneity among units to allow us to assign individuality to the units. The frailty model is a type of mixed effects models. The materials in this section are adapted from Deep et al. [2020].

In this section, we consider a situation that an event occurs multiple times on a unit. The event could be a warning event triggered by a certain working condition. The event could also be a terminating failure event for a component in the unit. Once it happens, the component will be replaced but it may fail again and the event occurs again for the unit. Obviously, due to specific environmental and inherent factors, the characteristics of the recurrent event sequences for different units are not identical. The conventional Cox PH model is used to model the terminating failure event that only occurs once for a unit. Andersen and Gill [1982] extended the model to address multiple events as follows:

$$h_{ij}(t|\mathbf{x}_{ij}) = \delta_{ij} h_0(t) \exp(\mathbf{x}_{ij}^T \gamma),$$

where $h_{ij}(t)$ is the *intensity* function for the jth observation of the ith unit and δ_{ij} is the indicator taking value 1 or 0 indicating if the event occurs. The intensity function can be roughly viewed as the "instantaneous" event occurrence risk at the specified time. The heterogeneity among the units can be characterized by the difference between the intensity for a specific unit and that for the population. Thereby, we distinguish two units by using an unobserved factor called *frailty*. The frailty is relative to the population and is unique to a unit. The intensity with the frailty term is now expressed as

$$h_{ij}(t|\mathbf{x}_{ij}, u_i) = \delta_{ij} h_0(t) u_i \exp(\mathbf{x}_{ij}^T \gamma), \tag{12.25}$$

where $u_i \sim f(U; \boldsymbol{\theta})$ is the frailty term for the ith unit acting multiplicatively on intensity function. The function $f(U; \boldsymbol{\theta})$ is a probability density function of the frailty term and $\boldsymbol{\theta}$ is the parameter for that function. The frailty model is actually a type of "mixed effects" model and $\boldsymbol{\theta}$ is also called "hyperparameters". Please note that the model in (12.25) implies that from each unit i, we have multiple event observations that are indexed by j. Thus, the observed data for an individual i now includes $\mathcal{D}_i = \{\mathbf{x}_{ij}, \delta_{ij}, t_{ij}\}$, $j = 1, \ldots, n_i$. Please note that u_i is unobservable. The parameters that we need to estimate for this model include $\{h_0(t), \gamma, \boldsymbol{\theta}\}$. The maximum likelihood method can be used to estimate these parameters. Specifically, the overall likelihood for the data $\boldsymbol{\mathcal{D}}$ of N units can be written as

$$L(h_0, \gamma, \boldsymbol{\theta}; \boldsymbol{\mathcal{D}}, U) = \prod_{i=1}^{N} \prod_{j=1}^{n_i} \left[h_0(t_{ij}) u_i \exp(\mathbf{x}_{ij}^T \gamma) \right]^{\delta_{ij}} \exp\left(-H_0(t)\, u_i \exp(\mathbf{x}_{ij}^T \gamma)\right), \tag{12.26}$$

where H_0 is the baseline cumulative intensity function.

Equation (12.26) contains the unobserved frailty term u_i. Thus, using conditional independence we can separate the (conditional) likelihood and frailty term

$$L(h_0, \gamma, \boldsymbol{\theta}; \mathcal{D}, U) = L_c(h_0, \gamma; \mathcal{D} \mid U) f(U). \tag{12.27}$$

To evaluate (12.27), we need to specify the probability density function for the frailty term and then integrate the nuisance parameter U over $f(U)$. Hougaard [2000] and Duchateau and Janssen [2007] discuss several distributions to model frailty. The most popular choice for modeling frailty in biostatistics as well as in reliability engineering is the Gamma distribution. If we let θ be the parameter for the one-parameter Gamma distribution, we can write $u_i \sim G(1/\theta, 1/\theta)$ and $E(u_i) = 1$ and $\text{var}(u_i) = \theta$.

$$f(u) = \frac{u^{1/\theta - 1} e^{-u/\theta}}{\theta^{1/\theta} + \Gamma(1/\theta)}.$$

The separation in the likelihood function (12.27) is very helpful for the estimation of parameters. In R, the package `frailtypack` provides very useful routines to implement the model parameter estimation for a frailty model (with gamma or log-normal frailty distribution) and Cox proportional hazard model.

The obtained estimates $\hat{h}_0, \hat{\boldsymbol{\theta}}, \hat{\gamma}$ using the historical data $\mathcal{D}$ represent the population characteristics. They can be used in an online updating scheme to update the frailty distribution for a new unit p based on the newly observed data $\mathcal{D}_p$ from it. With the updated frailty distribution, we can get a point estimate (e.g., the mean of the distribution) of the value of the frailty term for the new unit. Then we can use the frailty term to compute the intensity and hence the survival function for the new unit and eventually get an individualized event prediction. Comparing with the Cox PH model based event prediction, we are not only considering the covariates, but also addressing the individual characteristics of the new unit through the updated frailty term.

We can use a Bayesian updating scheme to update the distribution of the frailty term for the new unit. Specifically, using Bayes' formula, we can have the posterior distribution of the frailty term u_p as

$$f(u_p \mid \mathcal{D}_p) = \frac{f(\mathcal{D}_p \mid u_p) \pi(u_p)}{f(\mathcal{D}_p)}, \tag{12.28}$$

where $\mathcal{D}_p$ is the data from a new unit p, $f(u_p \mid \mathcal{D}_p)$ is the posterior distribution of u_p, $\pi(u_p)$ is the prior distribution of u_p, and $f(\mathcal{D}_p \mid u_p)$ is the data likelihood. We can naturally use the estimated population distribution of the frailty term as the prior distribution, i.e., $\pi(u_p) = f(U; \hat{\boldsymbol{\theta}})$.

Following (12.26), the conditional likelihood for the new unit can be written as

$$f(\mathcal{D}_p | u_p) = \prod_{j=1}^{n_p} \left[\hat{h}_0(t) u_p \exp(\mathbf{x}_{pj}^T \hat{\gamma}) \right]^{\delta_{pj}} \exp(-\widehat{H}_0(t) u_p \exp(\mathbf{x}_{pj}^T \hat{\gamma})) \quad (12.29)$$

and the posterior can be written as

$$f(u_p | \mathcal{D}_p) = \frac{\prod_{j=1}^{n_p} \left[\hat{h}_0(t) u_p \exp(\mathbf{x}_{pj}^T \hat{\gamma}) \right]^{\delta_{pj}} \exp(-\widehat{H}_0(t) u_p \exp(\mathbf{x}_{pj}^T \hat{\gamma})) \times f(u_p; \hat{\boldsymbol{\theta}})}{\int_0^{\infty} \prod_{j=1}^{n_p} \left[\hat{h}_0(t) u_p \exp(\mathbf{x}_{pj}^T \hat{\gamma}) \right]^{\delta_{pj}} \exp(-\widehat{H}_0(t) u_p \exp(\mathbf{x}_{pj}^T \hat{\gamma})) \times f(u_p; \hat{\boldsymbol{\theta}}) du}. \quad (12.30)$$

The posterior may or may not have a closed form expression depending upon the assumed distribution for the frailty term.

With the updated frailty distribution, we can use the mean of the distribution as a point estimation of the value of the frailty term for the new unit p.

$$\hat{u}_p = E(u_p | \mathcal{D}_p). \quad (12.31)$$

Equations (12.29), (12.30), and (12.31) provide the generic formula to online update the frailty term. Depending on the selected form of frailty distribution, the computation could be complex. Fortunately, for some popular choices of frailty distributions, we can obtain an explicit closed form expression for the frailty term or its approximation. We show the result for the Gamma frailty. Such a closed form expression can significantly speed up the calculation of the frailty term and make the on-line updating in real-time feasible.

For Gamma frailty, the posterior formulation in (12.28) has a closed form expression and the expectation of u_p is

$$\hat{u}_p = \frac{d_p + 1/\hat{\theta}}{1/\hat{\theta} + \widehat{H}_p(t)}, \quad (12.32)$$

where d_p is the number of events the unit p experienced, $\hat{\theta}$ is the estimated parameter for Gamma frailty, and $\widehat{H}_p(t)$ is conditional cumulative intensity of the unit until time t. The derivation of the above expression is provided in the appendix of this chapter.

For a general frailty distribution where the closed form expression is not available, we can use numerical methods such as sampling to obtain the posterior density function in (12.28). We draw samples and make inferences for the frailty term using its posterior distribution.

Once we have the updated frailty term, we can build the survival function for unit p at t^*. Substituting the estimates, the expression of the survival function becomes

$$\hat{S}_{pj}(t|t^*, \hat{u}_p, \hat{h}_0, \hat{\gamma}) = \hat{S}_0(t)^{\hat{u}_p \exp(\mathbf{x}_{pj}^T \hat{\gamma})},$$

where $\hat{u}_p$ is the updated value of the frailty term obtained in (12.31).

Example 12.6 To illustrate the intuition of the online updating and prediction scheme, we present a simple example. We adopt Gamma distribution for frailty in this example and compare the prediction result with the Cox PH model based method. For this example, we present two cases: (a) the new unit has constant time-to-events, and (b) the time-to-events are increasing as the time advances.

Evaluating the case of constant time-to-events (Figure 12.17), we find that as new event observations join the history, the frailty parameter is updated and we observe that the estimated remaining useful life adapts to the trend. The corresponding frailty value decreases reflecting that actual time-to-events are larger than the previously estimated values. The errors continue to decrease in subsequent predictions. We also observe that jump sizes reduce because variance in posterior reduces with more time-to-events observations. However, the

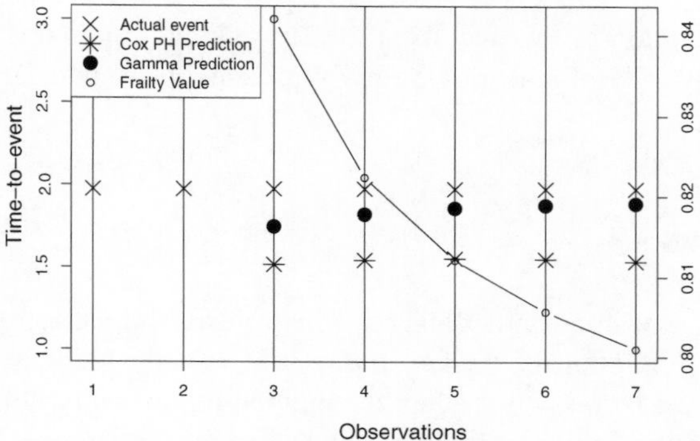

Figure 12.17 Prediction performance for a case with constant time-to-event data

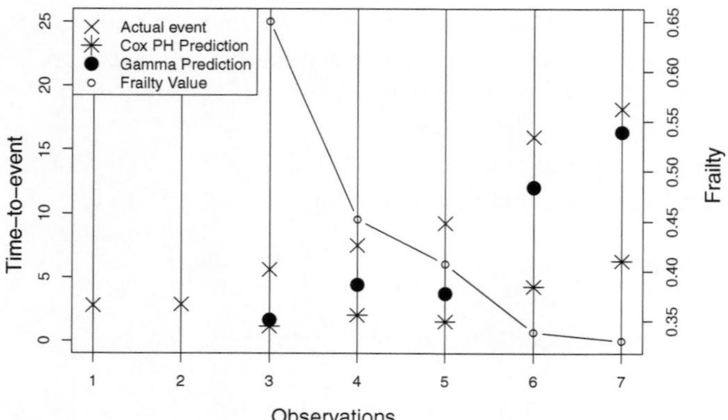

Figure 12.18 Prediction performance for a case with increasing time-to-event data

predictions in the Cox PH model based method do not involve any updating, and thereby the prediction remains fixed.

Likewise, considering the case of increasing time-to-events (Figure 12.18), the characteristics of the prediction are analogous to the previous case. Here our initial estimates are underestimating the actual time-to-events, and as more event observations are available we make remarkable improvements. There is also considerable decrease in estimates of the frailty term with each update, implying that the unit has a higher surviving probability than the previous estimate.

This example illustrates the dynamic updating nature of the frailty based event prediction method and its potential advantage over the Cox PH model based event prediction.

For more detailed description of this recurrent event prediction approach and a real world case study of the application of this approach, please refer to Deep et al. [2020].

Bibliographical Notes

Time-to-event data is a very important data type in industrial practices. Modeling, analysis, and prediction of time-to-event data are the focus of survival analysis in biomedical areas and reliability engineering in industrial applications. There are many good statistics books on this topic, for example, [Klein and Moeschberger, 2005; Hosmer Jr and Lemeshow, 1999; Moore, 2016; Meeker and Escobar, 2014; Elsayed, 2012]. The Cox PH model is comprehensively covered in Hosmer Jr et al. [2011], Therneau and Grambsch [2013]. The books covering the topic of the frailty model, which is the random effects model for time variables, include Duchateau and Janssen [2007], Hanagal [2019], and Wienke [2010]. An excellent book on recurrent events is Cook and Lawless [2007]. The joint model is an advanced statistical model to describe the association structure between longitudinal data and event times. The books on joint model include Rizopoulos [2012] and Elashoff et al. [2016] and reviews on this topic are given in Tsiatis and Davidian [2004] and Papageorgiou et al. [2019].

Exercises

1. (a). Estimate survival function through the Kaplan–Meier method for the data provided below. In the table, we have survival time (the actual survival time or censored survival time), denoted as T, for 14 units. δ is the censoring indicator, where 0/1 means censored/actual survival time.

Unit ID	1	2	3	4	5	6	7	8	9	10	11	12	13	14
T	2	3	3	7	5	3	9	8	8	7	2	10	15	12
δ	1	0	1	1	1	0	1	0	1	1	1	1	0	1

   (b). From the estimated survival function, what is the median survival life?
2. Suppose we are given the following survival data of five units, where T is the survival (actual or censored) time, δ is the indicator of censoring, and x is the fixed covariate for each unit.

unit id	T	δ	x
101	1	1	1
102	2	1	2
103	3	0	1
104	3	1	0
105	4	1	1

   If we want to use a Cox PH model $h_i(t) = h_0(t)\exp(x_i \cdot \gamma)$ to fit the data, please write down the partial-likelihood function of the data.
3. The R data file `chapter12_question3.csv` contains the survival times of 100 units. The units belong to type A if `x=0`, else they are of type 2. Considering that units are homogeneous, please fit a Cox PH model with `x` as predictor. Test for proportionality assumption (also give a figure). Plot

estimated survival functions for both types of units and obtain expected event time.

4. Consider the recurrent event setting. Assume that we have obtained model estimates from historical data. Particularly, we fit a piece-wise constant hazard without covariates in the form of (12.25). The estimates $\hat{h}(t)$ are denoted as `est_hazard` in the table. The variance parameter of the fitted Gamma frailty term is $\hat{\theta} = 0.5$.

t1	t2	est_hazard
0	1	0.50
1	2	0.50
2	3	1.00
3	4	1.00
4	5	0.75
5	6	0.75
6	7	0.50
7	8	1.00
8	9	1.00
9	10	2.00

Now, let us consider that we have a new unit which is expected to behave in a similar way to past units. We would like to continuously estimate and update the frailty parameter ($\hat{u}_m \sim \text{Gamma}(\hat{\theta})$) corresponding to this new unit. The data from this new unit is given below:

entry	event_time
1	2
2	3
3	4
4	6
5	8
6	10

Starting from ($t = 0$), obtain $\hat{u}_m$ for this unit just after each of the six realizations (use (12.32)). Comment on the trend of frailty value with respect to the population and the observed event times.

5. Using the data `chapter12_question5.csv` do the following:
   - Fit the joint model as in Example 12.5 considering all units except `JCI2`.

- Fit a time-varying Cox PH (Section 12.2.2) considering `x` and `r` as the covariates and using all units except `JCI2`.
- Now, we are interested in prediction of failure of `JCI2`. The actual time to failure in the dataset is $t = 9$. Use both of the fitted models to compute the mean remaining life at two instances: $t^* = \{2, 7\}$. Using the mean remaining life as a predictor of failure, compare and comment on the errors in prediction from both the models. (Please note that during the prediction from time-varying Cox PH model, it does not have the information of future covariates. Therefore, use the latest available set of covariates (i.e. available at t^*) and compute the future survival function as if it were a time-fixed Cox PH).

Appendix

We present the updating formula for the Gamma frailty term in this appendix.

Adapting the posterior (12.28) in expanded form with prior $G(1/\hat{\theta},1/\hat{\theta})$

$$f(u_p|\mathcal{D}_p)=\frac{\prod_{j=1}^{n_p}\left[\hat{h}_0(t)u_p\exp(\mathbf{x}_{pj}^T\hat{\gamma})\right]^{\delta_{pj}}\exp(-\widehat{H}_0(t)u_p\exp(\mathbf{x}_{pj}^T\hat{\gamma}))\times\dfrac{u_p^{1/\hat{\theta}-1}e^{-u_p/\hat{\theta}}}{\hat{\theta}^{1/\hat{\theta}}+\Gamma(1/\hat{\theta})}}{\int_0^\infty\prod_{j=1}^{n_p}\left[\hat{h}_0(t)u_p\exp(\mathbf{x}_{pj}^T\hat{\gamma})\right]^{\delta_{pj}}\exp(-\widehat{H}_0(t)u_p\exp(\mathbf{x}_{pj}^T\hat{\gamma}))\times\dfrac{u^{1/\hat{\theta}-1}e^{-u/\hat{\theta}}}{\hat{\theta}^{1/\hat{\theta}}+\Gamma(1/\hat{\theta})}du}.$$

Working out the integration of marginal likelihood in the denominator gives

$$L_{marg}=\frac{\Gamma(d_p+1/\hat{\theta})\prod_{j=1}^{n_p}\left[\hat{h}_0(t)\exp(\mathbf{x}_{pj}^T\hat{\gamma})\right]^{\delta_{pj}}}{\left(1/\hat{\theta}+\sum_{j=1}^{n_p}\widehat{H}_0(t)\exp(\mathbf{x}_{pj}^T\hat{\gamma})\right)^{d_p+1/\hat{\theta}}\hat{\theta}^{1/\hat{\theta}}\Gamma^{1/\hat{\theta}}}.$$

Carrying out algebraic manipulations and substituting $\widehat{H}_p(t_p)=\sum_{j=1}^{n_p}\widehat{H}_0(\mathrm{t})\exp(\mathbf{x}_{pj}^T\hat{\gamma})$ (called conditional cumulative hazard) we can obtain

$$f(u_p|\mathcal{D}_p)=\frac{u_p^{d_p+1/\hat{\theta}-1}\exp(-u_p(1/\hat{\theta}+\widehat{H}_p(t_p)))(1/\hat{\theta}+\widehat{H}_p(t_p))^{d_p+1/\hat{\theta}}}{\Gamma(d_p+1/\hat{\theta})}.$$

The above expression is Gamma density with parameters $(d_p+1/\hat{\theta},1/\hat{\theta}+\widehat{H}_p(t_p))$ therefore, the mean value of the posterior is in (12.32).

Appendix

Basics of Vectors, Matrices, and Linear Vector Space

A.1 Vectors and Linear Vector Space

Definition A.1 *[Vector]*: A vector is a list of numbers.

Each number in a vector is called an element of the vector. The number of elements in a vector is the dimension of a vector. If the elements are arranged in a row, it is called a row vector and if the elements are arranged in a column, it is called a column vector. We can view a vector as a point in space with magnitude (also called norm) and direction.

Definition A.2 *[Magnitude/norm of a vector]*: The magnitude or norm of a vector is the square root of the sum of squares of its elements, i.e., if $\mathbf{v} = \left(v_1\ v_2 \ldots, v_N\right)^T$, then $\|\mathbf{v}\| = \left(\sum_{i=1}^{N} v_i^2\right)^{1/2}$, where $\|\mathbf{v}\|$ represents the magnitude/norm of the vector $\mathbf{v}$.

Definition A.3 *[Direction of a vector]*: The direction of a vector $\mathbf{v}$ is represented by a vector with unit magnitude, $\mathbf{v}/\|\mathbf{v}\|$.

Some operations on vectors with the same dimension include addition, subtraction, and dot product. Let $\mathbf{v} = \left(v_1\ v_2 \ldots\ v_N\right)^T$ and $\mathbf{w} = \left(w_1\ w_2 \ldots\ w_N\right)^T$, then $\mathbf{v} + \mathbf{w} = \left(v_1 + w_1\ v_2 + w_2 \ldots\ v_N + w_N\right)^T$,

$\mathbf{v} - \mathbf{w} = \left(v_1 - w_1\ v_2 - w_2 \ldots\ v_N - w_N\right)^T$,

$a \cdot \mathbf{v} = \left(a \cdot v_1\ a \cdot v_2 \ldots\ a \cdot v_N\right)^T$, and $\mathbf{v} \cdot \mathbf{w} = \mathbf{v}^T \mathbf{w} = \sum_{i=1}^{N} (v_i \cdot w_i)$.

Industrial Data Analytics for Diagnosis and Prognosis: A Random Effects Modelling Approach, First Edition. Shiyu Zhou and Yong Chen.

Definition A.4 *[Orthogonality]*: Two vectors are orthogonal to one another if the dot product of those two vectors is equal to zero.

Definition A.5 *[Angle between two vectors]*: The angle θ between two vectors $\mathbf{v}$ and $\mathbf{w}$ is $\theta = \cos^{-1}[(\mathbf{v} \cdot \mathbf{w}) / \|\mathbf{v}\| \cdot \|\mathbf{w}\|]$.

Definition A.6 *[Linear independence]*: A group of vectors is linearly independent if no one of the vectors can be created by any linear combination of the other vectors in the group, where linear combination only involves scalar multiplication and addition operations of vectors.

Definition A.7 *[Vector space]*: A vector space is a set of vectors that is closed under finite vector addition and scalar multiplication.

A vector set is closed if the result of an arbitrary linear combination of an arbitrary group of vectors in this set is still in this set.

Definition A.8 *[Basis set of a vector space]*: A basis set of vectors is a linearly independent set of vectors that, when used in linear combination, can represent every vector in a given vector space.

The number of vectors in the basis set of a vector space is also called the dimension of the linear space.

A.2 Matrix and Its Operations

Definition A.9 *[Matrix]*: A matrix of dimension m by n is defined as a rectangular array of elements arranged in m rows and n columns, denoted by $\mathbf{A} = \begin{pmatrix} a_{11} & a_{12} & \cdots & a_{1n} \\ a_{21} & a_{22} & \cdots & a_{2n} \\ \vdots & \vdots & \ddots & \vdots \\ a_{m1} & a_{m2} & \cdots & a_{mn} \end{pmatrix}$, where a_{ij} is the element at the ith row and jth column of the matrix.

Obviously, a vector can be viewed as a special case of a matrix. Each column of a matrix is a column vector and each row of a matrix is a row vector. If we let the ith column of matrix $\mathbf{A}$ be the ith row of a new matrix, the new matrix is called the transpose of $\mathbf{A}$, often denoted by $\mathbf{A}^T$. Obviously $\mathbf{A}^T$ will have dimension of $n \times m$. For a $m \times n$ dimensional matrix $\mathbf{A}$, if $m = n$, we call the matrix a square matrix. If the columns of a square matrix are orthogonal vectors with unit magnitude, the matrix is called an orthogonal matrix. Sometimes we call a matrix a "tall" matrix if $m > n$ and a "fat" matrix if $m < n$. For a square matrix if we have $a_{ij} = a_{ji}$, for all $1 \le i \le m$ and $1 \le j \le m$, we call it a symmetric matrix. The sum of the diagonal elements of a square matrix is called a trace of the matrix, denoted by $\mathrm{tr}(\mathbf{A}) = \sum_{i=1}^{m} a_{ii}$.

If two matrices have the same dimension, then we can conduct addition and subtraction of these two matrices, which are simply the element-wise addition and subtraction. We can left- or right-multiply a matrix by a scalar through multiplying each element of the matrix by the scalar. However, multiplication of two matrices is not defined based on element-wise multiplications.

Definition A.10 *[Matrix multiplication]*: If $\mathbf{A}$ is a $m \times n$ matrix and $\mathbf{B}$ is a $n \times p$

matrix $\mathbf{A} = \begin{pmatrix} a_{11} & a_{12} & \cdots & a_{1n} \\ a_{21} & a_{22} & \cdots & a_{2n} \\ \vdots & \vdots & \ddots & \vdots \\ a_{m1} & a_{m2} & \cdots & a_{mn} \end{pmatrix}$, $\mathbf{B} = \begin{pmatrix} b_{11} & b_{12} & \cdots & b_{1p} \\ b_{21} & b_{22} & \cdots & b_{2p} \\ \vdots & \vdots & \ddots & \vdots \\ b_{n1} & b_{n2} & \cdots & b_{np} \end{pmatrix}$,

then the matrix product $\mathbf{C} = \mathbf{AB}$ is a $m \times p$ matrix and defined as

$$\mathbf{C} = \begin{pmatrix} c_{11} & c_{12} & \cdots & c_{1p} \\ c_{21} & c_{22} & \cdots & c_{2p} \\ \vdots & \vdots & \ddots & \vdots \\ c_{m1} & c_{m2} & \cdots & c_{mp} \end{pmatrix},$$

where $c_{ij} = \sum_{k=1}^{n} a_{ik} b_{kj}$, $1 \le i \le m$ and $1 \le j \le p$.

In fact, the (i, j)th element of the product of $\mathbf{AB}$ is the dot product of the ith row vector of $\mathbf{A}$ with the jth column vector of $\mathbf{B}$. The concept of matrix multiplication plays a critical role in linear algebra. Different from scalar multiplication, the matrix multiplication is not commutative, i.e., in general $\mathbf{AB} \neq \mathbf{BA}$. Matrix multiplication is distributive with respect to addition, i.e., $\mathbf{A}(\mathbf{B}+\mathbf{C}) = \mathbf{AB} + \mathbf{AC}$ and $(\mathbf{B}+\mathbf{C})\mathbf{A} = \mathbf{BA} + \mathbf{CA}$. Matrix multiplication is also associative, i.e., $(\mathbf{AB})\mathbf{C} = \mathbf{A}(\mathbf{BC})$.

Another multiplication operation of matrices we often use is the Kronecker product or direct product.

Definition A.11 *[Kronecker or direct product of two matrices]*: Let $\mathbf{A}$ be an $m \times n$ dimensional matrix with elements a_{ij}, $1 \le i \le m$ and $1 \le j \le n$ and $\mathbf{B}$ is a

$p \times q$ dimensional matrix, then the Kronecker product of these two matrices is a $mp \times nq$ dimensional matrix, given as $\mathbf{A} \otimes \mathbf{B} = \begin{pmatrix} a_{11}\mathbf{B} & a_{12}\mathbf{B} & \cdots & a_{1n}\mathbf{B} \\ a_{21}\mathbf{B} & a_{22}\mathbf{B} & \cdots & a_{2n}\mathbf{B} \\ \vdots & \vdots & \ddots & \vdots \\ a_{m1}\mathbf{B} & a_{m2}\mathbf{B} & \cdots & a_{mn}\mathbf{B} \end{pmatrix}$.

Definition A.12 *[Determinant of a square matrix]*: Let $\mathbf{A}$ be an m dimensional square matrix. The determinant of $\mathbf{A}$ denoted by $|\mathbf{A}|$ or $\det(\mathbf{A})$, is defined as $|\mathbf{A}| = \sum_{\mathbf{s} \in \mathbf{S}_m} \operatorname{sgn}(\mathbf{s}) \prod_{i=1}^{m} a_{is_i}$, where $\mathbf{S}_m$ is the set containing all $m!$ permutations of $\{1, 2, \ldots, m\}$, $\mathbf{s}$ is one of such permutations, s_i is the ith element of $\mathbf{s}$, and $\operatorname{sgn}(\mathbf{s})$ is positive if we need an even number of exchanges of elements of $\mathbf{s}$ to covert $\mathbf{s}$ into $\{1, 2, \ldots, m\}$; otherwise $\operatorname{sgn}(\mathbf{s})$ is negative.

Please note that the computational load for directly computing determinants using Definition A.12 is high for a high dimensional matrix as it requires the summation of $m!$ number of terms.

Definition A.13 *[Matrix inverse]*: The inverse of the square $m \times m$ matrix $\mathbf{A}$, denoted by $\mathbf{A}^{-1}$ is the matrix such that $\mathbf{A}\mathbf{A}^{-1} = \mathbf{I}$, where $\mathbf{I}$ is the $m \times m$ identity matrix with ones on the diagonal and zeros elsewhere.

Matrix inverse is a very important concept and has many interesting properties.

- The inverse of a square matrix may not exist. If it exists, it is unique and the matrix is called non-singular. Otherwise, the matrix is called singular.
- $\mathbf{A}\mathbf{A}^{-1} = \mathbf{A}^{-1}\mathbf{A} = \mathbf{I}$
- $(\mathbf{A}\mathbf{B})^{-1} = \mathbf{B}^{-1}\mathbf{A}^{-1}$
- $(\mathbf{A}^T)^{-1} = (\mathbf{A}^{-1})^T$
- $(\mathbf{A} + \mathbf{X}\mathbf{B}\mathbf{X}^T)^{-1} = \mathbf{A}^{-1} - \mathbf{A}^{-1}\mathbf{X}(\mathbf{B}^{-1} + \mathbf{X}^T\mathbf{A}^{-1}\mathbf{X})^{-1}\mathbf{X}^T\mathbf{A}^{-1}$, where $\mathbf{A}$ and $\mathbf{B}$ are square and invertible matrices but do not need to have the same dimension.

Definition A.14 *[Matrix rank]*: The rank of a matrix is defined as (1) the maximum number of linearly independent column vectors in the matrix, called column rank or (2) the maximum number of linearly independent row vectors in the matrix, called row rank. The column rank and row rank will always be the same.

A matrix is said to have full rank if its rank equals the lesser of the number of rows and columns. A matrix is said to be rank-deficient if it does not have full rank. The rank of a matrix can be obtained through its reduced row echelon form.

Definition A.15 *[Reduced row echelon form (RREF) of a matrix]*: A matrix is said to be in a reduced row echelon form if

- The first non-zero number in a row is 1 (we call it the leading 1).
- All rows of zeros (if any) are at the bottom part of the matrix.
- Each column that contains a leading 1 has zeros everywhere else.

Given an arbitrary matrix, its RREF can be obtained through elementary row operations, i.e., exchanging the locations of two rows, multiplying a row by a non-zero scalar, or adding a scalar multiple of one row to another row. It can be proven that the RREF of a matrix is unique and the number of non-zero rows of matrix is its rank.

Definition A.16 *[Range/image of a matrix]*: The range or image of a matrix $\mathbf{A}$ is the span (set of all possible linear combinations) of its column vectors.

Definition A.17 *[Null space of a matrix]*: The null space of matrix $\mathbf{A}$ is the set of all vectors $\mathbf{v}$ for which $\mathbf{Av} = \mathbf{0}$.

The null space is a linear space and its dimension is called the nullity of the matrix. It can be proven that the sum of nullity and the rank of the matrix is the number of columns of the matrix. Obviously, the nullity of a full column ranked matrix is zero.

Definition A.18 *[Eigenvalues and eigenvectors of a matrix]*: If a scalar λ and a non-zero vector $\mathbf{v}$ satisfy $\mathbf{Av} = \lambda\mathbf{v}$, then they are called eigenvalue and eigenvector of the matrix $\mathbf{A}$.

The eigenvalues of a matrix $\mathbf{A}$ can be obtained by solving the mth order characteristics equation $|\mathbf{A} - \lambda\mathbf{I}| = \mathbf{0}$ and with the eigenvalue, we can obtain the corresponding eigenvector by solving the linear system $(\mathbf{A} - \lambda\mathbf{I})\mathbf{v} = \mathbf{0}$ for $\mathbf{v}$. Please note that square matrices may have complex eigenvalues. There are many very interesting properties of eigenvalues and eigenvectors for a square matrix. Denote $\lambda_1, \lambda_2, \cdots, \lambda_m$ as the eigenvalues of a m dimensional matrix $\mathbf{A}$, we have

- $\mathrm{tr}(\mathbf{A}) = \sum_{i=1}^{m} \lambda_i$.
- $|\mathbf{A}| = \prod_{i=1}^{m} \lambda_i$.

- If a square matrix is nonsingular, all its eigenvalues are non-zero.
- The eigenvalues of a real symmetric matrix are all real.
- For a symmetric matrix, the eigenvectors corresponding to distinct eigenvalues are orthogonal to each other.
- For any real symmetric matrix, there exists an orthogonal matrix $\mathbf{U}$, such that $\mathbf{U}^T\mathbf{AU} = \text{diag}(\lambda_1, \lambda_2, \ldots, \lambda_m)$, where $\text{diag}(\lambda_1, \lambda_2, \ldots, \lambda_m)$ is a matrix with diagonal elements being $\lambda_1, \lambda_2, \ldots, \lambda_m$ and zeros elsewhere. Based on this relationship, we can also get the eigendecomposition of matrix $\mathbf{A}$ as $\mathbf{A} = \mathbf{U} \cdot \text{diag}(\lambda_1, \lambda_2, \ldots, \lambda_m) \cdot \mathbf{U}^T$.

Definition A.19 *[Positive definite and positive semi-definite matrix]*: A symmetric matrix $\mathbf{A}$ is called positive definite if $\mathbf{v}^T\mathbf{Av} > 0$ for any $\mathbf{v} \neq \mathbf{0}$, and positive semi-definite if $\mathbf{v}^T\mathbf{Av} \geq 0$ for any $\mathbf{v} \neq \mathbf{0}$.

A symmetric matrix is positive definite if and only if all its eigenvalues are positive. For positive semi-definite symmetric matrix, if its dimension is m and rank is r, then it has r positive eigenvalues and $m - r$ zero eigenvalues.

Definition A.20 *[Singular value decomposition of a matrix]*: For a $m \times n$ dimensional matrix $\mathbf{A}$, we can decompose it into a factorization of the form $\mathbf{U\Sigma V}^T$, where $\mathbf{U}$ is a $m \times m$ orthogonal matrix, $\mathbf{\Sigma}$ is a $m \times n$ rectangular diagonal matrix, and $\mathbf{V}$ is a $n \times n$ orthogonal matrix. The diagonal elements of $\mathbf{\Sigma}$ are non-negative real numbers, called singular values of $\mathbf{A}$.

Singular value decomposition generalizes the eigendecomposition of a square matrix to any $m \times n$ matrix. One important application of singular value decomposition is to construct the pseudoinverse, also called the Moore–Penrose inverse, of a $m \times n$ matrix, which is a generalization of the inverse of a matrix for a square matrix.

Definition A.21 *[Pseudoinverse of a matrix]*: For a $m \times n$ dimensional matrix $\mathbf{A}$ with singular value decomposition $\mathbf{A} = \mathbf{U\Sigma V}^T$, its pseudoinverse is $\mathbf{A}^- = \mathbf{V\Sigma}^-\mathbf{U}^T$, where $\mathbf{\Sigma}^-$ is the pseudoinverse of $\mathbf{\Sigma}$, which is formed by replacing every non-zero diagonal entry by its reciprocal and transposing the resulting matrix.

Some important properties of pseudoinverse of a matrix include the following

- if $\mathbf{A}$ is invertible, then $\mathbf{A}^- = \mathbf{A}^{-1}$.
- $\mathbf{AA}^-\mathbf{A} = \mathbf{A}$
- $\mathbf{A}^-\mathbf{AA}^- = \mathbf{A}^-$
- $\mathbf{AA}^-$ and $\mathbf{A}^-\mathbf{A}$ are symmetric.

References

Odd Aalen. Nonparametric estimation of partial transition probabilities in multiple decrement models. *The Annals of Statistics*, 6(3):534–545, 1978.

Alan Agresti. *Foundations of linear and generalized linear models*. John Wiley & Sons, 2015.

Mauricio A Álvarez and Neil D Lawrence. Computationally efficient convolved multiple output Gaussian processes. *The Journal of Machine Learning Research*, 12:1459–1500, 2011.

Per Kragh Andersen and Richard D Gill. Cox's regression model for counting processes: A large sample study. *The Annals of Statistics*, 10(4):1100–1120, 1982.

Theodore Wilbur Anderson. Asymptotic theory for principal component analysis. *Annals of Mathematical Statistics*, 34(1):122–148, 1963.

Theodore Wilbur Anderson. Statistical inference for covariance matrices with linear structure. In *Multivariate Analysis, II (Proceedings of the Second International Symposium, Dayton, Ohio, 1968)*, pages 55–66, 1969.

Theodore Wilbur Anderson. *An introduction to multivariate statistical analysis*. Wiley Series in Probability and Statistics. Wiley, 2003.

Daniel W Apley and Jianjun Shi. Diagnosis of multiple fixture faults in panel assembly. *Transactions-American Society of Mechanical Engineers Journal of Manufacturing Science and Engineering*, 120:793–801, 1998.

Piero Baraldi, Francesca Mangili, and Enrico Zio. A prognostics approach to nuclear component degradation modeling based on Gaussian process regression. *Progress in Nuclear Energy*, 78:141–154, 2015.

Kaveh Bastani, Babak Barazandeh, and Zhenyu James Kong. Fault diagnosis in multistation assembly systems using spatially correlated Bayesian learning algorithm. *Journal of Manufacturing Science and Engineering*, 140(3):031003, 2018.

Kaveh Bastani, Zhenyu Kong, Wenzhen Huang, and Yingqing Zhou. Compressive sensing-based optimal sensor placement and fault diagnosis for multi-station assembly processes. *IIE Transactions*, 48(5):462–474, 2016.

Industrial Data Analytics for Diagnosis and Prognosis: A Random Effects Modelling Approach, First Edition. Shiyu Zhou and Yong Chen.

Ilias Bilionis, Nicholas Zabaras, Bledar A Konomi, and Guang Lin. Multi-output separable Gaussian process: Towards an efficient, fully Bayesian paradigm for uncertainty quantification. *Journal of Computational Physics*, 241:212–239, 2013.

Christopher M Bishop. *Pattern recognition and machine learning*. Springer, 2006.

Johann F Böhme. Estimation of spectral parameters of correlated signals in wavefields. *Signal Processing*, 11(4):329–337, 1986.

Phillip Boyle and Marcus Frean. Dependent Gaussian processes. In *Advances in neural information processing systems*, pages 217–224, 2005.

Norman Breslow. Covariance analysis of censored survival data. *Biometrics*, 30(1):89–99, 1974.

Helen Brown and Robin Prescott. *Applied mixed models in medicine*. John Wiley & Sons, 2014.

Robert Grover Brown and Patrick YC Hwang. *Introduction to random signals and applied Kalman filtering*. Wiley New York, 1992.

VJ Carey and You-Gan Wang. *Mixed-effects models in S and S-PLUS*. Taylor & Francis, 2001.

Raymond J Carroll. *Transformation and weighting in regression*. CRC Press, 2017.

George Casella and Roger L Berger. *Statistical inference*. Duxbury Pacific Grove, CA, 2002.

Dariusz Ceglarek and Jianjun Shi. Fixture failure diagnosis for autobody assembly using pattern recognition. *Journal of Engineering for Industry*, 118(1):55–66, 1996.

Santanu Chakraborty, Nagi Gebraeel, Mark Lawley, and Hong Wan. Residual-life estimation for components with non-symmetric priors. *IIE Transactions*, 41(4):372–387, 2009.

Minho Chang and DC Gossard. Computational method for diagnosis of variation-related assembly problems. *International Journal of Production Research*, 36(11):2985–2995, 1998.

Jie Chen and Ron J Patton. *Robust model-based fault diagnosis for dynamic systems*. Springer Science & Business Media, 2012.

Yong Chen. Application of matroid theory for diagnosability study of coordinate sensing systems in discrete-part manufacturing processes. *Technometrics*, 48(3):386–398, 2006.

Woojin Cho, Youngrae Kim, and Jinkyoo Park. Hierarchical anomaly detection using a multioutput Gaussian process. *IEEE Transactions on Automation Science and Engineering*, 17(1):261–272, 2019.

Mario Cleves. *An introduction to survival analysis using Stata*. Stata Press, 2008.

Diana Cole. *Parameter redundancy and identifiability*. CRC Press, 2020.

Bianca M Colosimo and Enrique Del Castillo. *Bayesian process monitoring, control and optimization*. CRC Press, 2006.

Bianca M Colosimo, Paolo Cicorella, Massimo Pacella, and Marzia Blaco. From profile to surface monitoring: SPC for cylindrical surfaces via Gaussian processes. *Journal of Quality Technology*, 46(2):95–113, 2014.

Richard J Cook and Jerald Lawless. *The statistical analysis of recurrent events*. Statistics for Biology and Health. Springer New York, 2007.

David R Cox. Regression models and life-tables. *Journal of the Royal Statistical Society. Series B (Methodological)*, 34(2):187–220, 1972.

H d'Assumpcao. Some new signal processors for arrays of sensors. *IEEE Transactions on Information Theory*, 26(4):441–453, 1980.

Carl De Boor. *A practical guide to splines*. Springer-Verlag New York, 1978.

Akash Deep, Dharmaraj Veeramani, and Shiyu Zhou. Event prediction for individual unit based on recurrent event data collected in teleservice systems. *IEEE Transactions on Reliability*, 69(1):216–227, 2020.

Eugene Demidenko. *Mixed models: theory and applications*. John Wiley & Sons, 2005.

Richard E DeVor, Tsong-how Chang, and John William Sutherland. *Statistical quality design and control: contemporary concepts and methods*. Prentice Hall, 2007.

Peter Diggle, Peter J Diggle, Patrick Heagerty, Kung-Yee Liang, Patrick J Heagerty, and Scott Zeger. *Analysis of longitudinal data*. Oxford University Press, 2002.

Steven X Ding. *Model-based fault diagnosis techniques: design schemes, algorithms, and tools*. Springer Science & Business Media, 2008.

Yu Ding, Dariusz Ceglarek, and Jianjun Shi. Fault diagnosis of multistage manufacturing processes by using state space approach. *Journal of Manufacturing Science and Engineering*, 124(2):313–322, 2002a.

Yu Ding, Jianjun Shi, and Dariusz Ceglarek. Diagnosability analysis of multi-station manufacturing processes. *Journal of Dynamic Systems, Measurement, and Control*, 124(1):1–13, 2002b.

Yu Ding, Shiyu Zhou, and Yong Chen. A comparison of process variation estimators for in-process dimensional measurements and control. *Journal of Dynamic Systems, Measurement, and Control*, 127(1):69–79, 2005.

Dheeru Dua and Casey Graff. UCI machine learning repository, 2017. URL http://archive.ics.uci.edu/ml.

Luc Duchateau and Paul Janssen. *The frailty model*. Springer Science & Business Media, 2007.

Pham Luu Trung Duong, Hyunseok Park, and Nagarajan Raghavan. Application of multi-output Gaussian process regression for remaining useful life prediction of light emitting diodes. *Microelectronics Reliability*, 88:80–84, 2018.

Akram Eddahech, Olivier Briat, Eric Woirgard, and Jean-Michel Vinassa. Remaining useful life prediction of lithium batteries in calendar ageing for automotive applications. *Microelectronics Reliability*, 52(9):2438–2442, 2012.

Bradley Efron. The efficiency of cox's likelihood function for censored data. *Journal of the American Statistical Association*, 72(359):557–565, 1977.

Robert Elashoff, Gang li, and Ning Li. *Joint modeling of longitudinal and time-to-event data*. Chapman & Hall/CRC Monographs on Statistics and Applied Probability. CRC Press, 2016.

Elsayed A Elsayed. *Reliability engineering*. Wiley Series in Systems Engineering and Management. Wiley, 2012.

Ronald A Fisher. The correlation between relatives on the supposition of Mendelian inheritance. *Transactions of the Royal Society of Edinburgh*, 52(2):399–433, 1918.

Garrett M Fitzmaurice, Nan M Laird, and James H Ware. *Applied longitudinal analysis*. Wiley Series in Probability and Statistics - Applied Probability and Statistics Section Series. Wiley, 2004.

Garrett M Fitzmaurice, Nan M Laird, and James H Ware. *Applied longitudinal analysis*. John Wiley & Sons, 2012.

Bernhard K Flury. Two generalizations of the common principal component model. *Biometrika*, 74(1):59–69, 1987.

Chris Fraley and Adrian E Raftery. How many clusters? Which clustering method? Answers via model-based cluster analysis. *The Computer Journal*, 41(8):578–588, 1998.

Nicholas W Galwey. *Introduction to mixed modelling: beyond regression and analysis of variance*. John Wiley & Sons, 2014.

Zhiqiang Ge and Zhihuan Song. *Multivariate statistical process control: process monitoring methods and applications*. Springer Science & Business Media, 2012.

Nagi Gebraeel and Jing Pan. Prognostic degradation models for computing and updating residual life distributions in a time-varying environment. *IEEE Transactions on Reliability*, 57(4):539–550, 2008.

Nagi Z Gebraeel, Mark A Lawley, Rong Li, and Jennifer K Ryan. Residual-life distributions from component degradation signals: A Bayesian approach. *IIE Transactions*, 37(6):543–557, 2005.

Janos Gertler. *Fault detection and diagnosis in engineering systems*. CRC, Virginia, USA, 1998.

Erik W Grafarend. *Linear and nonlinear models: fixed effects, random effects, and mixed models*. de Gruyter, 2006.

Robert B Gramacy. *Surrogates: Gaussian process modeling, design, and optimization for the applied sciences*. CRC Press, 2020.

Mohinder S Grewal and Angus P Andrews. *Kalman filtering: theory and practice with MATLAB*. John Wiley & Sons, 2014.

David D Hanagal. *Modeling survival data using frailty models: second edition*. Industrial and Applied Mathematics. Springer Singapore, 2019.

Andrew C Harvey. *Forecasting, structural time series models and the Kalman filter*. Cambridge university press, 1990.

David Harville. Extension of the Gauss-Markov theorem to include the estimation of random effects. *The Annals of Statistics*, 384–395, 1976.

Trevor Hastie, Robert Tibshirani, and Jerome Friedman. *The elements of statistical learning: data mining, inference, and prediction*. Springer Science & Business Media, 2009.

Trevor Hastie, Robert Tibshirani, and Martin Wainwright. *Statistical learning with sparsity: the lasso and generalizations*. CRC press, 2015.

Haibo He and Edwardo A Garcia. Learning from imbalanced data. *IEEE Transactions on Knowledge and Data Engineering*, 21(9):1263–1284, 2009.

Donald Hedeker and Robert D Gibbons. *Longitudinal data analysis*. John Wiley & Sons, 2006.

Dave Higdon. Space and space-time modeling using process convolutions. In *Quantitative methods for current environmental issues*, pages 37–56. Springer, 2002.

David W Hosmer Jr and Stanley Lemeshow. *Applied survival analysis: time-to-event*. Wiley-Interscience, 1999.

David W Hosmer Jr, Stanley Lemeshow, and Susanne May. *Applied survival analysis: regression modeling of time-to-event data*. John Wiley & Sons, 2011.

Philip Hougaard. *Analysis of multivariate survival data*. Statistics for Biology and Health. Springer New York, 2000.

Alex Hrong-Tai Fai and Paul L Cornelius. Approximate f-tests of multiple degree of freedom hypotheses in generalized least squares analyses of unbalanced split-plot experiments. *Journal of Statistical Computation and Simulation*, 54(4):363–378, 1996.

Rolf Isermann. *Fault-diagnosis applications: model-based condition monitoring: actuators, drives, machinery, plants, sensors, and fault-tolerant systems*. Springer Science & Business Media, 2011.

Salman Jahani, Raed Kontar, Dharmaraj Veeramani, and Shiyu Zhou. Statistical monitoring of multiple profiles simultaneously using Gaussian processes. *Quality and Reliability Engineering International*, 34(8):1510–1529, 2018.

Salman Jahani, Raed Kontar, Shiyu Zhou, and Dharmaraj Veeramani. Remaining useful life prediction based on degradation signals using monotonic b-splines with infinite support. *IISE Transactions*, 52(5):537–554, 2020.

Andrew KS Jardine, Daming Lin, and Dragan Banjevic. A review on machinery diagnostics and prognostics implementing condition-based maintenance. *Mechanical Systems and Signal Processing*, 20(7):1483–1510, 2006.

Nong Jin and Shiyu Zhou. Data-driven variation source identification for manufacturing process using the eigenspace comparison method. *Naval Research Logistics (NRL)*, 53(5):383–396, 2006a.

Nong Jin and Shiyu Zhou. Signature construction and matching for fault diagnosis in manufacturing processes through fault space analysis. *IIE transactions*, 38(4):341–354, 2006b.

Richard Arnold Johnson and Dean W Wichern. *Applied multivariate statistical analysis*. Prentice Hall, Upper Saddle River, NJ, 2002.

Richard H Jones. *Longitudinal data with serial correlation: a state-space approach*. CRC Press, 1993.

Rudolph Emil Kalman. A new approach to linear filtering and prediction problems. *Journal of Basic Engineering*, 82(1):35–45, 1960.

Edward L Kaplan and Paul Meier. Nonparametric estimation from incomplete observations. *Journal of the American Statistical Association*, 53(282):457–481, 1958.

Heysem Kaya, Pmar Tüfekci, and Fikret S Gürgen. Local and global learning methods for predicting power of a combined gas & steam turbine. In *Proceedings of the International Conference on Emerging Trends in Computer and Electronics Engineering ICETCEE*, pages 13–18, 2012.

André I Khuri, Thomas Mathew, and Bimal K Sinha. *Statistical tests for mixed linear models*. John Wiley & Sons, 2011.

John P Klein and Melvin L Moeschberger. *Survival analysis: techniques for censored and truncated data*. Springer Science & Business Media, 2005.

Raed Kontar, Garvesh Raskutti, and Shiyu Zhou. Minimizing negative transfer of knowledge in multivariate Gaussian processes: A scalable and regularized approach. *IEEE Transactions on Pattern Analysis and Machine Intelligence*, DOI: 10.1109/TPAMI.2020.2987482, 2020.

Raed Kontar, Junbo Son, Shiyu Zhou, Chaitanya Sankavaram, Yilu Zhang, and Xinyu Du. Remaining useful life prediction based on the mixed effects model with mixture prior distribution. *IISE Transactions*, 49(7):682–697, 2017a.

Raed Kontar, Shiyu Zhou, and John Horst. Estimation and monitoring of key performance indicators of manufacturing systems using the multi-output Gaussian process. *International Journal of Production Research*, 55(8):2304–2319, 2017b.

Raed Kontar, Shiyu Zhou, Chaitanya Sankavaram, Xinyu Du, and Yilu Zhang. Nonparametric modeling and prognosis of condition monitoring signals using multivariate Gaussian convolution processes. *Technometrics*, 60(4):484–496, 2018.

WJ Krzanowski. Between-groups comparison of principal components. *Journal of the American Statistical Association*, 74(367):703–707, 1979.

WJ Krzanowski. Between-group comparison of principal components—some sampling results. *Journal of Statistical Computation and Simulation*, 15(2–3): 141–154, 1982.

Alexandra Kuznetsova, Per B Brockhoff, and Rune Haubo Bojesen Christensen. lmertest package: Tests in linear mixed effects models. *Journal of Statistical Software*, 82(13), 2017.

Nan M Laird and James H Ware. Random-effects models for longitudinal data. *Biometrics*, 38(4):963–974, 1982.

Jaesung Lee, Junbo Son, Shiyu Zhou, and Yong Chen. Variation source identification in manufacturing processes using Bayesian approach with sparse variance components prior. *IEEE Transactions on Automation Science and Engineering*, 17(3):1469–1485, 2020.

Frank L Lewis, Tong Heng Lee, Justin Chee Khiang Pang, and Zhao Yang Dong. *Intelligent diagnosis and prognosis of industrial networked systems*. CRC Press, 2011.

Fan Li and Jiuping Xu. A new prognostics method for state of health estimation of lithium-ion batteries based on a mixture of Gaussian process models and particle filter. *Microelectronics Reliability*, 55(7):1035–1045, 2015.

Xiaoyu Li, Changgui Yuan, Xiaohui Li, and Zhenpo Wang. State of health estimation for Li-ion battery using incremental capacity analysis and Gaussian process regression. *Energy*, 190:116467, 2020.

Yongxiang Li, Qiang Zhou, Xiaohu Huang, and Li Zeng. Pairwise estimation of multivariate Gaussian process models with replicated observations: Application to multivariate profile monitoring. *Technometrics*, 60(1):70–78, 2018.

Zhiguo Li and Shiyu Zhou. Robust method of multiple variation sources identification in manufacturing processes for quality improvement. *Journal of Manufacturing Science and Engineering*, 128(1):326–336, 2006.

Zhiguo Li, Shiyu Zhou, and Yu Ding. Pattern matching for variation-source identification in manufacturing processes in the presence of unstructured noise. *IIE Transactions*, 39(3):251–263, 2007.

Kaibo Liu, Nagi Z Gebraeel, and Jianjun Shi. A data-level fusion model for developing composite health indices for degradation modeling and prognostic analysis. *IEEE Transactions on Automation Science and Engineering*, 10(3):652–664, 2013.

Kailong Liu, Xiaosong Hu, Zhongbao Wei, Yi Li, and Yan Jiang. Modified Gaussian process regression models for cyclic capacity prediction of lithium-ion batteries. *IEEE Transactions on Transportation Electrification*, 5(4):1225–1236, 2019.

Yiqi Liu, Yongping Pan, Daoping Huang, and Qilin Wang. Fault prognosis of filamentous sludge bulking using an enhanced multi-output Gaussian processes regression. *Control Engineering Practice*, 62:46–54, 2017.

C Joseph Lu and William O Meeker. Using degradation measures to estimate a time-to-failure distribution. *Technometrics*, 35(2):161–174, 1993.

Bertil Matérn. *Spatial variation*. Springer Science & Business Media, 2013.

Charles E McCulloch and John M Neuhaus. *Generalized linear mixed models.* Wiley Online Library, 2001.

Geoffrey McLachlan and Thriyambakam Krishnan. *The EM algorithm and extensions.* John Wiley & Sons, 2007.

Geoffrey McLachlan and David Peel. *Finite mixture models.* John Wiley & Sons, 2004.

William Q Meeker and Luis A Escobar. *Statistical methods for reliability data.* John Wiley & Sons, 2014.

Russell B Millar. *Maximum likelihood estimation and inference: with examples in R, SAS and ADMB.* John Wiley & Sons, 2011.

Alan Miller. *Subset selection in regression.* CRC Press, 2002.

Geert Molenberghs and Geert Verbeke. *Linear mixed models for longitudinal data.* Springer, 2000.

Douglas C Montgomery. *Introduction to statistical quality control.* John Wiley & Sons (New York), 2009.

Dirk F Moore. *Applied survival analysis using R.* Springer, 2016.

Robb J Muirhead. *Aspects of multivariate statistical theory.* John Wiley & Sons, 2009.

Wayne Nelson. Theory and applications of hazard plotting for censored failure data. *Technometrics*, 14(4):945–966, 1972.

Gang Niu. *Data-driven technology for engineering systems health management.* Springer, 2017.

WJ Padgett and Meredith A Tomlinson. Inference from accelerated degradation and failure data based on Gaussian process models. *Lifetime Data Analysis*, 10(2):191–206, 2004.

Grigorios Papageorgiou, Katya Mauff, Anirudh Tomer, and Dimitris Rizopoulos. An overview of joint modeling of time-to-event and longitudinal outcomes. *Annual Review of Statistics and Its Application*, 2019.

Yudi Pawitan. *In all likelihood: statistical modelling and inference using likelihood.* Oxford science publications. OUP Oxford, 2001.

Karl Pearson. Liii. on lines and planes of closest fit to systems of points in space. *The London, Edinburgh, and Dublin Philosophical Magazine and Journal of Science*, 2(11):559–572, 1901.

Ying Peng, Ming Dong, and Ming Jian Zuo. Current status of machine prognostics in condition-based maintenance: A review. *The International Journal of Advanced Manufacturing Technology*, 50(1–4):297–313, 2010.

José Pinheiro and Douglas Bates. *Mixed-effects models in S and S-PLUS.* Springer Science & Business Media, 2006.

Adrian Pizzinga. *Restricted Kalman filtering: theory, methods, and application.* Springer Science & Business Media, 2012.

Peter ZG Qian, Huaiqing Wu, and CF Jeff Wu. Gaussian process models for computer experiments with qualitative and quantitative factors. *Technometrics*, 50(3):383–396, 2008.

S Joe Qin. Survey on data-driven industrial process monitoring and diagnosis. *Annual Reviews in Control*, 36(2):220–234, 2012.

BLS Prakasa Rao. *Identifiability in stochastic models: characterization of probability distributions*. Academic Press, 1992.

Calyampudi Radhakrishna Rao. Estimation of variance and covariance components - MINQUE theory. *Journal of Multivariate Analysis*, 1(3):257–275, 1971.

Calyampudi Radhakrishna Rao and Jürgen Kleffe. *Estimation of variance components and applications*. North Holland, 1988.

Carl Edward Rasmussen and Christopher KI Williams. *Gaussian process for machine learning*. MIT press, 2006.

R Core Team. "R: A Language and Environment for Statistical Computing", R Foundation for Statistical Computing, Vienna, Austria, 2020.

Marvin Rausand and Høyland Arnljot. *System reliability theory: models, statistical methods, and applications*. John Wiley & Sons, 2004.

Alvin C Rencher. *Methods of multivariate analysis*. John Wiley & Sons, 2003.

Robert R Richardson, Michael A Osborne, and David A Howey. Gaussian process regression for forecasting battery state of health. *Journal of Power Sources*, 357:209–219, 2017.

Dimitris Rizopoulos. *Joint models for longitudinal and time-to-event data: with applications in R*. Chapman & Hall/CRC Biostatistics Series. CRC Press, 2012.

Dimitris Rizopoulos, Laura A Hatfield, Bradley P Carlin, and Johanna JM Takkenberg. Combining dynamic predictions from joint models for longitudinal and time-to-event data using Bayesian model averaging. *Journal of the American Statistical Association*, 109(508):1385–1397, 2014.

Qiang Rong, Darek Ceglarek, and Jianjun Shi. Dimensional fault diagnosis for compliant beam structure assemblies. *Journal of Manufacturing Science and Engineering*, 122(4):773–780, 2000.

Thomas J Santner, Brian J Williams, and William I Notz. *The design and analysis of computer experiments*. Springer, 2003.

Bernhard Scholkopf and Alexander J Smola. *Learning with kernels: support vector machines, regularization, optimization, and beyond*. MIT press, 2001.

James R Schott. Common principal component subspaces in two groups. *Biometrika*, 75(2):229–236, 1988.

James R Schott. Some tests for common principal component subspaces in several groups. *Biometrika*, 78(4):771–777, 1991.

Shayle R Searle, George Casella, and Charles E McCulloch. *Variance components*. John Wiley & Sons, 2009.

Jian Qing Shi and Taeryon Choi. *Gaussian process regression analysis for functional data*. CRC Press, 2011.

Jianjun Shi and Shiyu Zhou. Quality control and improvement for multistage systems: A survey. *IIE Transactions*, 41(9):744–753, 2009.

Galit Shmueli, Peter C Bruce, Inbal Yahav, Nitin R Patel, and Kenneth C Lichtendahl Jr. *Data mining for business analytics: concepts, techniques, and applications in R*. John Wiley & Sons, 2017.

Xiao-Sheng Si, Wenbin Wang, Chang-Hua Hu, and Dong-Hua Zhou. Remaining useful life estimation–a review on the statistical data driven approaches. *European Journal of Operational Research*, 213(1):1–14, 2011.

Xiao-Sheng Si, Zheng-Xin Zhang, and Chang-Hua Hu. *Data-driven remaining useful life prognosis techniques*. Beijing, China: National Defense Industry Press and Springer-Verlag GmbH, 2017.

Steven M Sidik, Harold F Leibecki, and John M Bozek. *Cycles till failure of silver-zinc cells with completing failures modes: preliminary data analysis*. American Statistical Association Annual Meeting, Houston, Texas, 1980.

Dan Simon. *Optimal state estimation: Kalman, H infinity, and nonlinear approaches*. John Wiley & Sons, 2006.

Judith D Singer, John B Willett, and John B Willett *Applied longitudinal data analysis: modeling change and event occurrence*. Oxford university press, 2003.

Junbo Son, Yilu Zhang, Chaitanya Sankavaram, and Shiyu Zhou. Rul prediction for individual units based on condition monitoring signals with a change point. *IEEE Transactions on Reliability*, 64(1):182–196, 2015a.

Junbo Son, Shiyu Zhou, Chaitanya Sankavaram, Xinyu Du, and Yilu Zhang. Remaining useful life prediction based on noisy condition monitoring signals using constrained Kalman filter. *Reliability Engineering & System Safety*, 152:38–50, 2016.

Junbo Son, Qiang Zhou, Shiyu Zhou, and Mutasim Salman. Prediction of the failure interval with maximum power based on the remaining useful life distribution. *IIE Transactions*, 47(10):1072–1087, 2015b.

Petre Stoica and Arye Nehorai. On the concentrated stochastic likelihood function in array signal processing. *Circuits, Systems and Signal Processing*, 14(5):669–674, 1995.

Daniel O Stram and Jae Won Lee. Variance components testing in the longitudinal mixed effects model. *Biometrics*, 1171–1177, 1994.

Piyush Tagade, Krishnan S Hariharan, Sanoop Ramachandran, Ashish Khandelwal, Arunava Naha, Subramanya Mayya Kolake, and Seong Ho Han. Deep Gaussian process regression for lithium-ion battery health prognosis and degradation mode diagnosis. *Journal of Power Sources*, 445:227281, 2020.

Heidar A Talebi, Farzaneh Abdollahi, Rajni V Patel, and Khashayar Khorasani. *Neural network-based state estimation of nonlinear systems: application to fault detection and isolation*. Springer, 2009.

Monika Tanwar and Nagarajan Raghavan. Lubricating oil remaining useful life prediction using multi-output Gaussian process regression. *IEEE Access*, 8:128897–128907, 2020.

Grace Tapia, Alaa H Elwany, and Huiyan Sang. Prediction of porosity in metal-based additive manufacturing using spatial Gaussian process models. *Additive Manufacturing*, 12:282–290, 2016.

Toon W Taris. *A primer in longitudinal data analysis*. Sage, 2000.

Terry M Therneau and Patricia M Grambsch. *Modeling survival data: extending the Cox model*. Statistics for Biology and Health. Springer New York, 2013.

Robert Tibshirani. Regression shrinkage and selection via the lasso. *Journal of the Royal Statistical Society: Series B (Methodological)*, 58(1):267–288, 1996.

Michael E Tipping and Christopher M Bishop. Probabilistic principal component analysis. *Journal of the Royal Statistical Society: Series B (Statistical Methodology)*, 61(3):611–622, 1999.

Anastasios A Tsiatis and Marie Davidian. Joint modeling of longitudinal and time-to-event data: An overview. *Statistica Sinica*, 809–834, 2004.

Kwok L Tsui, Nan Chen, Qiang Zhou, Yizhen Hai, and Wenbin Wang. Prognostics and health management: A review on data driven approaches. *Mathematical Problems in Engineering*, 2015, 2015.

Pınar Tüfekci. Prediction of full load electrical power output of a base load operated combined cycle power plant using machine learning methods. *International Journal of Electrical Power & Energy Systems*, 60:126–140, 2014.

Fernando Tusell. Kalman filtering in r. *Journal of Statistical Software*, 39(2):1–27, 2011.

George Z Vachtsevanos. *Intelligent fault diagnosis and prognosis for engineering systems*. John Wiley & Sons, 2006.

George Vachtsevanos, Frank Lewis, Michael Roemer, Andrew Hess and Biqing Wu. *Intelligent fault diagnosis and prognosis for engineering systems*. Wiley, Hoboken, NJ, 2006.

Sergio Valle, Weihua Li, and S Joe Qin. Selection of the number of principal components: the variance of the reconstruction error criterion with a comparison to other methods. *Industrial & Engineering Chemistry Research*, 38(11):4389–4401, 1999.

Geert Verbeke. Linear mixed models for longitudinal data. In *Linear mixed models in practice*, pages 63–153. Springer, 1997.

Rui Wang, Linmiao Zhang, and Nan Chen. Spatial correlated data monitoring in semiconductor manufacturing using Gaussian process model. *IEEE Transactions on Semiconductor Manufacturing*, 32(1):104–111, 2018.

Hadley Wickham. *ggplot2: elegant graphics for data analysis*. Springer, 2016.

Andreas Wienke. *Frailty models in survival analysis*. Chapman & Hall/CRC Biostatistics Series. CRC Press, 2010.

Graham Williams. *Data mining with Rattle and R: the art of excavating data for knowledge discovery*. Springer Science & Business Media, 2011.

Andrew Gordon Wilson, David A Knowles, and Zoubin Ghahramani. Gaussian process regression networks. arXiv preprint arXiv:1110.4411, 2011.

Marcin Witczak. *Modelling and estimation strategies for fault diagnosis of non-linear systems: from analytical to soft computing approaches*. Springer Science & Business Media, 2007.

CF Jeff Wu and Michael S Hamada. *Experiments: planning, analysis, and optimization*. John Wiley & Sons, 2011.

Hulin Wu and Jin-Ting Zhang. *Nonparametric regression methods for longitudinal data analysis: mixed-effects modeling approaches*. John Wiley & Sons, 2006.

Lang Wu. *Mixed effects models for complex data*. CRC Press, 2009.

Michael S Wulfsohn and Anastasios A Tsiatis. A joint model for survival and longitudinal data measured with error. *Biometrics*, 330–339, 1997.

Linmiao Zhang, Kaibo Wang, and Nan Chen. Monitoring wafers' geometric quality using an additive Gaussian process model. *IIE Transactions*, 48(1):1–15, 2016.

Yang Zhang, Zhen He, Chi Zhang, and William H Woodall. Control charts for monitoring linear profiles with within-profile correlation using Gaussian process models. *Quality and Reliability Engineering International*, 30(4):487–501, 2014.

Zhengxin Zhang, Xiaosheng Si, Changhua Hu, and Xiangyu Kong. Degradation modeling–based remaining useful life estimation: A review on approaches for systems with heterogeneity. *Proceedings of the Institution of Mechanical Engineers, Part O: Journal of Risk and Reliability*, 229(4):343–355, 2015.

Jing Zhao and Shiliang Sun. Variational dependent multi-output Gaussian process dynamical systems. *The Journal of Machine Learning Research*, 17(1):4134–4169, 2016.

Qiang Zhou, Peter ZG Qian, and Shiyu Zhou. A simple approach to emulation for computer models with qualitative and quantitative factors. *Technometrics*, 53(3):266–273, 2011.

Qiang Zhou, Junbo Son, Shiyu Zhou, Xiaofeng Mao, and Mutasim Salman. Remaining useful life prediction of individual units subject to hard failure. *IIE Transactions*, 46(10):1017–1030, 2014.

Shiyu Zhou, Yong Chen, and Jiayun Shi. Root cause estimation and statistical testing for quality improvement of multistage manufacturing processes. *IEEE Transactions on Automation Science and Engineering*, 1(1):73–83, 2004.

Shiyu Zhou, Yu Ding, Yong Chen, and Jianjun Shi. Diagnosability study of multistage manufacturing processes based on linear mixed-effects models. *Technometrics*, 45(4):312–325, 2003.

Alain Zuur, Elena N Ieno, Neil Walker, Anatoly A Saveliev, and Graham M Smith. *Mixed effects models and extensions in ecology with R*. Springer Science & Business Media, 2009.

Index

Industrial Data Analytics for Diagnosis and Prognosis: A Random Effects Modelling Approach, First Edition. Shiyu Zhou and Yong Chen.

Printed and bound by CPI Group (UK) Ltd, Croydon, CR0 4YY